Environmental Econometrics
Using Stata

Environmental Econometrics Using Stata

Christopher F. Baum
Department of Economics and School of Social Work
Boston College

Stan Hurn
School of Economics and Finance
Queensland University of Technology

A Stata Press Publication
StataCorp LLC
College Station, Texas

Copyright © 2021 StataCorp LLC
All rights reserved. First edition 2021

Published by Stata Press, 4905 Lakeway Drive, College Station, Texas 77845
Typeset in $\LaTeX\,2_\varepsilon$
Printed in the United States of America
10 9 8 7 6 5 4 3 2 1

Print ISBN-10: 1-59718-355-5
Print ISBN-13: 978-1-59718-355-0
ePub ISBN-10: 1-59718-356-3
ePub ISBN-13: 978-1-59718-356-7
Mobi ISBN-10: 1-59718-357-1
Mobi ISBN-13: 978-1-59718-357-4

Library of Congress Control Number: 2021934557

No part of this book may be reproduced, stored in a retrieval system, or transcribed, in any form or by any means—electronic, mechanical, photocopy, recording, or otherwise—without the prior written permission of StataCorp LLC.

Stata, **STaTa**, Stata Press, Mata, **mata**, and NetCourse are registered trademarks of StataCorp LLC.

Stata and Stata Press are registered trademarks with the World Intellectual Property Organization of the United Nations.

NetCourseNow is a trademark of StataCorp LLC.

$\LaTeX\,2_\varepsilon$ is a trademark of the American Mathematical Society.

Contents

List of figures	xv
List of tables	xix
Preface	xxi
Acknowledgments	xxv
Notation and typography	xxvii

1 Introduction   1
- 1.1 Features of the data 1
  - 1.1.1 Periodicity 2
  - 1.1.2 Nonlinearity 3
  - 1.1.3 Structural breaks and nonstationarity 5
  - 1.1.4 Time-varying volatility 6
  - Types of data 7

2 Linear regression models   11
- 2.1 Air pollution in Santiago, Chile 11
- 2.2 Linear regression and OLS estimation 14
- 2.3 Interpreting and assessing the regression model 17
  - 2.3.1 Goodness of fit 17
  - Tests of significance 18
  - 2.3.2 Residual diagnostics 20
  - Homoskedasticity 21
  - Serial independence 23
  - Normality 24
- 2.4 Estimating standard errors 26

3	**Beyond ordinary least squares**		**33**
	3.1	Distribution of particulate matter	33
	3.2	Properties of estimators	35
		Consistency	35
		Asymptotic normality	36
		Asymptotic efficiency	38
	3.3	Maximum likelihood and the linear model	38
	3.4	Hypothesis testing	43
		Likelihood-ratio test	43
		Wald test	44
		LM test	44
	3.5	Method-of-moments estimators and the linear model	44
	3.6	Testing for exogeneity	48
4	**Introducing dynamics**		**55**
	4.1	Load-weighted electricity prices	55
	4.2	Specifying and fitting dynamic time-series models	58
		AR models	59
		Moving-average models	60
		ARMA models	60
	4.3	Exploring the properties of dynamic models	61
	4.4	ARMA models for load-weighted electricity price	65
	4.5	Seasonal ARMA models	71
5	**Multivariate time-series models**		**77**
	5.1	CO_2 emissions and growth	77
	5.2	The VARMA model	79
	5.3	The VAR model	80
	5.4	Analyzing the dynamics of a VAR	85
		5.4.1 Granger causality testing	85

		5.4.2	Impulse–responses	87
			Vector moving-average form	87
			Orthogonalized impulses	88
		5.4.3	Forecast-error variance decomposition	92
	5.5	SVARs		94
		5.5.1	Short-run restrictions	95
		5.5.2	Long-run restrictions	98

6 Testing for nonstationarity — 105

	6.1	Per capita CO_2 emissions	105
	6.2	Unit roots	108
	6.3	First-generation unit-root tests	112
		6.3.1 Dickey–Fuller tests	112
		6.3.2 Phillips–Perron tests	116
	6.4	Second-generation unit-root tests	117
		6.4.1 KPSS test	117
		6.4.2 Elliott–Rothenberg–Stock DFGLS test	119
	6.5	Structural breaks	121
		6.5.1 Known breakpoint	121
		6.5.2 Single-break unit-root tests	123
		6.5.3 Double-break unit-root tests	124

7 Modeling nonstationary variables — 129

	7.1	The crush spread	129
	7.2	Illustrating equilibrium relationships	131
	7.3	The VECM	133
	7.4	Fitting VECMs	135
		7.4.1 Single-equation methods	135
		7.4.2 System estimation	137
	7.5	Testing for cointegration	140
	7.6	Cointegration and structural breaks	143

8	**Forecasting**		**151**
	8.1	Forecasting wind speed	151
	8.2	Introductory terminology	153
	8.3	Recursive forecasting in time-series models	154
		8.3.1 Single-equation forecasts	155
		8.3.2 Multiple-equation forecasts	156
		8.3.3 Properties of recursive forecasts	157
	8.4	Forecast evaluation	158
	8.5	Daily forecasts of wind speed for Santiago	160
	8.6	Forecasting with logarithmic dependent variables	166
		8.6.1 Staying in the linear regression framework	169
		8.6.2 Generalized linear models	171
9	**Structural time-series models**		**175**
	9.1	Sea level and global temperature	175
	9.2	The Kalman filter	177
	9.3	Vector autoregressive moving-average models in state-space form	179
	9.4	Unobserved component time-series models	184
		9.4.1 Trends	184
		9.4.2 Seasonals	188
		9.4.3 Cycles	189
	9.5	A bivariate model of sea level and global temperature	191
10	**Nonlinear time-series models**		**197**
	10.1	Sunspot data	198
	10.2	Testing	200
	10.3	Bilinear time-series models	203
	10.4	Threshold autoregressive models	208
	10.5	Smooth transition models	212
	10.6	Markov switching models	220
11	**Modeling time-varying variance**		**229**
	11.1	Evaluating environmental risk	229

	11.2	The generalized autoregressive conditional heteroskedasticity model .	231
	11.3	Alternative distributional assumptions	237
	11.4	Asymmetries	239
	11.5	Motivating multivariate volatility models	242
	11.6	Multivariate volatility models	245
		11.6.1 The vech model	246
		11.6.2 The dynamic conditional correlation model	248
12	**Longitudinal data models**		**255**
	12.1	The pollution haven hypothesis	255
	12.2	Data organization	257
		12.2.1 Wide and long forms of panel data	258
		12.2.2 Reshaping the data	259
	12.3	The pooled model	262
	12.4	Fixed effects and random effects	264
		12.4.1 Individual FEs	265
		12.4.2 Two-way FE	268
		12.4.3 REs	270
		12.4.4 The Hausman test in a panel context	272
		12.4.5 Correlated RE	274
	12.5	Dynamic panel-data models	279
13	**Spatial models**		**283**
	13.1	Regulatory compliance	283
	13.2	The spatial weighting matrix	286
		13.2.1 Specification	286
		Distance weights	287
		Contiguity weights	288
		13.2.2 Construction	288
	13.3	Exploratory data analysis	292

13.4	Spatial models	294
	Spatial lag model	295
	Spatial error model	296
13.5	Fitting spatial models by maximum likelihood	297
	Spatial lag model	297
	Spatial error model	299
13.6	Estimating spillover effects	300
13.7	Model selection	303

14 Discrete dependent variables — 309

14.1	Humpback whales	309
14.2	The data	311
14.3	Binary dependent variables	316
14.3.1	Linear probability model	316
14.3.2	Binomial logit and probit models	318
14.4	Ordered dependent variables	326
14.5	Censored dependent variables	331

15 Fractional integration — 339

15.1	Mean sea levels and global temperature	339
15.2	Autocorrelations and long memory	340
15.3	Testing for long memory	343
15.4	Estimating d in the frequency domain	346
15.5	Maximum likelihood estimation of the ARFIMA model	351
15.6	Fractional cointegration	354

A Using Stata — 361

A.1	File management	362
A.1.1	Locating important directories: adopath	362
A.1.2	Organization of do-, ado-, and data files	364
A.1.3	Editing Stata do- and ado-files	364
A.2	Basic data management	365
A.2.1	Data types	365

	A.2.2	Getting your data into Stata	367
		Handling text files	368
		The import delimited command	368
		Accessing data stored in spreadsheets	369
		Importing data from other package formats	370
	A.2.3	Other data issues	371
		Protecting the data in memory	371
		Missing data handling	371
		Recoding missing values: the mvdecode and mvencode commands	372
	A.2.4	String-to-numeric conversion and vice versa	372
A.3	General programming hints		373
		Variable names	373
		Observation numbering: _n and _N	373
		The varlist	373
		The numlist	373
		The if exp and in range qualifiers	374
		Local macros	374
		Global macros	375
		Scalars	375
		Matrices	376
		Looping	377
		The generate command	378
		The egen command	378
		Computation for by-groups	379
A.4	A smorgasbord of important topics		380
		Date and time handling	380
		Time-series operators	382
A.5	Factor variables and operators		383
A.6	Circular variables		384

References	**385**
Author index	**403**
Subject index	**409**

Figures

1.1	Daily average Brisbane temperature	2
1.2	Distribution of daily average Brisbane temperature	3
1.3	Monthly sunspot numbers	4
1.4	Monthly Southern Oscillation index	5
1.5	Global temperature anomalies	6
1.6	Brisbane temperature volatility	7
1.7	Cross-country log CO_2 emissions per capita	8
1.8	Spatial distribution of GDP	9
2.1	Weather stations in Santiago, Chile	12
2.2	PM2.5 concentration, Cerrillos station	13
2.3	Regression residuals	21
2.4	Histogram of regression residuals	25
3.1	Distribution of PM2.5 concentration, Santiago	34
3.2	Illustrating the consistency property	36
4.1	Load-weighted electricity price	58
4.2	ACFs of ARMA models	62
4.3	PACFs of ARMA models	64
4.4	ACFs of ARMA models	65
4.5	ACFs of residuals of ARMA models	70
4.6	ACFs of residuals of seasonal ARMA models	74
5.1	Ratio of CO_2 emissions to real gross domestic product (GDP)	78
5.2	CO_2 emissions and real GDP	79

5.3	Impulse–response functions	91
5.4	Impulse–responses from CO_2 SVAR	101
6.1	Per capita CO_2 emissions	106
6.2	AR(2) model with two real roots	110
6.3	AR(2) model with complex roots	111
6.4	AR(2) model with one unit root	112
6.5	Structural breaks	122
6.6	Double structural break unit-root test	126
7.1	Crush spread	130
7.2	Illustrating equilibrium adjustment	132
7.3	Engle–Granger residuals	141
7.4	Gregory–Hansen residuals	146
8.1	Periodicity in wind speed	153
8.2	Forecast wind speed	162
8.3	Distribution of wind speed	166
8.4	Distribution of wind speed	171
9.1	Mean annual global sea level and temperature	176
9.2	Global temperature with smooth trend	187
9.3	Histogram of irregular component	188
9.4	Long-term forecasts of sea level	194
10.1	Annual sunspot numbers	199
10.2	ACF and PACF of sunspot numbers	199
10.3	Bilinear time series	204
10.4	Bilinear model of sunspots	207
10.5	Actual and fitted values of SETAR model	212
10.6	Threshold functions	214
10.7	Actual and fitted values from the LSTAR model	220
10.8	Smoothed probabilities from Markov switching model	224

11.1	DJSI	230
11.2	Excess returns on DJSI World	232
11.3	Conditional variance of returns on DJSI World Index	235
11.4	Conditional variance forecasts	236
11.5	Conditional variance gap	242
11.6	Conditional covariance and time-varying β	245
11.7	DVECH time-varying β	248
11.8	DVECH time-varying β	252
12.1	Costs of abatement and PHH	256
13.1	Choropleth map of fishing compliance indicators	285
13.2	Visualizing a spatial weighting matrix	291
13.3	Moran's scatterplot	293
14.1	Annual humpback whale catch	310
14.2	Age distribution of respondents	315
14.3	Sensitivity and specificity versus cutoff	326
14.4	Scatterplot of willingness to pay	332
15.1	Global mean sea level and temperature	340
15.2	Autocorrelations of an $I(0)$ process and $I(d)$ process	342
15.3	Autocorrelations of mean sea level and temperature	343
15.4	Global sea ice extent	356
15.5	Residuals from fractional cointegration analysis	358

Tables

2.1 Implementing robust, HAC, and clustered standard errors for the simple regression model of PM2.5 in Santiago, Chile 30

4.1 Summary of the Queensland load-weighted electricity price broken down by month. The data are daily for the period January 1, 2000 to August 9, 2013. 57

9.1 Common trend specifications in unobserved component time-series models 185

11.1 GARCH(1,1) estimates of excess returns on the DJSI World Index under different distributional assumptions 238

11.2 Estimates of the CAPM for DJSI indices for the period February 1, 2004 to June 17, 2017 243

13.1 Summary of the cross-sectional variables from 51 countries used to capture spatial interactions in compliance with international environmental agreements 292

A.1 Numeric data types 365

Preface

There is no doubt that the environment is one of the greatest challenges faced by policymakers today. The key issues addressed by environmental sciences are often empirical. In many instances, very detailed, sizable datasets are available. Researchers in this field, including those in academe, research bodies, and government agencies, should have a solid understanding of the econometric tools best suited for analysis of these data.

Of course, there exist complex and expensive physical models of the environment that deal with many of the problems addressed in this book, such as pollution, temperature, greenhouse gas emissions, and sea levels to name but a few. However, it is becoming increasingly clear, through the increased involvement of econometricians in environmental issues that reduced-form models have a role to play not only in modeling environmental phenomena but also in producing point and density forecasts. In short, successful environmental modeling does not necessarily require a structural model, but it does require that the econometrics underlying the reduced-form approaches is competently done. This provides the essential raison d'être for the book.

This book is designed to introduce environmental researchers to a broad range of econometric techniques that can be effectively applied to environmental data. The study of environmental issues is inherently interdisciplinary, encompassing the physical sciences, economics, sociology, political science, and public health. Researchers in these fields are likely to have some statistical training, and an understanding of basic statistical concepts is presumed. The development of modern econometrics, coupled with increasing computational capability to process sizable datasets, has broadened our ability to study environmental data using powerful analytical and graphical tools.

Although our focus is on applied econometric techniques appropriate for the analysis of environmental data, we expect this book to be widely used. We believe that the potential audience includes economists at the undergraduate, graduate, and professional levels in academia, research institutes, consulting firms, government agencies, and international organizations. Our approach provides a gentle introduction to the most widely used econometric tools, which should serve to address the needs of those who may have only seen econometrics at an undergraduate level, such as those in public policy programs. We not only emphasize how to fit models in Stata but also highlight the need for using a wide range of diagnostic tests to validate the results of estimation and subsequent policy conclusions. This emphasis on careful, reproducible research should be appreciated by academic and non-academic researchers who are seeking to produce credible, defensible conclusions about key issues in environmental science.

Although appendix A provides a brief guide to using Stata effectively, thus book assumes that the reader is familiar with Stata's command line interface and elementary concepts of Stata programming such as do-files and data management facilities. An understanding of basic linear regression techniques will also be helpful but is not essential, because the book covers the basic building blocks of modern econometrics. More advanced econometric methods are also introduced, interspersing presentation of the underlying theory with clear examples of their employment on environmental data. In contrast with many existing econometric textbooks that deal mainly with the theoretical properties of estimators and test statistics, this book addresses the implementation issues that arise in the computational aspects of applied econometrics. The computer code that is provided will also help to bridge the gap between theory and practice so that the reader, as a result, can build on the code and tailor it to more challenging applications.

Organization

Although not specifically designated as such, the material presented in this book falls naturally into two parts. Chapters 1 to 8 provide a first course in applied environmental econometrics. These chapters cover the basic building blocks upon which the rest of the book is based, including the usual regression framework taught in standard econometric courses but always related to the modeling of environmental data. Chapter 2 describes the workhorse of applied econometrics, the linear regression model, while chapter 3 covers additional important estimation methods beyond the simple least-squares method. Chapter 4 extends the single-equation model to include dynamic components, while chapter 5 considers multiple time-series models, particularly vector autoregression and structural vector autoregression. The next two chapters develop the tools to deal with nonstationary data. Chapter 6 presents a range of tests for nonstationarity, known as unit-root tests. Chapter 7 discusses the extension of nonstationarity to deal with multiple time series and the idea of cointegrated systems. The last chapter in the first part of the book is chapter 8, which deals with forecasting methods and evaluation of forecast accuracy. Our philosophy is to make the treatment accessible by avoiding, wherever possible, the use of matrix algebra and potentially confusing notation. Where the use of this kind of notation is unavoidable, our intention is to provide as much intuition as possible.

The second part of the book comprises chapters 9 to 15 and relates to important econometric methods that may be of particular interest to empirical environmental studies. These chapters could form the core of a second course in applied environmental econometrics, and the level of difficulty steps up slightly. The first three chapters deal with techniques aimed at dealing with the nonlinear behavior that characterizes many environmental data series. Consequently, chapter 9 presents unobserved component models that decompose a given time series into its unobserved components, chapter 10 covers models that exhibit fundamental nonlinearity in mean, such as threshold models and Markov switching models, and chapter 11 deals with models that are nonlinear in variance and display what is known as volatility clustering. The remaining four chapters are perhaps best described as topics in applied environmental econometrics. Chapter 12 illustrates selected models for longitudinal data, taking advantage of the multiple measurements of environmental series, such as climate conditions. Chapter 13 is concerned with models for data that are measured at different geographical locations and covers the estimation of models that are characterized by spatial effects. Chapter 14 presents a selection of limited dependent variable models, focusing on the modeling of willingness to pay for environmental preservation and mitigation. Chapter 15 presents models of fractional integration and cointegration, which were first studied in hydrology and biological processes.

The Stata code and datasets to reproduce all the examples in the book are available from a companion website. One of the features of this book is that each chapter has several nontrivial exercises that not only reinforce the material covered in the chapter but also extend it. Code to solve the exercises at the end of each chapter is also available.

Acknowledgments

Creating a manuscript of this scope and magnitude is a daunting task, and there are many people to whom we are indebted. In particular, we would like to thank David Drukker, Enrique Pinzon, Bill Rising, and other StataCorp staff for their support. In addition, we wish to thank colleagues and students who have provided useful feedback during the writing of the book. These include Adam Clements, Kenneth Lindsay, Annastiina Silvennoinen, Timo Teräsvirta, Lina Xu, and Nicholas Johnson. In addition, we would like to acknowledge the financial support of the National Centre for Econometric Research (http://www.ncer.edu.au), based in the Queensland University of Technology Business School for generous financial support that enabled visits to Brisbane and Boston that greatly aided the completion of the project.

It is fair to say that writing this book was an immense task, and the biggest debt of gratitude we owe, therefore, is to our respective families.

Christopher F. Baum
Stan Hurn

March 2021

Notation and typography

In this book, we assume that you are somewhat familiar with Stata, so you know how to input data, use previously created datasets, create new variables, run regressions, and the like. For those readers who are completely unfamiliar with Stata but would like to use the book, a useful companion volume for reference is Baum (2006).

The book is specifically designed for you to learn by doing, so it is expected that you will read it while sitting at a computer so that you can try using the sequences of commands contained in the book to replicate the results. In this way, you will be able to generalize these sequences to suit your own needs.

Generally, the `typewriter font` is used to refer to Stata commands, syntax, and variables. A "dot" prompt followed by a command indicates that you can type verbatim what is displayed after the dot (in context) to replicate the results in the book.

Italic font is used for words that are not supposed to be typed; instead, you are to substitute another word or words for them. For example, the instruction type `by`(*groupvar*) means that you should replace "*groupvar*" with the actual name of the group variable.

All the datasets and do-files for this book are freely available for you to download. You can also download all the community-contributed commands described in this book. At the Stata dot prompt, type

```
. net from http://www.stata-press.com/data/eeus/
  (output omitted)
. net install eeus-ado
  (output omitted)
. net get eeus-data
  (output omitted)
. net get eeus-do
  (output omitted)
```

After installing the files, type `spinst_eeus` to obtain all the community-contributed commands used in the book's examples. You should check the messages produced by the `spinst_eeus` command. If there are any error messages, follow the instructions at the bottom of the output to complete the download.

In a net-aware Stata, you can also load the dataset by specifying the complete URL of the dataset. For example,

```
. use http://www.stata-press.com/data/eeus/sunspots
```

This text complements the material in the PDF Stata manuals but does not replace it. Stata manuals are often referred to using [R], [P], etc. For example, [R] **summarize** refers to the *Stata Reference Manual* entry for `summarize`, and [P] **syntax** refers to the entry for `syntax` in the *Stata Programming Manual*.

1 Introduction

Some of the most important research issues facing our society concern the environment. What are the effects of global warming on sea levels and agriculture? To what degree are observed variations in global temperature and CO_2 deviating from those predictable from data spanning several centuries? Do we face greater extremes in temperature, precipitation, and severe climactic events such as Atlantic hurricanes and Pacific cyclones as a consequence of human activity? What are the effects of environmental pollution on public health in major cities? How does increased development of arid regions, from the American Southwest to Beijing, affect water supplies? What impact will green energy initiatives have on the demand for fossil fuels? What are the unintended consequences of environmental regulation of fuels used in electricity generation?

All of these questions can be addressed with a wealth of environmental data, often available over long time spans, many locations, and at an increasing frequency. But the mere availability of data does not lead to substantive and justifiable conclusions in the environmental field. Those who have taken an elementary statistics course are well aware of the dangers of applying *post hoc, ergo propter hoc* logic—observing a phenomenon and attributing its cause to a prior notion of causality. Likewise, those who have used linear regression techniques are well aware of the dangers of misspecification bias due to omitted variables or measurement error. Consequently, researchers must study environmental data using the appropriate statistical tools. Many of those tools have been developed in recent decades in econometric theory. In this book, we focus on those tools that are of particular relevance to environmental researchers, juxtaposing discussion of their theoretical background with relevant applications from readily available environmental data.

1.1 Features of the data

Several statistical issues facing environmental researchers are related to the particular characteristics of environmental data. These data are often time series or multiple time series, so their analysis requires a good understanding of the econometric techniques that have been developed in that field. These techniques extend well beyond simple linear models to nonlinear methods.

1.1.1 Periodicity

Consider the time-series plot of average daily temperature in Brisbane, Australia for the period January 2000 to December 2007 shown in figure 1.1.

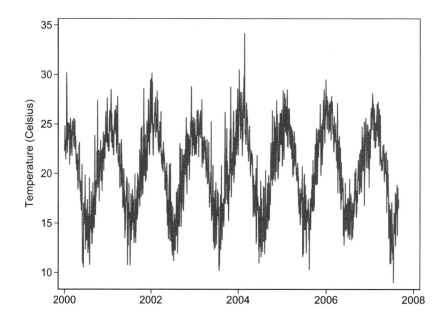

Figure 1.1. Plot of the average daily temperature in Brisbane, Australia

The regular periodicity related to seasonal fluctuations is evident in these data. We can take these same data and consider their distribution in the form of a histogram of temperature values in figure 1.2. The distribution of temperature is clearly bimodal, with a peak around 16 degrees centigrade reflecting mean low temperatures as well as the higher peak around 24 degrees centigrade reflecting mean high temperatures. In some instances, it is useful to proceed by decomposing a time series into its constituent features, following the classical decomposition into trend, seasonal, cyclical, and irregular components. These techniques are discussed in chapter 9.

1.1.2 Nonlinearity

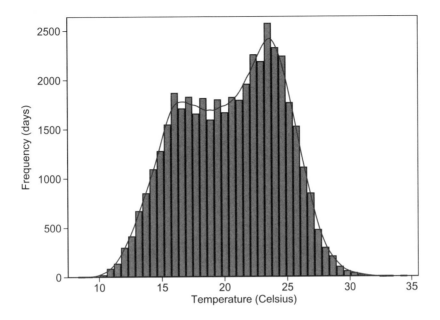

Figure 1.2. Histogram of daily average temperature in Brisbane, Australia

1.1.2 Nonlinearity

Environmental data may be periodic, but it is also possible that this periodicity is irregular, giving rise to nonlinearity. Consider the time series of the number of sunspots observed per month for the period from January 1749 to June 2016. The monthly time series is plotted in figure 1.3. It has been established that important variations in solar radiation occur over the sunspot cycle, which has a periodicity of approximately 11 years. With hundreds of years of data available, time-series techniques can be applied to study this cyclical behavior and its variability. In the case of sunspots, variability is apparent in terms of not only the length of the cycle but also the amplitude of the cycle. Notice, for example, periods in the early 1800s and around 1900, where the amplitude of the cycle appears considerably damped compared with other periods.

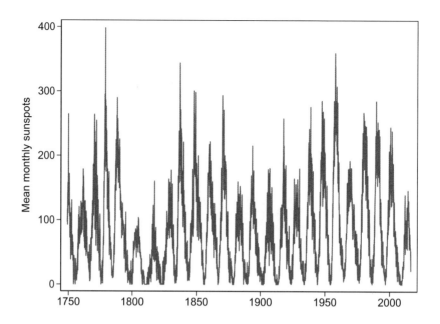

Figure 1.3. Plot of average monthly sunspots from January 1749 to June 2016

Another irregular cycle of crucial importance to countries in the Pacific region is the appearance and strength of the El Niño Southern Oscillation, which is the variations in temperature and precipitation caused by fluctuating atmospheric pressure in the eastern and western Pacific. These fluctuations can be summarized by the Southern Oscillation Index, which is computed as 10 times the standardized value of the difference in mean sea level air pressure between Tahiti and Darwin, Australia and illustrated in figure 1.4. On this scale, a low value is associated with the El Niño phenomenon, leading to drought and brushfires in Australia and Indonesia, with torrential rains in places like Southern California, Peru, and Chile. In contrast, a large positive value would be associated with an extreme La Niña event, leading to droughts in the eastern Pacific and heavy rainfall in the western Pacific. The importance of these phenomena linked to the Southern Oscillation go beyond the tropical Pacific, with severe events causing climactic effects in the United States and Canada.

1.1.3 Structural breaks and nonstationarity 5

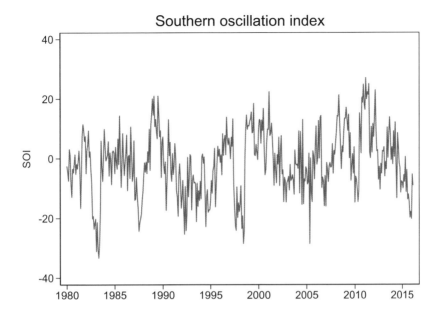

Figure 1.4. Plot of the monthly Southern Oscillation Index

In these data, we see features absent from the sunspot data. Not only is the cycle irregular (note the extraordinary El Niño episode in 1998), but there is also a longer-term cycle evident in its mean value over the period. Analysis of these nonlinearities calls for nonlinear modeling methods. These models include the threshold and switching classes of nonlinear time-series models, which are discussed in chapter 10.

1.1.3 Structural breaks and nonstationarity

In figure 1.5, two series are illustrated representing global temperature anomalies, defined as deviations in degrees Celsius from the 1961–1990 reference period. The term temperature anomaly means a departure from a reference value or long-term average. A positive anomaly indicates that the observed temperature was warmer than the reference value, while a negative anomaly indicates that the observed temperature was cooler than the reference value. The HadCRUT4 series is a global temperature dataset developed by the Climatic Research Unit at the University of East Anglia in conjunction with the Hadley Centre in the UK Meteorological Office. The Giss Surface Temperature Analysis is provided by the NASA Goddard Institute for Space Studies[1] with details on its construction in Hansen et al. (2010). In this figure, it appears that global temperature was relatively constant through the early 1900s, with an upward trend through the 1940s leveling off in the 1950s, with a sharp increase in recent years.

---

1. http://data.giss.nasa.gov/gistemp/

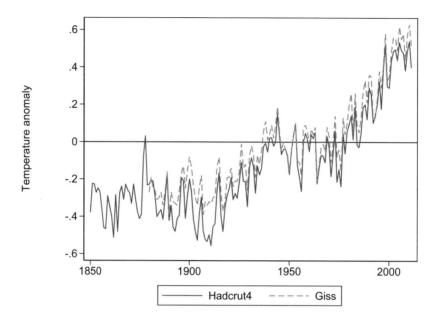

Figure 1.5. The HadCRUT4 and Giss global temperature anomalies, defined relative to the 1961–1990 reference period

The dynamics evident in these processes require dealing with both deterministic trends and stochastic trends, which are important aspects of effective econometric modeling. This has prompted the development of sophisticated techniques to identify and address these challenges. Although the theoretical justification for nonstationary behavior may be more compelling for economic and financial series, similar concerns may be held for time series in the environmental domain. Testing for nonstationarity behavior is dealt with in chapter 6. Of particular importance will be the ability to distinguish between nonstationary behavior and those environmental time series that are simply subject to structural breaks.

Econometric modeling with two or more nonstationary time series is the subject matter of chapter 7. Of particular interest here is the question of whether two or more nonstationary time series can be combined so as to produce a stationary series. This concept, known as cointegration, is an important component in the toolbox of the applied environmental econometrician and is crucial for effective dynamic modeling.

1.1.4 Time-varying volatility

A time series may exhibit a relatively constant mean, but its volatility (or variance) may be time varying, indicating greater or lesser degrees of uncertainty. Econometric models for the modeling of time series and their time-varying volatility are potentially

1.1.4 Time-varying volatility

important in a growing trend in the literature aimed at using option pricing models, developed in the financial econometrics literature, to price derivatives written on environmental variables such as temperature, rainfall, and the wind. As an example of time-varying volatility, the estimated standard deviation of average temperature in Brisbane, Australia for each of the 365 days of the year over more than 130 years is plotted in figure 1.6. The seasonal pattern in temperature variability over the calendar year is clearly evident.[2]

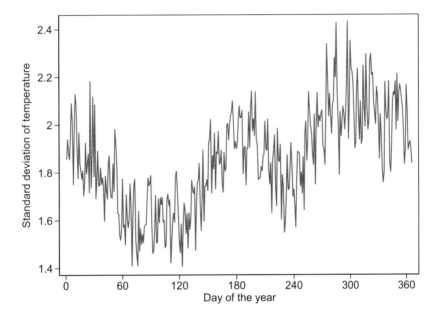

Figure 1.6. Standard deviation of daily Brisbane average temperature, 1887–2007

Types of data

Longitudinal data

Environmental data are often available in longitudinal (panel) format, that is, regularly spaced measurements from several sites, such as daily extreme temperatures for a set of 50 cities. These data have clear advantages because we can answer questions with longitudinal data that cannot be addressed with pure time-series or pure cross-sectional data. At the same time, modeling these data is more complex because unobserved heterogeneity must be accounted for. Unobserved time-invariant heterogeneity relates to the feature that each unit in the panel has unique characteristics that may not be quantifiable. Extensions of simple models for longitudinal data involve dynamic panel structures, which allow for persistence in the outcome variable, seemingly unrelated

---

2. For simplicity, observations relating to the 29th of February were deleted from the sample.

regression models, and hierarchical models, in which the data have more than two dimensions.

Figure 1.7 plots annual data for per capita CO_2 and SO_2 emissions for three countries. We see nearly parallel growth in CO_2 emissions in the United States and France in the last half of the 19th century, a period when Great Britain was already heavily industrialized. The effects of the Great Depression are evident in the UK series, while the impact of the two world wars are clearly visible in the French series. Data such as these can be used to answer questions relating to the so-called environmental Kuznets curve, a proposed nonlinear relationship between emissions and income.

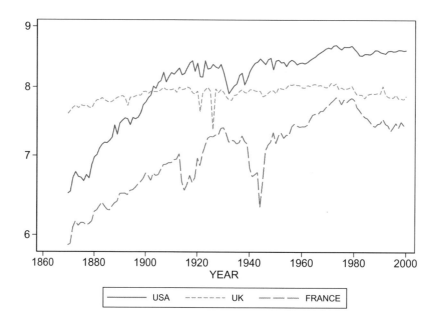

Figure 1.7. Log CO_2 emissions per capita, selected countries

Spatial relationships

With the increasing availability of environmental monitoring of sea levels, temperatures, and other climactic conditions at several stations, we may want to take advantage of spatial econometric techniques that account for the spatiotemporal relationships related to the underlying geography. Spillover effects of environmental phenomena may reflect proximity, such as the drift of radioactive isotopes after the Chernobyl disaster. They may also be related to the progress of flooding in which the crest of floodwaters inexorably travels downstream.

1.1.4 Time-varying volatility

Climate plays an important part in economic development, and this link is particularly evident in countries that are located close to the equator. This phenomenon is partly due to more frequent weather extremes in these areas but also because these countries tend to be more dependent on sectors that are climate sensitive, such as agriculture and tourism. Figure 1.8 uses 2016 data to illustrate world per capita gross domestic product (GDP) measured in millions of U.S. dollars and based on estimated population. This figure is what is known as a choropleth map, in which areas are shaded or patterned in proportion to the measurement of the statistical variable being displayed, in this case per capita GDP. Choropleth maps provide an easy way to visualize how a measurement varies across a geographic area and hence illustrate the spatial dimension of the problem. For example, on examining figure 1.8, a tentative hypothesis might be that countries adjacent to the equator suffer from low standards of living.

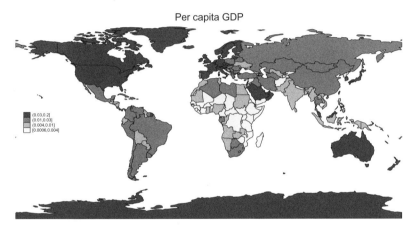

Figure 1.8. Choropleth map of world per capita GDP. Darker areas indicate countries with higher per capita GDP.

Discrete data

Survey data often give rise to variables that are qualitative or not continuous. For example, surveys aimed at ascertaining the willingness of individuals to pay for environmental conservation programs naturally give rise to binary (yes or no) responses or responses in which there is an ordering of preferences. Dependent variables that are binary, ordered, or filtered in some other way such that the data are not continuously observed require a special set of econometric methods.

Summary

In summary, environmental data come in many flavors. The research questions that they may be used to answer are likely to require the use of different econometric methods depending on the hypotheses to be tested and the nature of the available data. In this book, we provide an overview of a broad set of econometric tools and guidance in their use.

2 Linear regression models

This chapter presents the classical linear regression model. Although we provide a review of the mechanics of the regression model, the exposition is brief because the reader is assumed to be familiar with the material that would be covered in a standard undergraduate course in econometrics. The exposition of the chapter starts with the ordinary least-squares (OLS) estimator of the classical linear regression model. It then discusses the interpretation and analysis of the fitted model in terms of statistical tests of significance and residual diagnostics. Throughout this chapter, the dependent variable will be taken to be stationary, and the explanatory variables will be assumed to have finite means and variances that are not explicit functions of time. Formal details relating to stationarity and modeling with nonstationary variables are discussed in chapters 6 and 7. Aspects of relaxing assumptions on the regressors are addressed in chapters 3 and 4.

2.1 Air pollution in Santiago, Chile

The major dataset used in this chapter relates to the air pollution problem in Santiago, Chile. The data are daily observations on air pollution and other climatic variables for the period January 1, 2009 to December 31, 2014. The air pollutant most significantly related to premature mortality is a particularly fine particle that is less than 2.5 micrometers (millionths of a metre) in diameter. These fine pollutants are known as PM2.5 particles and are caused primarily by vehicle emissions, particularly diesel trucks, and by the use of wood-burning heaters. The World Health Organization (WHO), recognizing the adverse effects on human respiratory and cardiovascular systems, has established guidelines on acceptable levels of PM2.5. The background concentration of PM2.5 particles has been estimated to be 3–5 micrograms per cubic metre ($\mu g/m^3$) and the long-term guideline for exposure to PM2.5 has been set at an annual average concentration of 10 $\mu g/m^3$. The short-term air quality guideline is set as a daily average of 25 $\mu g/m^3$. The WHO recommends that countries with areas not meeting the 24-hour air quality guideline undertake immediate remedial action to achieve reductions to these levels.

In 2014, the OECD ranked Chile as the country with the highest air pollution among its 36 members. In Chile, at least 60% of the inhabitants are exposed to PM2.5 concentrations over 15 $\mu g/m^3$ (Cifuentes 2010). In fact, it is estimated by the Chilean Ministerio del Medio Ambiente (2011) that approximately 4,000 premature deaths in Chile are due to chronic exposure to this component of pollution. There are also significant broader economic consequences of air pollution. In 2013, the World Bank estimated that lost work-related income due to air pollution was 225 billion U.S. dollars. In Chile, the net economic benefit of effectively regulating PM2.5 is estimated to be 7.1 billion U.S. dollars. Santiago, the capital and largest city of Chile, where 41% of the country's total population resides, has a severe problem with PM2.5 pollution and is ranked fourth in terms of cities with the worst air quality on the continent according to the WHO. The problem is exacerbated by geography. Santiago lies in a bowl-shaped valley surrounded by the Andes to the east and the Chilean Coastal Range to the west. During winter, a layer of warm air holds the colder air close to the ground and causes high levels of smog and air pollution to be trapped, a phenomenon known as thermal inversion.

Hourly average PM2.5 concentrations for 11 monitoring stations in Santiago are shown in figure 2.1 for the period 2011 to 2014. Daily PM2.5 concentration measured by the Cerrillos weather station January 1, 2009 to December 31, 2014 will be the focus of the discussion in this chapter. This data series is plotted in figure 2.2. As is evident from the figures, lengthy periods of dangerously high levels of PM2.5 pollution are experienced every year. These depressing numbers mean that models of PM2.5 concentration have an important role to play in helping Chilean authorities to engage with the extent of the problem and in informing environmental policy.

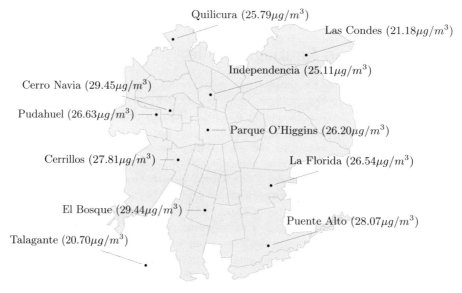

Figure 2.1. The 11 monitoring stations in Santiago, Chile. Values in parentheses indicate hourly average PM2.5 for the period 2011 to 2014 for each commune corresponding to the monitoring station.

2.1 *Air pollution in Santiago, Chile* 13

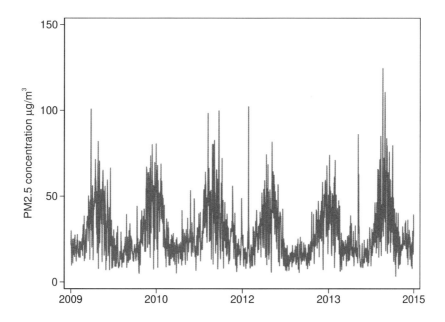

Figure 2.2. The concentration of PM2.5 particles measured in $\mu g/m^3$ at the Cerrillos weather station in Santiago, Chile. The data are daily for the period January 1, 2009 to December 31, 2014.

In addition to the concentration of PM2.5 particles, `pm25`, the dataset that will be used in this chapter also includes several other variables relating to air pollution. These are carbon monoxide levels (`co`), temperature (`temperature`), and wind speed (`vv`). Summary statistics for the data are as follows:

```
. use http://www.stata-press.com/data/eeus/pm_daily
. tabstat pm25 co temperature vv, statistics(N mean med sd min max)
> columns(statistics)

    variable |        N       mean        p50         sd        min        max
-------------+------------------------------------------------------------------
        pm25 |     2191   28.03872   23.50375   15.54959   3.333333   124.5833
          co |     2191   .7933811   .5746167   .6089451         .1    4.18425
 temperature |     2191   16.55248   16.71708   4.903541   4.833333   27.90542
          vv |     2191   2.129484   2.071917    .806073   .5186667   4.284083
```

Wind direction (`dv`) is also potentially important, but because it is measured in compass degrees (0°–360°) it is an example of a circular statistic (Cox 2009). Although wind direction is a continuous variable, it would make no sense to use its value (in terms of degrees of a circle) without recognizing the innate nature of the measurement. For instance, the simple average of a wind direction of 10° and 350° is 180°, representing almost precisely the opposite direction! The wind direction variable in the Cerrillos dataset has been treated by applying Nicholas Cox's (2004) command `circsummarize` from the `circular` package, available from the Statistical Software Components Archive, and the resultant variable `winddir` contains the compass direction from which the wind blows. As can be gleaned from the output, the prevailing wind in Santiago is from the south-southwest.

```
. tabstat dv, by(winddir) statistics(N min max)
Summary for variables: dv
     by categories of: winddir

 winddir |        N        min        max
---------+---------------------------------
     NNE |      104   4.709051   45.92259
     ENE |       54   46.63645   89.90435
     ESE |      103   93.25986   135.9989
     SSE |      317   136.0407   180.9678
     SSW |     1381   181.0414   225.9577
     WSW |      215   226.0288   270.0622
     WNW |       10   273.6462   314.1476
     NNW |        7    318.629   358.7118
---------+---------------------------------
   Total |     2191   4.709051   358.7118
```

2.2 Linear regression and OLS estimation

The fundamental aim of the linear regression model is to explain the movements in the dependent variable as a linear function of a set of explanatory variables. The population linear regression function is given by

$$y_t = \beta_0 + \beta_1 x_{1t} + \beta_2 x_{2t} + \cdots + \beta_K x_{Kt} + u_t \tag{2.1}$$
$$y_t = \boldsymbol{\beta}' \mathbf{x}_t + u_t$$

in which y_t is the dependent variable, $\boldsymbol{\beta} = [\beta_0 \ \beta_1 \ldots \beta_K]'$ is a $(K+1) \times 1$ vector of parameters, and $\mathbf{x}_t = [1 \ x_{1t} \ldots x_{Kt}]'$ is a $(K+1) \times 1$ vector of explanatory variables including the constant. For current purposes, both y_t and the regressors ($\mathbf{x}_t$) are assumed to have finite means and variances that are not explicit functions of time.

The disturbances, u_t, represent the movements in y_t that are not explained by the model. There are three important assumptions that the disturbances are assumed to satisfy:

$$E(u_t | \mathbf{x}_t) = 0 \tag{2.2}$$
$$E(u_t^2 | \mathbf{x}_t) = \sigma_u^2 \tag{2.3}$$
$$E(u_t u_s | \mathbf{x}_t \mathbf{x}_s) = 0, \quad t \neq s \tag{2.4}$$

Taken together, these assumptions imply that conditional on the regressors ($\mathbf{x}_t$) the disturbance terms are a sequence of serially uncorrelated random variables with zero mean and finite variance. If it is additionally assumed either that they are drawn from a normal distribution or that all their higher-order moments are identical, then the

2.2 Linear regression and OLS estimation

disturbance terms are said to be independently and identically distributed with mean zero and constant variance, σ_u^2.

The assumption in (2.2) is known as the zero-conditional mean assumption and has two important implications: that the unconditional expectation of the disturbance term is zero $[E(u_t) = 0]$ and that the regressors are orthogonal to the disturbance term $[E(u_t \mathbf{x}_t) = 0]$.[1] The zero-conditional mean assumption can be problematic because in most practical situations, some or all the regressors are themselves random variables, a point that is particularly important in the context of the environment in which observed outcomes are generated by a complex set of environmental interactions.

The sample counterpart of the population regression function is

$$y_t = \widehat{\boldsymbol{\beta}}' \mathbf{x}_t + \widehat{u}_t$$

where $\widehat{\boldsymbol{\beta}}$ is the estimator of the vector of population parameters $\boldsymbol{\beta}$ and $\widehat{u}_t$ is known as the residual. The three major estimation procedures that are used in the book are OLS, maximum likelihood, and the generalized method of moments. This chapter discusses OLS, and chapter 3 presents the other methods.

The OLS estimator of the parameters of the model is chosen so as to minimize the sum of squared disturbances given by

$$\text{RSS}(\boldsymbol{\beta}) = \sum_{t=1}^{T} u_t^2 = \sum_{t=1}^{T} (y_t - \boldsymbol{\beta}' \mathbf{x}_t)^2$$

Differentiating RSS with respect to $\boldsymbol{\beta}$ yields

$$\frac{\partial \text{RSS}}{\partial \boldsymbol{\beta}} = -2 \sum_{t=1}^{T} (y_t - \boldsymbol{\beta}' \mathbf{x}_t) \mathbf{x}_t \qquad (2.5)$$

The OLS estimator $\widehat{\boldsymbol{\beta}} = [\ \widehat{\beta}_0\ \widehat{\beta}_1\ \ldots\ \widehat{\beta}_K\ ]'$ of the parameters of the multiple regression model is obtained by setting the first-order conditions in (2.5) equal to zero and solving the resultant system of equations, which are collectively known as the *normal equations*. The solution is given by

$$\widehat{\boldsymbol{\beta}} = \left(\sum_{t=1}^{T} \mathbf{x}_t \mathbf{x}_t' \right)^{-1} \left(\sum_{t=1}^{T} \mathbf{x}_t y_t \right) \qquad (2.6)$$

Once the estimates $\{\widehat{\beta}_0, \widehat{\beta}_1, \ldots \widehat{\beta}_K\}$ are available, the estimated variance of the disturbances may be computed as

$$\widehat{\sigma}_u^2 = \frac{1}{T} \sum_{t=1}^{T} \widehat{u}_t^2 = \frac{1}{T} \sum_{t=1}^{T} \left(y_t - \widehat{\boldsymbol{\beta}}' \mathbf{x}_t \right)^2 \qquad (2.7)$$

---

1. For more details and derivations of these results, see Wooldridge (2016, 62–63, 76–77) or Hayashi (2000, 7–10).

Note that in computing $\widehat{\sigma}_u^2$ in (2.7), it is common to express the denominator in terms of the degrees of freedom, $T - K - 1$, instead of merely T, where K is the number of explanatory variables excluding the constant. In a small sample, the estimate given by (2.7) will be biased downward.

The precision of the OLS estimator is summarized by its covariance matrix, which, if the assumptions made on the regression disturbance term in (2.2)–(2.3) are satisfied, is given by

$$\text{VCE}\left(\widehat{\boldsymbol{\beta}}\right) = \left(\sum_{t=1}^{T} \mathbf{x}_t \mathbf{x}_t'\right)^{-1} \widehat{\sigma}_u^2 \tag{2.8}$$

in which the explicit conditioning on $\mathbf{x}_t$ has been suppressed for notational convenience. The standard errors of the parameters are the square roots of the diagonal entries of this matrix; the smaller the standard error, the more precise is the least-squares estimator of the population parameter. Standard errors play an important role in statistical tests of significance on the parameters of the regression model, as will become apparent shortly.

The Stata command **regress** (see [R] **regress**) computes the OLS estimators of the classical linear regression model. We fit the linear regression model for daily PM2.5 levels using carbon monoxide, temperature, wind direction, and wind speed as explanatory variables. Instead of simply allowing wind direction to enter the regression model as a single explanatory variable, the effect of the wind in each quadrant of the compass may be included in the regression, using Stata's factor-variable notation.[2]

```
. regress pm25 co temperature vv i.winddir

      Source |       SS           df       MS      Number of obs   =     2,191
-------------+----------------------------------   F(10, 2180)     =    538.37
       Model |  376903.035        10  37690.3035   Prob > F        =    0.0000
    Residual |  152616.779     2,180  70.007697   R-squared       =    0.7118
-------------+----------------------------------   Adj R-squared   =    0.7105
       Total |  529519.814     2,190  241.789869   Root MSE        =    8.3671

        pm25 |      Coef.   Std. Err.      t    P>|t|     [95% Conf. Interval]
-------------+----------------------------------------------------------------
          co |   22.02843   .4524855    48.68   0.000     21.14108    22.91578
 temperature |   .2932477   .0464253     6.32   0.000     .2022052    .3842903
          vv |  -1.088717   .3375641    -3.23   0.001    -1.750698   -.4267362
             |
     winddir |
         ENE |  -5.184654   1.457544    -3.56   0.000    -8.042975   -2.326333
         ESE |  -7.446572   1.268848    -5.87   0.000    -9.934849   -4.958294
         SSE |  -6.024663   1.017916    -5.92   0.000    -8.02085    -4.028475
         SSW |  -5.888658       .8539   -6.90   0.000    -7.563201   -4.214115
         WSW |  -5.926742   1.009067    -5.87   0.000    -7.905576   -3.947908
         WNW |  -5.368964   2.788711    -1.93   0.054    -10.83777    .0998464
         NNW |   -4.22259   3.286866    -1.28   0.199    -10.66831    2.223128
             |
       _cons |   13.70695   1.367645    10.02   0.000     11.02493    16.38898
```

---

2. See [U] **11.4.3 Factor variables** and appendix A.

Carbon monoxide is known to have a positive correlation with the concentration of PM2.5 because both pollutants are products of incomplete combustion and are major components of biomass smoke. Some studies have used CO as a proxy for PM2.5 given that it is easier to measure (see, for example, McCracken et al. [2013]). The sign and the strength of the relationship between CO and PM2.5 in this regression therefore comes as no surprise, but potential problems caused by the inclusion of CO in the regression are discussed in chapter 3. The relationships between fine particulate matter and meteorological variables have also received some attention. Tai, Mickley, and Jacob (2010) find that temperature, wind speed, wind direction, and precipitation all have a significant impact on PM2.5 and therefore imply that climate change will impact on air quality. The positive influence of temperature on PM2.5 concentration is similar to that found by Tai, Mickley, and Jacob (2010). The negative influence of wind speed, as a contrast with the excluded direction north-northeast, accords with intuition. Six of the possible seven wind directions enter the equation as having significantly different effects than a wind from the north-northeast.

2.3 Interpreting and assessing the regression model

2.3.1 Goodness of fit

The standard error of the regression is

$$\widehat{\sigma}_u = \sqrt{\frac{1}{T}\sum_{t=1}^{T}\widehat{u}_t^2} \qquad (2.9)$$

which is simply the large-sample standard deviation of the OLS residuals as given in (2.7). Stata uses the divisor $(T - K - 1)$ in the computation of the standard error of the regression. This adjustment reflects the fact that $K + 1$ degrees of freedom in the sample are expended by the prior estimation of the parameters $\{\widehat{\beta}_0, \widehat{\beta}_1, \ldots \widehat{\beta}_K\}$. In the `regress` output, the standard error of regression is labeled as `Root MSE` and in this case is 8.3671.

On its own, the `Root MSE` is difficult to interpret. A natural measure of the explanatory power of a fitted model is given by the proportion of the variation in the dependent variable explained by the model. This measure is given by the coefficient of determination

$$R^2 = \frac{\text{Explained sum of squares}}{\text{Total sum of squares}} = \frac{\sum_{t=1}^{T}(y_t - \overline{y})^2 - \sum_{t=1}^{T}\widehat{u}_t^2}{\sum_{t=1}^{T}(y_t - \overline{y})^2}$$

The coefficient of determination satisfies the inequality $0 \leq R^2 \leq 1$. Values close to unity suggest a good model fit, and values close to zero represent a poor fit. In this example, the regressors account for 71% of the variation in weekly PM2.5 values in Santiago.

A potential drawback with R^2 is that it never decreases when another variable is added to the model. From a statistical point of view, we should include those variables that significantly improve the explanatory power of the model. This is achieved by penalizing the R^2 statistic for the addition of extra parameters. The adjusted coefficient of determination, labeled by Stata as `Adj R-squared`, is given by

$$\overline{R}^2 = 1 - (1 - R^2)\frac{T-1}{T-K-1}$$

in which the adjustment factor $(T-1)/(T-K-1)$ becomes larger as K, the number of regressors, grows. This correction therefore represents a degrees-of-freedom correction to penalize the addition of additional variables that do not significantly increase R^2. From the algebra, $\overline{R}^2$ is always less than R^2, in this case marginally so at 71%.

Tests of significance

The aim of the regression model is to explain the variation in the dependent variable around its sample mean $\overline{y}$, using information on the explanatory variables $x_{1t}, x_{2t}, \ldots, x_{Kt}$. Consequently, for this information to be considered useful, the coefficients $\beta_1, \beta_2, \ldots, \beta_K$ associated with the explanatory variables must not all be zero. To investigate the validity of the model, we perform tests on the estimated point and interval values of these parameters individually and jointly.

To test the importance of a single explanatory variable in the regression equation, the associated parameter estimate is tested to see if it can be distinguished from zero using a two-tailed t test. The null and alternative hypotheses are, respectively,

$$H_0: \quad \beta_k = 0 \quad (x_k \text{ does not contribute to explaining } y_t)$$
$$H_1: \quad \beta_k \neq 0 \quad (x_k \text{ does contribute to explaining } y_t)$$

The t statistic to perform this test is defined as

$$t = \frac{\widehat{\beta}_k}{\text{se}\left(\widehat{\beta}_k\right)} \qquad (2.10)$$

where $\widehat{\beta}_k$ is the estimated coefficient of β_k and $\text{se}(\widehat{\beta}_k)$ is the corresponding estimated standard error. Under the null hypothesis, this test statistic follows the t distribution with $T - K - 1$ degrees of freedom, denoted t_{T-K-1}. In most practical situations, the distribution of t is simply taken to be the normal distribution because in large samples, where $T - K - 1 > 30$, the t distribution is indistinguishable from a normal distribution. The null hypothesis is rejected at the α significance level if the probability of rejecting a true null hypothesis, commonly known as the p-value, is smaller than this chosen level of significance. The decision rule is

$$p\text{-value} < \alpha: \quad \text{Reject } H_0 \text{ at the } \alpha \text{ level of significance}$$
$$p\text{-value} > \alpha: \quad \text{Fail to reject } H_0 \text{ at the } \alpha \text{ level of significance}$$

2.3.1 Goodness of fit

It is common to choose $\alpha = 0.05$ as the significance level, which means that the probability of rejecting the null hypothesis when it is true[3] is 5%.

In the output of `regress`, Stata provides the simple t tests of significance of the explanatory variables together with the p-value of the test in the column labeled P > |t|. The results of the PM2.5 regression show that carbon monoxide levels, temperature, and wind speed are all statistically significant in explaining the concentration of PM2.5 particles. For wind direction, five of the seven directions which appear in the daily data are found to be statistically significant: the west-northwest direction with a p-value of 0.054 and the north-northwest direction with a p-value of 0.199 are not statistically significant.

The t test in (2.10) is designed to determine the importance of an explanatory variable by considering whether its coefficient is zero. More general tests can be performed to test for any arbitrary value of β_1 by using the t statistic

$$t = \frac{\widehat{\beta}_1 - \beta_1}{\text{se}\left(\widehat{\beta}_1\right)}$$

This can be carried out in Stata by using the `test` (see [R] **test**) command. For instance, a test of the null hypothesis that the coefficient on wind speed, vv, is exactly -3, is

```
. test vv = -3
 ( 1)  vv = -3
       F(  1,  2180) =    32.06
            Prob > F =    0.0000
```

The result is given as an F statistic with one numerator degree of freedom, which is merely the square of the associated t statistic. The F and t forms of the statistic will have the same p-value and lead to the same conclusion, namely, that this restriction is not acceptable to the data.

We often want to test multiple restrictions on the coefficient vector, which leads to a joint hypothesis test. It is important to note that a joint test is not the box score of individual tests; for instance, we may have a model in which each regressor has a t statistic below the critical value, but the overall model is quite informative. Thus, we must consider how such a joint test can be performed.

In the case of the multiple regression model with K explanatory variables, a joint test of the significance of all the explanatory variables is based on the null and alternative hypotheses

$$H_0: \beta_1 = \beta_2 = \cdots = \beta_K = 0$$
$$H_1: \text{at least one } \beta_k \text{ is not zero}$$

---

3. Rejecting a true null hypothesis is known as a type 1 error. Failing to reject the null hypothesis when it is false is referred to as a type 2 error.

Notice that this test does not include the intercept parameter β_0, so the total number of restrictions is K. Two tests of the null hypothesis are the large-sample χ^2 test

$$\chi_K^2 = \frac{R^2}{(1-R^2)/(T-K-1)}$$

which is distributed as χ^2 with K degrees of freedom. Applying a finite-sample correction to the χ^2 statistic and dividing by K yields the F test

$$F_{K,T-K-1} = \frac{R^2/K}{(1-R^2)/(T-K-1)}$$

which, under the null, follows an F distribution with K degrees of freedom in the numerator and $(T-K-1)$ degrees of freedom in the denominator. This is often termed the analysis of variance F test because it is based on the analysis of variance decomposition of y_t into that part explained by the model and that part left unexplained, the residual sum of squares (RSS). When you use `regress`, the $F(K, T-K-1)$ statistic is provided along with its degrees of freedom and p-value.

Note that subset F tests, where the significance of some of the regressors, such as a set of indicator variables, is jointly tested, will often be used. This can be done after a regression in Stata using the `test` or `testparm` command (see [R] test), where the regressors to be challenged are listed on the command line. The null hypothesis is that each regressors' coefficient cannot be distinguished from zero. For example, to test the joint significance of wind direction, the `testparm` command can be used:

```
. testparm i.winddir
 ( 1)  2.winddir = 0
 ( 2)  3.winddir = 0
 ( 3)  4.winddir = 0
 ( 4)  5.winddir = 0
 ( 5)  6.winddir = 0
 ( 6)  7.winddir = 0
 ( 7)  8.winddir = 0
       F(  7,  2180) =    7.44
            Prob > F =    0.0000
```

The null hypothesis that all the coefficients are zero is strongly rejected despite the fact that two of the coefficients are indistinguishable from zero. When a qualitative factor is expressed as a set of indicators, as wind direction is in this model, the F test should always be used to determine whether the factor adds to the explanatory power of the model.

2.3.2 Residual diagnostics

Having explored the basic specification and estimation of a linear regression, we now focus on the evaluation of the fitted model beyond the simple assessment of the statistical significance of coefficients and its goodness of fit. Several residual diagnostics are

2.3.2 Residual diagnostics

commonly used to evaluate the validity of the basic assumptions made on the regression disturbances.

The residual series `uhat` from the basic regression of PM2.5 on carbon monoxide, temperature, wind speed, and wind direction is computed with the `predict` command (see [R] **predict**) using its `residuals` option, and plotted in figure 2.3.

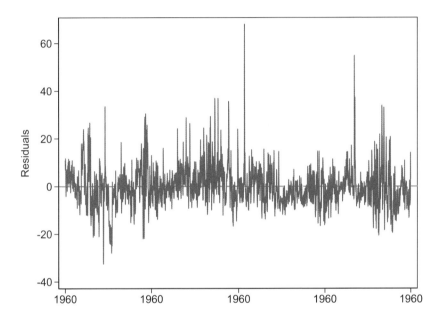

Figure 2.3. Regression residuals from daily PM2.5 data

Homoskedasticity

The most commonly used diagnostic tests for heteroskedasticity in the disturbance process are those of Breusch and Pagan (1979) and their variant by Cook and Weisberg (1983). These tests may be applied when using cross-sectional data or when modeling time-varying volatility, a topic that is dealt with in chapter 11. These tests are based on an auxiliary regression in which the squared residuals are regressed on any variables that might be considered to influence the presence of heteroskedasticity. Their null and alternative hypotheses are, respectively,

$$
\begin{aligned}
H_0: &\quad \text{Homoskedasticity:} \quad (\sigma_u^2 \text{ is constant}) \\
H_1: &\quad \text{Heteroskedasticity:} \quad (\sigma_u^2 \text{ is time-varying})
\end{aligned}
$$

Rejection of the null hypothesis of homoskedasticity will bias the standard errors of the estimated coefficients of the regression model and any hypothesis tests based on the variance–covariance matrix. This will lead to false inference concerning the significance of the explanatory variables.

Under the null hypothesis, the regression of squared residuals on any set of explanatory variables should have no explanatory power. The test statistic is computed as TR^2, where R^2 comes from this auxiliary regression, and it is distributed as χ^2_M under the null hypothesis, where M is the number of variables included in the auxiliary regression. The original version of the Breusch–Pagan/Cook–Weisberg test assumes that the regression disturbances are normally distributed. After a regression, the command estat hettest (see [R] estat hettest) performs this test using the fitted values ($\widehat{y}_t$) as the sole auxiliary regressor. The command may also be given as estat hettest varlist, where a list of possibly influential variables is given. The variables need not be restricted to those included in the regression equation. For example, heteroskedasticity is often related to scale, so a measure of scale could be included even if it does not explicitly appear in the regression.

In White's (1980) form of the test, the variables are chosen as the original regressors, their squares and cross-products, dropping redundant elements (for instance, the square of an indicator). That test may be invoked with estat imtest, white. One drawback of the White test is that if there are many regressors, the test uses many degrees of freedom, with a potential loss of power.

```
. estat hettest

Breusch-Pagan / Cook-Weisberg test for heteroskedasticity
         Ho: Constant variance
         Variables: fitted values of pm25

         chi2(1)      =     223.60
         Prob > chi2  =     0.0000

. estat imtest, white

White's test for Ho: homoskedasticity
         against Ha: unrestricted heteroskedasticity

         chi2(37)     =     175.88
         Prob > chi2  =     0.0000

Cameron & Trivedi's decomposition of IM-test
```

Source	chi2	df	p
Heteroskedasticity	175.88	37	0.0000
Skewness	28.07	10	0.0018
Kurtosis	5.05	1	0.0246
Total	208.99	48	0.0000

The tests strongly reject the null hypothesis of homoskedasticity.

2.3.2 Residual diagnostics

Serial independence

A test for serial correlation, or autocorrelation, is important when using time-series data. The aim of the test is to detect if the disturbance process is autocorrelated, that is, if successive errors are systematically related, with the disturbance term related to previous disturbance terms. The null and alternative hypotheses are, respectively,

$$H_0: \quad u_t \text{ are independently distributed}$$
$$H_1: \quad u_t \text{ exhibit serial correlation}$$

Rejection of the null hypothesis should be taken seriously because it often reflects a model specification problem such as the omission of relevant variables or inadequate dynamics. Furthermore, the presence of positive first-order serial correlation will bias the standard errors of estimated coefficients downward, providing a false indication of their significance. Serial correlation may appear between u_t, u_{t-1}, or first-order autocorrelation, but it may also appear at higher orders. For instance, in daily data, there may be neglected day-of-week effects that show up as correlations between u_t and u_{t-7}. Thus, useful tests for autocorrelation should consider the possibility that higher-order autocorrelations may be significant.

A test for a general form of the null hypothesis was developed by Box and Jenkins (1970) and refined by Ljung and Box (1978): the Q statistic, or portmanteau test, as provided by Stata's `wntestq` command (see [TS] **wntestq**). This test may be applied to any time series, but if applied to regression residuals, the model must contain strictly exogenous regressors. This rules out, for instance, the presence of predetermined regressors such as lagged dependent variables. An extension to this test was developed by Breusch (1978) and Godfrey (1978). The residuals are regressed on p of their own lags as well as the original regressors. The regressors are, of course, uncorrelated with the residuals at time t but may well be correlated with lagged residual series, adding power to the test. The Breusch–Godfrey test for p lags has the null that the first p autocorrelations of the residuals are jointly zero. It may be performed after a regression in Stata using the command `estat bgodfrey` with the option `lag(p)`. To test for the presence of higher-order serial correlation, given that there is serial correlation at the first, second, etc. order, the appropriate test is that of Cumby and Huizinga (1992). Their test, which generalizes the Breusch–Pagan test, is available as `actest` (Baum and Schaffer 2004) from the Statistical Software Components Archive.

```
. estat bgodfrey, lag(1/6)
Breusch-Godfrey LM test for autocorrelation
```

lags(p)	chi2	df	Prob > chi2
1	979.047	1	0.0000
2	983.743	2	0.0000
3	990.682	3	0.0000
4	997.999	4	0.0000
5	1000.796	5	0.0000
6	1015.840	6	0.0000

H0: no serial correlation

```
. actest, lags(6) robust
Cumby-Huizinga test for autocorrelation
  H0: variable is MA process up to order q
  HA: serial correlation present at specified lags >q
```

H0: q=0 (serially uncorrelated) HA: s.c. present at range specified				H0: q=specified lag-1 HA: s.c. present at lag specified			
lags	chi2	df	p-val	lag	chi2	df	p-val
1 - 1	291.939	1	0.0000	1	291.939	1	0.0000
1 - 2	295.185	2	0.0000	2	84.747	1	0.0000
1 - 3	296.178	3	0.0000	3	39.027	1	0.0000
1 - 4	297.673	4	0.0000	4	26.554	1	0.0000
1 - 5	301.415	5	0.0000	5	26.505	1	0.0000
1 - 6	308.392	6	0.0000	6	27.709	1	0.0000

Test allows predetermined regressors/instruments
Test robust to heteroskedasticity

Once more, the results of the tests are unequivocal. The regression residuals are autocorrelated, indicating that the model is missing something systematic in the trajectory of the PM2.5 series.

Normality

The assumption that the regression disturbances are normally distributed has not been relied upon and is not strictly necessary in terms of establishing the properties of the OLS estimators. Notwithstanding this fact, empirical studies will often provide tests of the normality of the regression residuals, $\hat{u}_t$. Figure 2.4 plots a histogram of the regression residuals with the best fitting normal distribution overlaid. Initial inspection suggests that the normality assumption is not satisfied in this particular case.

2.3.2 Residual diagnostics

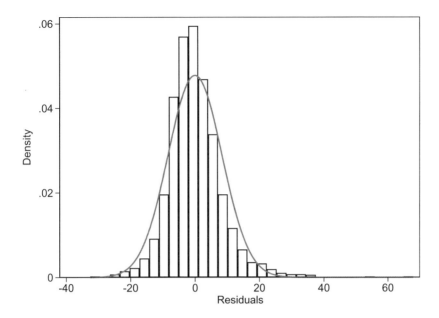

Figure 2.4. Histogram of the regression residuals from the PM2.5 regression with the best fitting normal distribution overlaid.

Two commonly used tests are the Jarque–Bera and the Doornik–Hansen tests.[4] The null and alternative hypotheses for these tests are, respectively,

$$H_0: \quad u_t \text{ is normally distributed}$$
$$H_1: \quad u_t \text{ is not normally distributed}$$

The Jarque and Bera (1987) test statistic is

$$\text{JB} = T\left(\frac{\text{SK}^2}{6} + \frac{(\text{KT} - 3)^2}{24}\right)$$

where T is the sample size. The sample skewness and kurtosis of the least-squares residuals, denoted by SK and KT, respectively, are given by

$$\text{SK} = \frac{1}{T}\sum_{t=1}^{T}\left(\frac{\widehat{u}_t}{\widehat{\sigma}_u}\right)^3 \qquad \text{KT} = \frac{1}{T}\sum_{t=1}^{T}\left(\frac{\widehat{u}_t}{\widehat{\sigma}_u}\right)^4$$

and $\widehat{\sigma}$ is the standard error of the regression in (2.9). The Jarque and Bera (1987) statistic is distributed as χ^2 with 2 degrees of freedom under the null hypothesis. The

---

4. Various alternative tests appear in the literature, such as those implemented by the Stata commands **swilk** and **sfrancia** (see [R] **swilk**). If you are using these tests, be careful to ensure that their constraints on sample size are satisfied.

Doornik and Hansen (2008) statistic can be applied to a single series or to a set of series to test for multivariate normality. In the context of a single series, the univariate skewness and kurtosis values are computed and transformed into standard normal variates, as described in `mvtest normality` (see [MV] **mvtest normality**).

```
. predict double uhat, resid
. quietly summarize uhat, detail
. scalar sk = r(skewness)^2/6
. scalar kt = (r(kurtosis)-3)^2/24
. scalar JB = r(N)*(sk+kt)
. display as text "JB test statistic   = " as res JB
JB test statistic   =  2177.0342
. display as text "Prob > chi2   =  "  as res chi2tail(2, JB)
Prob > chi2 = 0
. mvtest normality uhat
Test for multivariate normality
        Doornik-Hansen                  chi2(2) =   365.980   Prob>chi2 =   0.0000
```

Both of the tests indicate a strong rejection of the normality of the underlying disturbance process. A normal quantile plot (`qnorm`) is also recommended as a diagnostic tool. Given the clarity of these test results, the plot is not presented here.

2.4 Estimating standard errors

Violation of the assumptions of the classical linear regression model can have serious implications for the estimated parameters and their standard errors. In particular, if the assumptions of homoskedasticity (2.3), independence (2.4), or both are violated, the point estimates of the OLS estimators remain consistent, but it is always the case that the classical standard errors are no longer consistent and must be adjusted to account for these violations.

Equation (2.6), defining the solution to the OLS problem, can be decomposed as

$$\widehat{\boldsymbol{\beta}} = \left(\sum_{t=1}^{T} \mathbf{x}_t \mathbf{x}_t'\right)^{-1} \sum_{t=1}^{T} \mathbf{x}_t y_t = \boldsymbol{\beta} + \left(\sum_{t=1}^{T} \mathbf{x}_t \mathbf{x}_t'\right)^{-1} \sum_{t=1}^{T} \mathbf{x}_t u_t$$

where the last term is obtained by substituting for y_t from (2.1). Conditional on $\mathbf{x}_t$, the covariance matrix of $\widehat{\boldsymbol{\beta}}$ is given by

$$\text{VCE}\left(\widehat{\boldsymbol{\beta}}\right) = \text{var}\left\{\left(\sum_{t=1}^{T} \mathbf{x}_t \mathbf{x}_t'\right)^{-1} \sum_{t=1}^{T} \mathbf{x}_t u_t\right\}$$

This equation may be rewritten as

$$\text{VCE}\left(\widehat{\boldsymbol{\beta}}\right) = E\left(\sum_{t=1}^{T} \mathbf{x}_t \mathbf{x}_t'\right)^{-1} E\left\{\left(\sum_{t=1}^{T} \mathbf{x}_t u_t\right)\left(\sum_{t=1}^{T} \mathbf{x}_t u_t\right)'\right\} E\left(\sum_{t=1}^{T} \mathbf{x}_t \mathbf{x}_t'\right)^{-1} \quad (2.11)$$

2.4 Estimating standard errors

The term in the square brackets on the right-hand side of (2.11) may be expressed as the double summation

$$\left(\sum_{t=1}^{T} \mathbf{x}_t u_t\right)\left(\sum_{t=1}^{T} \mathbf{x}_t u_t\right)' = E\left(\sum_{t=1}^{T}\sum_{s=1}^{T} \mathbf{x}_t u_t u_s \mathbf{x}_s'\right)$$

This double summation contains exactly T elements in which the indices t and s coincide so that the following decomposition is possible:

$$E\left(\sum_{t=1}^{T}\sum_{s=1}^{T} \mathbf{x}_t u_t u_s \mathbf{x}_s'\right) = \left\{\sum_{t=1}^{T} E\left(u_t^2 \mathbf{x}_t \mathbf{x}_t'\right) + \sum_{t=1}^{T}\sum_{\substack{s=1\\s\neq t}}^{T} E\left(\mathbf{x}_t u_t u_s \mathbf{x}_s'\right)\right\} \quad (2.12)$$

There are now four cases to consider.

Case 1: u_t independent and identically distributed

If u_t satisfies homoskedasticity $[E(u_t^2|\mathbf{x}_t) = \sigma^2]$ and independence $[E(u_t u_s|\mathbf{x}_t \mathbf{x}_s) = 0$ for $t \neq s]$ then the cross-product terms on the right-hand side of expression (2.12) disappear, and the estimated covariance matrix of $\widehat{\boldsymbol{\beta}}$ is given by the familiar expression

$$\widehat{\text{VCE}}\left(\widehat{\boldsymbol{\beta}}\right) = \left(\sum_{t=1}^{T} \mathbf{x}_t \mathbf{x}_t'\right)^{-1} \widehat{\sigma}^2 \left(\sum_{t=1}^{T} \mathbf{x}_t \mathbf{x}_t'\right)\left(\sum_{t=1}^{T} \mathbf{x}_t \mathbf{x}_t'\right)^{-1} = \widehat{\sigma}^2 \left(\sum_{t=1}^{T} \mathbf{x}_t \mathbf{x}_t'\right)^{-1}$$

This expression is the now familiar form for the $\text{VCE}(\widehat{\boldsymbol{\beta}})$ given in (2.8).

Case 2: u_t independent but not identically distributed

If u_t is heteroskedastic $[E(u_t^2|\mathbf{x}_t) = \sigma_t^2]$ but still independent $[E(u_t u_s|\mathbf{x}_t \mathbf{x}_s) = 0$ for $t \neq s]$ then all the cross-product terms on the right-hand side of (2.12) disappear once more, but the simplification is not quite as neat. White (1980) derives a heteroskedasticity-consistent covariance matrix estimator that provides consistent estimates of the covariance matrix of the OLS estimator. The White estimator is given by

$$\widehat{\text{VCE}}\left(\widehat{\boldsymbol{\beta}}\right) = \left(\sum_{t=1}^{T} \mathbf{x}_t \mathbf{x}_t'\right)^{-1} \left(\sum_{t=1}^{T} \widehat{u}_t^2 \mathbf{x}_t \mathbf{x}_t'\right)\left(\sum_{t=1}^{T} \mathbf{x}_t \mathbf{x}_t'\right)^{-1}$$

White's robust standard errors are available in Stata by using the vce(robust) option after regress or most other estimation commands.

Case 3: u_t nonidentically and nonindependently distributed

In this case, u_t is assumed to be both heteroskedastic $[E(u_t^2|\mathbf{x}_t) = \sigma_t^2]$ and not independent $[E(u_t u_s|\mathbf{x}_t \mathbf{x}_s) \neq 0$ for $t \neq s]$. In this case, the cross-product terms on the right-hand side of (2.12) do not disappear. The relevant expression for the covariance matrix of $\widehat{\boldsymbol{\beta}}$ becomes

$$\widehat{\text{VCE}}\left(\widehat{\boldsymbol{\beta}}\right) = \left(\sum_{t=1}^{T} \mathbf{x}_t \mathbf{x}_t'\right)^{-1} E\left(\sum_{t=1}^{T}\sum_{s=1}^{T} \mathbf{x}_t \widehat{u}_t \widehat{u}_s \mathbf{x}_s'\right)\left(\sum_{t=1}^{T} \mathbf{x}_t \mathbf{x}_t'\right)^{-1}$$

The problem with implementing this expression as it stands is that implementing more terms in the summation as $T \to \infty$ will decrease the precision of each term in the expansion and hence also the precision of the estimator. On the other hand, truncating the sum too soon will yield an inconsistent estimator of the covariance matrix.

The most common method used to estimate the troublesome double summation is that suggested by Newey and West (1987). The double summation can be expressed in a slightly different form by reindexing the sum so that the summation is taken over the product $u_j u_k$ such that terms in which j and k are a fixed distance apart are collected together. Truncating the maximum number of lagged terms to include in the estimator at P, the term on the left-hand side of (2.12) becomes

$$\left(\sum_{t=1}^{T}\sum_{s=1}^{T} \mathbf{x}_t \widehat{u}_t \widehat{u}_s \mathbf{x}_s'\right) = \sum_{p=0}^{P}\sum_{k=1}^{T-p} \omega_p \widehat{u}_k \widehat{u}_{k+p} (\mathbf{x}_k \mathbf{x}_{k-p}' + \mathbf{x}_{k-p} \mathbf{x}_t')$$

in which the final term involving the vector $\mathbf{x}_t$ is required because, unlike the scalar case, $\mathbf{x}_k \mathbf{x}_{k-p}' \neq \mathbf{x}_{k-p} \mathbf{x}_t'$. The ω_p are weights chosen to ensure the optimal performance of the estimator. Newey and West suggest the weights (ω_p) and the maximum lag (P) be chosen as

$$\omega_n = 1 - \frac{p}{P+1}$$
$$P = \text{int}\left(\frac{3}{4} T^{1/3}\right)$$

in which int($\cdot$) refers to the integer part of the argument. This weighting scheme is known as the Bartlett kernel and is guaranteed to produce a positive definite VCE estimate.

Standard errors computed this way to allow for both heteroskedastic and nonindependent regression disturbances, known as heteroskedasticity- and autocorrelation-consistent (HAC) standard errors, are available in Stata by using the newey command (see [TS] **newey**) rather than regress. Stata's ivregress command (see [R] **ivregress**) also implements HAC standard errors through its vce() option.

Case 4: u_t correlated within clusters and not identically distributed

In many empirical applications, disturbances are often considered to be correlated within clusters of observations but independently distributed across clusters. These clusters may be motivated by geography; for example, cities in the same region may be subject

2.4 Estimating standard errors

to the same climatic patterns. Alternatively, the clusters may be motivated by behavior, such as individuals living in the same neighborhood or firms operating in the same industry. The cluster–robust VCE allows arbitrary correlations of disturbances within clusters as well as arbitrary heteroskedasticity across clusters.[5] The within-cluster correlation of errors can arise if the errors are not independently and identically distributed but rather contain a common shock component as well as an idiosyncratic component.

The cluster–robust VCE of u_t can then be expressed as

$$\boldsymbol{\Sigma}_u = \begin{pmatrix} \boldsymbol{\Sigma}_1 & & & & 0 \\ & \ddots & & & \\ & & \boldsymbol{\Sigma}_g & & \\ & & & \ddots & \\ 0 & & & & \boldsymbol{\Sigma}_G \end{pmatrix}$$

In this notation, $\boldsymbol{\Sigma}_g$ represents an intracluster covariance matrix. For cluster (group) g with τ observations, $\boldsymbol{\Sigma}_g$ will be the $\tau \times \tau$ matrix

$$\boldsymbol{\Sigma}_g = \begin{bmatrix} \sigma_1^2 & \cdots & \sigma_{1\tau} \\ \vdots & \ddots & \vdots \\ \sigma_{\tau 1} & \cdots & \sigma_\tau^2 \end{bmatrix}$$

Zero covariance between observations in the G different clusters gives the covariance matrix $\boldsymbol{\Sigma}_u$ a block-diagonal form.

If the within-cluster correlations are meaningful, ignoring them leads to inconsistent estimates of the VCE. Ignoring the intracluster correlation will bias the standard errors downward, so we might expect cluster–robust standard errors to be larger than their OLS counterparts if these correlations are meaningful. Because the vce(robust) estimate of the VCE ignores these correlations, its estimate of the VCE is not valid in the presence of clustering.

The cluster–robust estimate of the VCE takes the form

$$\widehat{\text{VCE}}\left(\widehat{\boldsymbol{\beta}}\right) = \left(\sum_{t=1}^T \mathbf{x}_t \mathbf{x}_t'\right)^{-1} \left(\sum_{j=1}^G \sum_{t=1}^T \sum_{s=1}^T \mathbf{x}_t \widehat{u}_t \widehat{u}_s \mathbf{x}_s' I_j\right) \left(\sum_{t=1}^T \mathbf{x}_t \mathbf{x}_t'\right)^{-1}$$

where I_j is an indicator function that takes the value 1 if the current u_t is in group j and zero otherwise. Stata's vce(cluster *varname*) option, available on most estimation commands including regress, allows such an error structure to be accounted for. Like the vce(robust) option, which it encompasses, application of the cluster option does not affect the point estimates but only modifies the estimated VCE of the estimated

---

5. In the context of panel, or longitudinal, data, which are dealt with in chapter 12, the natural clusters are the individual units, but the natural clustering is harder to determine in time-series data.

parameters. The vce(cluster *varname*) option requires that a group- or cluster-membership variable be specified that indicates how the observations are grouped.

The various estimates of precision are now computed for the linear regression using the daily Cerrillos data on PM2.5 concentration, temperature, wind speed, and wind direction. The results are reported in table 2.1. For the newey estimator, seven lags are specified for these daily data. For the cluster–robust VCE, clusters correspond to each calendar month.

Table 2.1. Implementing robust, HAC, and clustered standard errors for the simple regression model of PM2.5 in Santiago, Chile

	OLS	Robust	HAC	Cluster
co	22.028	22.028	22.028	22.028
	(0.452)	(0.608)	(0.905)	(0.949)
temperature	0.293	0.293	0.293	0.293
	(0.046)	(0.046)	(0.067)	(0.084)
vv	−1.089	−1.089	−1.089	−1.089
	(0.338)	(0.328)	(0.443)	(0.382)
ENE	−5.185	−5.185	−5.185	−5.185
	(1.458)	(1.895)	(2.076)	(2.223)
ESE	−7.447	−7.447	−7.447	−7.447
	(1.269)	(1.522)	(2.123)	(1.866)
SSE	−6.025	−6.025	−6.025	−6.025
	(1.018)	(1.231)	(1.807)	(0.989)
SSW	−5.889	−5.889	−5.889	−5.889
	(0.854)	(1.019)	(1.729)	(1.333)
WSW	−5.927	−5.927	−5.927	−5.927
	(1.009)	(1.170)	(1.825)	(1.338)
WNW	−5.369	−5.369	−5.369	−5.369
	(2.789)	(3.571)	(3.959)	(3.787)
NNW	−4.223	−4.223	−4.223	−4.223
	(3.287)	(4.708)	(4.689)	(4.702)
R^2	0.712	0.712		0.712

All four forms of the fitted model display the same point estimates of the coefficients and the same R^2 measure. Relaxing the assumption of independence, as we do in using newey and the cluster–robust VCE estimator, leads to much larger standard errors (lower t statistics) for the estimated coefficients, suggesting that the classical standard errors and robust standard errors are biased downward. These results are strongly suggestive that the residuals obtained from the linear model are not independently and identically distributed conditional on the explanatory variables, $\mathbf{x}_t$. This theme will be revisited in more detail in chapter 4.

Exercises

1. `spatialfish_small.dta` contains cross-sectional data on the compliance of 51 signatory countries with the voluntary obligations of Article 7 of the 1995 United Nations Code of Conduct for Responsible Fisheries. The dependent variable is `all_mean`, which ranges from 0 to 10, with higher scores indicating better compliance, and is made up of score variables that relate to the `behavior` and `intention` of signatory countries. Explanatory variables include the logarithm of per capita gross domestic product (`gpd`), distance from the fishing zone (`cost`), the degree of competition in the fishing zone (`compet`), the logarithm of the value of fish exports relative to gross domestic product (`export`), the fraction of catch in the country's exclusion zone (`eezcatch`), an index of biodiversity measuring the health of the environment (`bio`), and total involvement in 443 other international agreements (`treaty`). The dummy variable `eu_dum` indicates whether a country is a member of the European Union (EU).

    a. Rename the variable `all_mean` as `overall`, and rename `distance` as `cost`.
    b. Provide summary statistics (number of observations, mean, standard deviation, maximum values, and minimum values) for `overall`, `behavior`, and `intention` by whether they are members of the EU. Examine whether there is any evidence to suggest that EU members are more compliant with the Code of Conduct than non-EU members by performing a one-sample t test on `overall`, `behavior`, and `intention`.
    c. Run a regression of `overall` on `gdp`, `cost`, `compet`, `export`, `eezcatch`, `bio`, `treaty`, and `eu_dum`. Interpret your results, and comment on whether they accord with your intuition.
    d. For the regression model in exercise 1c, test for the significance of `eu_dum`. What do you conclude? Is this conclusion altered by changing the dependent variable to `behavior` or `intention`?
    e. Compute the residuals from the regression using `overall` as the dependent variable. Draw a histogram of the residuals with a normal distribution overlaid. Provide a formal test of normality.
    f. Test the residuals for heteroskedasticity using the Breusch–Pagan test.

2. `stern&kaufmann2014.dta` contains annual time-series data on global temperature anomalies (the variables `hadcrut4` and `gissv3`) and other variables relating to climate change for the period 1850 to 2011. The term temperature anomaly means a departure from the reference value or long-term average, in this case the 1951–1980 mean temperature. The Stata `notes` command will display notes on the individual series. Global temperature is determined by the balance between solar radiation (incoming) and infrared radiation (outgoing). This balance is disturbed by both natural `rfnat` and anthropogenic `rfanth` emissions of gases in a process known as radiative forcing.

    a. Plot the global temperature anomalies, and comment on the properties of the data.

b. Regress the `hadcrut4` temperature anomaly on a constant, a deterministic trend, and the radiative forcing variables `rfnat` and `rfanth`. Interpret your results.

c. Provide a time-series plot of the actual and fitted values of the regression.

d. Test the residuals for autocorrelation (of up to two lags) and heteroskedasticity. What do you conclude?

e. Now, regress the `hadcrut4` temperature anomaly on a constant, a deterministic trend, and natural radiative forcing `rfnat`, but instead of using a summary measure for anthropogenic forcing, substitute the greenhouse cases, carbon dioxide (`rfco2`), methane (`rfch4`), sulfur dioxide (`rfsox`), and black carbon (`rfbc`). Interpret your results. Which model do you prefer?

f. Test the residuals of the model in exercise 2e for autocorrelation (of up to two lags) and heteroskedasticity. What do you conclude? Adjust the standard errors of the model in exercise 2e in the light of your results, and comment on your results.

3 Beyond ordinary least squares

This chapter introduces two other approaches to estimating the parameters of the linear model. Estimation by maximum likelihood and generalized method of moments (GMM) will be used throughout the book, so a brief introduction to each of these methods is given here in the context of linear regression. The real power of these methods stems from their ability to handle nonlinear estimation problems, but the basic concepts relating to their implementation can be explained straightforwardly in terms of linear regression.

Under certain technical conditions, the maximum likelihood estimator possesses several important and desirable statistical properties, which makes it the benchmark against which all other estimators are measured. In addition, the principle of maximum likelihood provides the framework for three important types of tests in econometrics, namely, the likelihood-ratio (LR), the Wald (W), and the Lagrange multiplier (LM) tests. Consequently, a brief treatment of maximum likelihood is important. Method of moments estimators have become increasingly important recently because the informational requirements on which they are based are less demanding than for maximum likelihood. In addition, the use of moment-based estimators guards against incorrectly designating variables as exogenous.

3.1 Distribution of particulate matter

The discussion in chapter 2 concerning the concentration of PM2.5 measured in Santiago, Chile, focused on the time-series properties of the data. Consider now the link between the observed sample data and the probability distribution from which they are drawn. The most often used distributional assumption is the normal distribution. Let the observed sample of PM2.5 be denoted y_t for notational simplicity. Then, assuming normality means that

$$f(y_t; \mu, \sigma^2) = \frac{1}{\sqrt{2\pi\sigma^2}} \exp\left(-\frac{(y_t - \mu)^2}{2\sigma^2}\right)$$

A better approximation to the distribution of these strictly positive data may be given by the lognormal distribution given by

$$f(y_t; \mu, \sigma^2) = \frac{1}{\sqrt{2\pi\sigma^2} y_t} \exp\left(-\frac{(\log y_t - \mu)^2}{2\sigma^2}\right)$$

in which case it is the logarithm of y_t that follows a normal distribution. This distribution has the added advantage that it guarantees positive values.

The histogram of daily PM2.5 concentration measured at the Cerrillos weather station in Santiago, Chile is shown in figure 3.1, overlaid with the best-fitting normal and lognormal distributions. The normal distribution is not a particularly good fit and assigns a fair amount of density into the negative region, where there are no data. The lognormal distribution does a much better job of picking up the skewness and heavy right tail of the PM2.5 data, although it does understate the peak of the distribution somewhat.

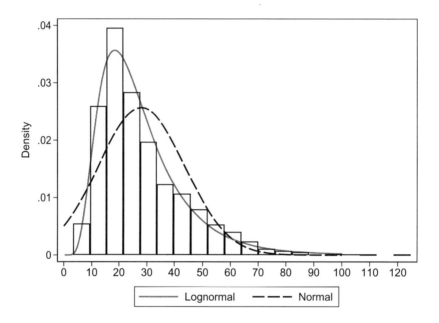

Figure 3.1. The distribution of daily PM2.5 particles measured in $\mu g/m^3$ at the Cerrillos weather station in Santiago, Chile. The data are daily for the period January 1, 2009 to December 31, 2014.

The two estimation methods that will be introduced in this chapter, namely, GMM and maximum likelihood estimation, take differing views of the importance of the distribution of the data. For example, it is obvious from figure 3.1 that the lognormal distribution provides a better fit of the distribution of particulate matter than the normal distribution. But the question of whether the lognormal distribution is in fact the correct distribution to use is unanswered. Often, theory provides little or no guidance on the specification of the distribution or relevant variables. Consequently distributional assumptions about the underlying stochastic processes are by necessity ad hoc. The particular advantage of method of moments estimators is that they only require knowledge of the set of moment conditions characterizing the model and do not require the entire distribution of the dependent variable to be known. In contrast, maximum likelihood estimation requires that the distribution of the dependent variable be specified and can be computed for all possible values of the parameters of the model.

3.2 Properties of estimators

The properties of estimators fall into two categories. In small samples, they are known as finite sample or exact properties. Typically, these properties are difficult to establish except for simple linear cases. Consequently, attention is often focused on the limiting properties of the estimator because the sample size increases indefinitely—the asymptotic properties. Given the sample sizes typically encountered in environmental econometrics, the focus will be on the latter.[1] The discussion here will be relatively informal. For more details, see Hamilton (1994), Hayashi (2000), or Martin, Hurn, and Harris (2013).

Recall from chapter 2 that the ordinary least-squares estimator can be decomposed to give

$$\widehat{\boldsymbol{\beta}} = \left(\sum_{t=1}^{T} \mathbf{x}_t \mathbf{x}_t'\right)^{-1} \sum_{t=1}^{T} \mathbf{x}_t y_t = \boldsymbol{\beta} + \left(\sum_{t=1}^{T} \mathbf{x}_t \mathbf{x}_t'\right)^{-1} \sum_{t=1}^{T} \mathbf{x}_t u_t \qquad (3.1)$$

This expression for $\widehat{\boldsymbol{\beta}}$ is a useful one for discussing the properties of the estimators.

Consistency

Consistency refers to the distance between $\widehat{\boldsymbol{\beta}}$ and $\boldsymbol{\beta}$ as the sample size T increases. A consistent estimator is one in which more information, in terms of a larger sample size, results in it being closer to the true population parameter $\boldsymbol{\beta}$. It is important to understand that even though $\widehat{\boldsymbol{\beta}}$ approaches, $\boldsymbol{\beta}$ it never quite reaches $\boldsymbol{\beta}$ deterministically because $\widehat{\boldsymbol{\beta}}$ is a random variable and not a deterministic process. This behavior as T grows is known as convergence in probability, and the constant to which the estimator converges in probability is found by taking the probability limit or "plim" of the estimator. For $\widehat{\boldsymbol{\beta}}$, consistency is established by taking the probability limit of the right-hand side of (3.1) as follows:

$$\text{plim}\left(\widehat{\boldsymbol{\beta}}\right) = \boldsymbol{\beta} + \left(\text{plim}\ \frac{1}{T}\sum_{t=1}^{T} \mathbf{x}_t \mathbf{x}_t'\right)^{-1} \left(\text{plim}\ \frac{1}{T}\sum_{t=1}^{T} \mathbf{x}_t u_t\right) = \boldsymbol{\beta}$$

The last step requires that the assumption $E(u_t|\mathbf{x}_t) = 0$ is satisfied and also some technical conditions on $\mathbf{x}_t$ that ensure that in the limit, the covariance matrix of explanatory variables, $\{\text{plim}\ (1/T)\sum_{t=1}^{T} \mathbf{x}_t \mathbf{x}_t'\}$, exists and is a positive definite matrix.[2]

Intuitively, the property of consistency requires two conditions be satisfied: the value of the estimator $\widehat{\boldsymbol{\beta}}$ must approach the true parameter value as T increases (known as asymptotic unbiasedness), and the variance of $\widehat{\boldsymbol{\beta}}$ around the true parameter value must

---

1. Another important distinction is whether $\mathbf{x}_t$ contains lagged values of y_t, as will be the case in chapter 4. If, as is currently assumed, $\mathbf{x}_t$ does not contain lagged dependent variables, then exact theory is possible. If $\mathbf{x}_t$ does contain lags of y_t, only the asymptotic properties of the estimators apply.
2. Positive definiteness is the matrix version of the requirement that a variance must be positive.

converge to zero as T increases. Consistency might be thought of as the minimum requirement for a useful estimator. Consider the two-variable linear regression model

$$y_t = \beta_0 + \beta_1 x_t + u_t$$

with $\beta_0 = \beta_1 = 1$ and $u_t \sim N(0,1)$. The explanatory variable is generated as $x_t \sim N(0,1)$ and then treated as fixed in repeated sampling. Estimation of β_1 using simulated data in samples of size 25 and 50 shows that the probability mass of the sampling distribution is centered on the true value of 1 and that the variance of the distribution gets smaller as sample size increases. The distributions shown in figure 3.2 are computed from 10,000 repetitions.

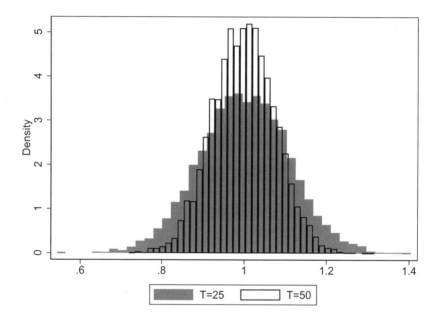

Figure 3.2. Illustrating the consistency of the maximum likelihood estimator of the slope parameter of the two-variable linear regression model with true slope parameter $\beta = 1$ and with $x_t \sim N(0,1)$. The distributions are computed from 10,000 repetitions in samples of size 25 and 50, respectively.

Asymptotic normality

The behavior of the sampling distribution of an estimator as the sample size becomes infinitely large is complicated by the fact that, for consistent estimators, the variance of the distribution tends to become infinitely small as $T \to \infty$. Consequently, it is usual to work with scaled versions of the distributions of the form $\sqrt{T}(\widehat{\boldsymbol{\beta}} - \boldsymbol{\beta})$, in which the distribution is centered to have mean zero. Recall that for a consistent estimator,

3.2　Properties of estimators

the variance of the estimator decreases to zero as $T \to \infty$. To examine the limiting distribution of the estimator, therefore, the use of the scale factor $\sqrt{T}$ is exactly what is needed to ensure that the variance of $\widehat{\boldsymbol{\beta}}$ is constant as T grows. In other words, from (3.1), it is usual to consider the distribution of

$$\sqrt{T}\left(\widehat{\boldsymbol{\beta}} - \boldsymbol{\beta}\right) = \left(\frac{1}{T}\sum_{t=1}^{T}\mathbf{x}_t\mathbf{x}_t'\right)^{-1}\frac{1}{\sqrt{T}}\sum_{t=1}^{T}\mathbf{x}_t u_t \tag{3.2}$$

We know already that $\{\text{plim }(1/T)\sum_{t=1}^{T}\mathbf{x}_t\mathbf{x}_t'\}$ is a positive-definite matrix. Central limit theorems establish the conditions under which $\{(1/\sqrt{T})\sum_{t=1}^{T}\mathbf{x}_t u_t\}$ will converge to a normal distribution, a process known as convergence in distribution. Provided that $E(u_t|\mathbf{x}_t) = 0$ and $E(u_t^2|\mathbf{x}_t) = \sigma_u^2$, an appropriate central limit theorem establishes that

$$\frac{1}{\sqrt{T}}\sum_{t=1}^{T}\mathbf{x}_t u_t \xrightarrow{d} N\left\{0, \sigma_u^2\left(\text{plim }\frac{1}{T}\sum_{t=1}^{T}\mathbf{x}_t\mathbf{x}_t'\right)\right\} \tag{3.3}$$

where $\xrightarrow{d}$ is read as "converges in distribution to".

Combining the two results about the limiting behavior of the two terms on the right-hand side of (3.2) is accomplished using Slutsky's theorem. Essentially, the limiting distribution of $\sqrt{T}(\widehat{\boldsymbol{\beta}} - \boldsymbol{\beta})$ will be a normal distribution but with a variance that accounts for the term $\{\text{plim }(1/T)\sum_{t=1}^{T}\mathbf{x}_t\mathbf{x}_t'\}^{-1}$, which is a constant matrix scaling the limiting normal distribution given in (3.3).

The end result is that

$$\sqrt{T}\left(\widehat{\boldsymbol{\beta}} - \boldsymbol{\beta}\right) \xrightarrow{d} N\left\{0, \sigma_u^2\left(\text{plim }\frac{1}{T}\sum_{t=1}^{T}\mathbf{x}_t\mathbf{x}_t'\right)^{-1}\right\}$$

or

$$\widehat{\boldsymbol{\beta}} \stackrel{a}{\sim} N\left\{\boldsymbol{\beta}, \sigma_u^2\left(\sum_{t=1}^{T}\mathbf{x}_t\mathbf{x}_t'\right)^{-1}\right\} \tag{3.4}$$

where $\stackrel{a}{\sim}$ is read as "is asymptotically distributed as". Note that the form of the variance of the asymptotic distribution in (3.4) is identical to the expression given in (2.8) for the VCE($\widehat{\boldsymbol{\beta}}$). This limiting normal distribution forms the basis of hypothesis tests on the parameters of the regression model given in chapter 2.

Asymptotic efficiency

An asymptotically efficient estimator is a consistent estimator whose variance is smaller than that of any other consistent estimator. Suppose we have two estimators $\widehat{\boldsymbol{\beta}}$ and $\widetilde{\boldsymbol{\beta}}$ where

$$\sqrt{T}\left(\widehat{\boldsymbol{\beta}} - \boldsymbol{\beta}\right) \xrightarrow{d} (0, \gamma_1), \qquad \sqrt{T}\left(\widetilde{\boldsymbol{\beta}} - \boldsymbol{\beta}\right) \xrightarrow{d} (0, \gamma_2)$$

If $\gamma_1 < \gamma_2$, then $\widehat{\boldsymbol{\beta}}$ is asymptotically efficient relative to $\widetilde{\boldsymbol{\beta}}$. To establish that $\widehat{\boldsymbol{\beta}}$ is asymptotically efficient relative to all other consistent estimators requires showing that its asymptotic variance attains the Cramér–Rao lower bound. The Cramér–Rao inequality establishes a lower bound for the variance–covariance matrix of any unbiased estimator. It is this search for an asymptotically efficient estimator that leads directly to the method of maximum likelihood.

3.3 Maximum likelihood and the linear model

Under certain technical conditions, the maximum likelihood estimator possesses several important and desirable statistical properties, which makes it the benchmark against which all other estimators are measured. In particular, subject to the regularity conditions being met, the maximum likelihood estimator not only is consistent and asymptotically normally distributed but it also automatically attains the Cramér–Rao lower bound and is therefore asymptotically efficient. The maximum likelihood estimator encompasses many other estimators often used in econometrics, including the ordinary least-squares estimator. These estimators then automatically inherit the properties of maximum likelihood estimators.

The distributions shown in figure 3.1 require the maximum likelihood estimates of parameters $\boldsymbol{\theta} = \{\mu, \sigma^2\}$. The first step in the process is to assume that the functional form of the joint probability density function of the variable, denoted $f(y_t)$, is known. The standard interpretation of the joint probability density is that $f(\cdot)$ is a function of y_t for given parameters, in this case $f(y_t; \mu, \sigma^2)$. To estimate the optimal parameters, however, this interpretation has been reversed so that $f(\cdot)$ is taken as a function of the parameters of the model for a given y_t, or $f(\mu, \sigma^2; y_t)$. The motivation behind this change in the interpretation of the arguments of the joint probability density function is to regard $\{y_1, y_2, \ldots, y_T\}$ as realized data that are no longer random. The optimal choices of the parameter values are therefore those that are "most likely" to have generated the observed data. This process of changing the arguments of the joint probability distribution of the data is the basis of maximum likelihood estimation.

The general process for estimating the parameters of a model, $\boldsymbol{\theta}$, by maximum likelihood are as follows, based on the assumption that the observations $\{y_1, y_2, \ldots, y_T\}$ are independent and identically distributed:

3.3 Maximum likelihood and the linear model

Step 1: Propose a functional form of the joint probability density function of the observed dependent variable, $\{y_1, y_2, \ldots, y_T\}$, denoted by $f(y_t)$. For notational simplicity, the explicit conditioning on the parameters is dropped.

Step 2: Reinterpret the joint probability function in terms of a likelihood function in which the parameters of the joint probability density function are taken as its arguments. Making use of the fact that the data are independent and identically distributed yields the likelihood function

$$L(\boldsymbol{\theta}) = \prod_{t=1}^{T} f(y_t)$$

or, after taking logarithms, the log-likelihood function

$$\log L(\boldsymbol{\theta}) = \sum_{t=1}^{T} \log f(y_t)$$

Step 3: The maximum likelihood estimator is now defined as

$$\widehat{\boldsymbol{\theta}} = \arg\max_{\boldsymbol{\theta}} \log L(\boldsymbol{\theta})$$

which, by the rules of calculus, is found where

$$\left. \frac{\partial \log L(\boldsymbol{\theta})}{\partial \boldsymbol{\theta}} \right|_{\boldsymbol{\theta}=\widehat{\boldsymbol{\theta}}} = 0$$

The independent and identically distributed assumption makes the construction of the likelihood in step 2 particularly easy, being the product of the likelihoods of the individual observations. Log-likelihood functions for dependent data will be introduced in chapter 4.

Consider the linear model given by

$$y_t = \boldsymbol{\beta}'\mathbf{x}_t + u_t \qquad u_t \sim \text{i.i.d. } N(0, \sigma^2)$$

where, in addition to the properties listed in (2.2)–(2.4), the disturbance term u_t is also assumed to be normally distributed. Although the distributional assumption is made about u_t, it follows immediately that the joint probability density function of y_t is given by

$$f(y_t) = \frac{1}{\sqrt{2\pi\sigma^2}} \exp\left(-\frac{(y_t - \boldsymbol{\beta}'\mathbf{x}_t)^2}{2\sigma^2}\right)$$

The construction of the likelihood function for a sample of $t = 1, 2, \ldots, T$ observations, $\{y_1, y_2, \ldots, y_T\}$, relies on the independent and identically distributed assumption so that

$$L(\boldsymbol{\beta}, \sigma^2) = \prod_{t=1}^{T} \frac{1}{\sqrt{2\pi\sigma^2}} \exp\left(-\frac{(y_t - \boldsymbol{\beta}'\mathbf{x}_t)^2}{2\sigma^2}\right)$$

in which the arguments of the likelihood function are now the parameters of the model with y_t taken as given and where conditioning on $\mathbf{x}_t$ has been suppressed for notational convenience. The product in the likelihood function means that it is easier to work with log-likelihood function, which is given by

$$\log L_T(\boldsymbol{\beta}, \sigma^2) = -\frac{T}{2}\ln(2\pi) - \frac{T}{2}\log \sigma^2 - \frac{1}{2\sigma^2}\sum_{t=1}^{T}(y_t - \boldsymbol{\beta}'\mathbf{x}_t)^2 \qquad (3.5)$$

The maximum of the log-likelihood function is found using calculus. The maximum likelihood estimator of $\boldsymbol{\beta}$ and σ^2, denoted $\widehat{\boldsymbol{\beta}}$ and $\widehat{\sigma}^2$, is obtained by setting the gradients of the log-likelihood function with respect to these parameters equal to zero and solving the resulting $K+1$ first-order conditions so that the maximum likelihood estimates satisfy

$$\left.\frac{\partial \log L_T}{\partial \boldsymbol{\beta}}\right|_{\boldsymbol{\beta}=\widehat{\boldsymbol{\beta}}} = 0$$

$$\left.\frac{\partial \log L_T}{\partial \sigma^2}\right|_{\sigma^2=\widehat{\sigma}^2} = 0$$

Differentiating $\log L_T(\boldsymbol{\beta}, \sigma^2)$ in (3.5) with respect to $\boldsymbol{\beta}$ and σ^2 yields

$$\frac{\partial \ln L_T(\boldsymbol{\beta}, \sigma)}{\partial \boldsymbol{\beta}} = \frac{1}{\sigma^2}\sum_{t=1}^{T}(y_t - \boldsymbol{\beta}'\mathbf{x}_t)\mathbf{x}_t \qquad (3.6)$$

$$\frac{\partial \ln L_T(\boldsymbol{\beta}, \sigma)}{\partial \sigma^2} = -\frac{1}{2\sigma^2} + \frac{1}{2\sigma^4}\sum_{t=1}^{T}(y_t - \boldsymbol{\beta}'\mathbf{x}_t)^2 \qquad (3.7)$$

It is immediately apparent that setting the K equations in (3.6) to zero gives the normal equations from which the ordinary least-squares estimator is derived. Furthermore, solving (3.7) for $\widehat{\sigma}^2$ yields an identical result to (2.7). These results establish that the maximum likelihood estimator and the ordinary least-squares estimator are identical in the case of the single-equation linear regression model with normally distributed disturbance terms for a large sample.

The degree of confidence in the estimated parameters $\boldsymbol{\beta}$ and σ^2 depends on the degree of curvature of the log-likelihood function. If the log-likelihood function is flat in the region of the optimal parameters, then the estimates of the parameters will be relatively imprecise by comparison with a situation where the log-likelihood function exhibits stronger curvature near the optimum. The curvature of the log-likelihood function is given by the matrix of second derivatives with respect to the parameters, known as the Hessian matrix. Intuitively, the inverse of the Hessian matrix is a measure of the degree of precision of the maximum likelihood estimator.

3.3 Maximum likelihood and the linear model

In the case of the simple linear regression model, the Hessian matrix is

$$\mathbf{H} = \begin{bmatrix} -\dfrac{1}{\sigma^2} \sum_{t=1}^{T} \mathbf{x}_t \mathbf{x}_t' & 0 \\ 0 & \dfrac{1}{2\sigma^4} - \dfrac{1}{2\sigma^6} \sum_{t=1}^{T} \left(y_t - \widehat{\boldsymbol{\beta}}' \mathbf{x}_t \right)^2 \end{bmatrix}$$

The covariance matrix of the maximum likelihood estimator, known as the information matrix, is given by the expected value of the negative of the inverse Hessian matrix evaluated at the maximum likelihood estimator. Recognizing that

$$E\left(\sum_{t=1}^{T} (y_t - \boldsymbol{\beta}' \mathbf{x}_t)^2 \right) = T\sigma^2$$

when evaluated at the maximum likelihood estimator, the information matrix is

$$E\left\{ H\left(\widehat{\boldsymbol{\theta}}\right)^{-1} \right\} = \begin{bmatrix} \widehat{\sigma}^2 \left(\sum_{t=1}^{T} \mathbf{x}_t \mathbf{x}_t' \right)^{-1} & 0 \\ 0 & \dfrac{\widehat{\sigma}^2}{T} \end{bmatrix}$$

which is also the covariance matrix of the maximum likelihood estimator. The square roots of the elements on the main diagonal are the standard errors of $\widehat{\boldsymbol{\beta}}$ and $\widehat{\sigma}^2$. It is immediately apparent that the variance and therefore the standard error of the maximum likelihood estimator, $\widehat{\boldsymbol{\beta}}$, is identical to that of the ordinary least-squares estimator.

There is another common way to estimate the covariance matrix of the maximum likelihood estimator. The information matrix equality (see, for example, Martin, Hurn, and Harris [2013, 58]), states that in the limit as $T \to \infty$

$$-E\left\{ H\left(\widehat{\boldsymbol{\theta}}\right) \right\} = E\left\{ J\left(\widehat{\boldsymbol{\theta}}\right) \right\}$$

where $H(\widehat{\boldsymbol{\theta}})$ is the Hessian matrix evaluated at the maximum likelihood estimator and $J(\widehat{\boldsymbol{\theta}})$ is the outer product of the gradient matrix given by

$$\begin{aligned} J\left(\widehat{\boldsymbol{\theta}}\right) &= \frac{1}{T} \sum_{t=1}^{T} \frac{\partial \log L_t(\boldsymbol{\theta})}{\partial \boldsymbol{\theta}} \frac{\partial \log L_t(\boldsymbol{\theta})}{\partial \boldsymbol{\theta}'} \bigg|_{\boldsymbol{\theta} = \widehat{\boldsymbol{\theta}}} \\ &= \frac{1}{T} \sum_{t=1}^{T} g_t\left(\widehat{\boldsymbol{\theta}}\right) g_t'\left(\widehat{\boldsymbol{\theta}}\right) \end{aligned}$$

This notation emphasizes that the gradient, $g_t(\widehat{\boldsymbol{\theta}})$, is computed at every observation T. In some instances, this estimate of the covariance matrix is numerically more stable than the computation of the Hessian, a point that is returned to in chapter 4.

A straightforward way to fit models by maximum likelihood in Stata is to use the mlexp command (see [R] **mlexp**). This command fits any model for which the log likelihood for an individual observation may simply be summed over observations to compute the total log likelihood. Standard errors based on the Hessian matrix are denoted vce(oim) where the argument refers to the negative of the expected value of the inverse Hessian as the observed information matrix. Standard errors based on the outer product of the gradient matrix are denoted vce(opg).

```
. use http://www.stata-press.com/data/eeus/pm_daily
. mlexp (ln(normalden(pm25, {xb:co temperature dv vv _cons}, exp({lnsigma})))),
> nolog
initial:       log likelihood =    -<inf>  (could not be evaluated)
feasible:      log likelihood = -91017.685
rescale:       log likelihood = -13006.813
rescale eq:    log likelihood = -9751.8271

Maximum likelihood estimation
Log likelihood = -7771.9515                    Number of obs    =     2,191
```

	Coef.	Std. Err.	z	P>|z|	[95% Conf.	Interval]
xb						
co	21.2735	.4229553	50.30	0.000	20.44452	22.10247
temperature	.2895487	.0464886	6.23	0.000	.1984328	.3806647
dv	-.0183468	.0037936	-4.84	0.000	-.0257821	-.0109114
vv	-.8764131	.3290759	-2.66	0.008	-1.52139	-.2314362
_cons	11.71787	1.322461	8.86	0.000	9.125893	14.30984
/lnsigma	2.128278	.0151065	140.88	0.000	2.09867	2.157886

```
. nlcom exp(_b[/lnsigma])
       _nl_1:  exp(_b[/lnsigma])
```

	Coef.	Std. Err.	z	P>|z|	[95% Conf.	Interval]
_nl_1	8.400389	.1269005	66.20	0.000	8.151669	8.649109

The parameter values and standard errors are identical to those obtained using ordinary least squares. Instead of estimating σ, the parameter $\log(\sigma)$ is estimated. This ensures that $\sigma = \exp\{\log(\sigma)\}$ is always positive. Using transformations like this one is not crucial in this simple linear problem, but this strategy becomes an important part of estimation in complex nonlinear problems. Once $\log(\sigma)$ is returned by the estimation, the nonlinear combination command nlcom (see [R] **nlcom**) can be used to obtain an estimate of σ and its standard error. The standard error is computed using the delta method. The maximum likelihood estimate of σ is slightly different from that of the ordinary least-squares estimate, given in the output as Root MSE, even once rounding is accounted for. This slight difference is due to the degrees of freedom correction used to compute the ordinary least-squares estimate.

Finally, it is worth noting that the properties of maximum likelihood estimators rely on the assumption that the joint probability density $f(y_t; \boldsymbol{\theta})$ is correctly specified. If the

3.4 Hypothesis testing

model is misspecified, the maximum likelihood estimator is known as a quasi–maximum likelihood estimator. Note that the quasi–maximum likelihood estimator will still be consistent if the assumed density is a member of the linear exponential family (including the Gaussian and gamma distributions) and the functional form for the conditional mean of the distribution is correctly specified (Cameron and Trivedi 2010, 322). In this instance, however, the standard errors based on either the Hessian matrix or the outer product of the gradient matrix will no longer be efficient. Instead, inference is based on a robust version of the covariance matrix, the so-called sandwich estimator, given by

$$\text{VCE}\left(\widehat{\boldsymbol{\theta}}\right) = H\left(\widehat{\boldsymbol{\theta}}\right)^{-1} J\left(\widehat{\boldsymbol{\theta}}\right) H\left(\widehat{\boldsymbol{\theta}}\right)^{-1}$$

Standard errors that are based on the sandwich estimator after using `mlexp` are denoted `vce(robust)`.

3.4 Hypothesis testing

In the estimation problem, $\widehat{\boldsymbol{\theta}}$ is the value of $\boldsymbol{\theta}$ that maximizes the log-likelihood function. In the test problem, the focus is on determining if the population parameter has a certain hypothesized value, $\boldsymbol{\theta}_0$. If this value differs from $\widehat{\boldsymbol{\theta}}$, then by definition, it must correspond to a lower value of the log-likelihood function and the crucial question is then the significance of this decrease. The general statement of a hypothesis test is of the form

$$H_0 : \boldsymbol{\theta} = \boldsymbol{\theta}_0, \quad H_1 : \boldsymbol{\theta} \neq \boldsymbol{\theta}_0$$

where H_0 and H_1 are known as the null and alternative hypotheses, respectively. If estimation is under the null hypothesis, the estimated parameters are denoted $\widehat{\boldsymbol{\theta}}_0$ with covariance matrix, $\text{VCE}(\widehat{\boldsymbol{\theta}}_0)$. If estimation is under the alternative hypothesis, the estimated parameters are denoted $\widehat{\boldsymbol{\theta}}_1$ with an associated covariance matrix, $\text{VCE}(\widehat{\boldsymbol{\theta}}_1)$.

There are three important types of test based on the maximum likelihood principle, namely, the LR, Wald (W), and LM tests. Each of these tests is now discussed in turn.

Likelihood-ratio test

The LR test measures the distance between the value of the log-likelihood under the null hypothesis [$\log L(\widehat{\boldsymbol{\theta}}_0)$] and the alternative hypothesis [$\log L(\widehat{\boldsymbol{\theta}}_1)$]. Suppose that there are M restrictions imposed by the null hypothesis; then, the test statistic is given by

$$\text{LR} = -2\left\{\log L\left(\widehat{\boldsymbol{\theta}}_0\right) - \log L\left(\widehat{\boldsymbol{\theta}}_1\right)\right\} \sim \chi_M^2$$

Implementing the LR test therefore requires fitting the model under both the null and alternative hypotheses.

Wald test

The Wald test is based on the distance between the parameter vector under the null hypothesis (θ_0) and the estimates under the alternative hypothesis ($\widehat{\theta}_1$) weighted by the inverse of the covariance evaluated under the alternative hypothesis $[\Omega(\widehat{\theta}_1)]$, which provides a measure of the curvature of log-likelihood function. In particular, sharper log-likelihood functions that correspond to relatively smaller variances provide tighter inference. The test statistic is given by

$$W = \left(\widehat{\theta}_1 - \theta_0\right)' \left\{\text{VCE}\left(\widehat{\theta}_1\right)\right\}^{-1} \left(\widehat{\theta}_1 - \theta_0\right) \sim \chi^2_M$$

The form of the test is a generalized version of a squared t test. Implementing the Wald test requires fitting the model under the alternative hypothesis only.

LM test

The LM test is based on the distance between the gradient of the log-likelihood function under the null hypothesis $[G(\widehat{\theta}_0)]$ and the gradient of the log-likelihood function under the alternative hypothesis $[G(\widehat{\theta}_1) = 0]$, this time scaled by the covariance matrix $\Omega(\widehat{\theta}_0)$, which is also the inverse of the variance of $G(\widehat{\theta}_0)$. The test statistic is given by

$$\text{LM} = G\left(\widehat{\theta}_0\right)' \left\{\text{VCE}\left(\widehat{\theta}_0\right)\right\} G\left(\widehat{\theta}_0\right) \sim \chi^2_M$$

Implementing the LM tests requires estimation of the model under the null hypothesis only.

A convenient form for computing the LM statistic in linear models is when the estimate of the covariance matrix $\text{VCE}(\widehat{\theta}_0)$ is based on the outer product of the gradient matrix rather than the Hessian matrix. In these circumstances, the LM test may be implemented using an auxiliary regression based on the residuals from the estimation of the original, restricted linear model. The details are beyond the scope of this book, but it is worth recognizing that the popular tests for autocorrelation and heteroskedasticity discussed in chapter 2 are in fact LM tests.

3.5 Method-of-moments estimators and the linear model

The importance of method-of-moments estimators in econometrics stems mainly from the fact that the desirable properties of maximum likelihood estimators rely on the assumption that the joint probability density $f(y_t; \theta)$ is correctly specified. Apart from the special case referred to in terms of the quasi–maximum likelihood estimator, the misspecification of the density function will generally result in the maximum likelihood estimator being inconsistent. Method-of-moments estimators do not require the specification of the entire joint density of the dependent variable. They are based on specifying a set of conditions whose expected values are equal to zero when evaluated at the true parameter vectors.

3.5 Method-of-moments estimators and the linear model

Consider, for example, the exogeneity assumption in the linear model given by $E(u_t \mathbf{x}_t) = 0$ in (2.2), which states that the regressors are orthogonal to the disturbances. If this orthogonality result is justified, then $E(u_t \mathbf{x}_t) = 0$ may be interpreted as a set of population moment conditions. Replacing the population expectations with sample counterparts leads to the system of equations

$$\frac{1}{T}\sum_{t=1}^{T}(y_t - \boldsymbol{\beta}'\mathbf{x}_t)\mathbf{x}_t = 0$$

which are in fact the normal equations for the ordinary least-squares estimator. In other words, when the exogeneity condition holds, ordinary least squares may be interpreted as a simple method-of-moments estimator.

Developing this line of argument to the situation in which exogeneity does not hold so that $E(u_t \mathbf{x}_t) \neq 0$, suppose there is a set of at least $K+1$ instrumental variables $\mathbf{z}_t$ that satisfy two conditions:

$$E(u_t \mathbf{z}_t) = 0, \qquad E(\mathbf{z}_t \mathbf{x}_t) \neq 0$$

In this situation, the population moment conditions are $E(u_t \mathbf{z}_t) = 0$ with sample counterparts given by

$$\frac{1}{T}\sum_{t=1}^{T}(y_t - \boldsymbol{\beta}'\mathbf{x}_t)\mathbf{z}_t = 0$$

Solving these equations gives the instrumental-variables estimator of the linear regression model

$$\widetilde{\boldsymbol{\beta}} = \left(\sum_{t=1}^{T}\mathbf{z}_t \mathbf{x}_t'\right)^{-1}\left(\sum_{t=1}^{T}\mathbf{z}_t y_t\right)$$

The covariance matrix of the instrumental-variables estimator is

$$\widetilde{\text{VCE}}\left(\widetilde{\boldsymbol{\beta}}\right) = \widetilde{\sigma}^2 \left(\sum_{t=1}^{T}\mathbf{z}_t \mathbf{x}_t'\right)^{-1}\left(\sum_{t=1}^{T}\mathbf{x}_t \mathbf{x}_t'\right)\left(\sum_{t=1}^{T}\mathbf{z}_t \mathbf{x}_t'\right)^{-1} \tag{3.8}$$

The expression reduces to the ordinary least-squares estimate of the covariance matrix when $\mathbf{z}_t$ is identically $\mathbf{x}_t$, that is, when there is no violation of the assumption of exogeneity.

Method-of-moments estimators are obtained using a system of equations where the number of moment equations is as least as large as the number of unknown parameters. If the number of moment conditions and number of parameters are equal, the problem is said to be exactly identified. If there are more moment conditions than parameters, it is said to be overidentified and the estimation method is referred to as the GMM (Hansen 1982). In this situation, the problem is formulated as a numerical minimization problem in which the GMM parameter estimates satisfy

$$\widetilde{\boldsymbol{\beta}} = \arg\min_{\boldsymbol{\beta}} Q(\boldsymbol{\beta}) = \left(\frac{1}{T}\sum_{t=1}^{T}\mathbf{z}_t u_t\right)' \mathbf{W} \left(\frac{1}{T}\sum_{t=1}^{T}\mathbf{z}_t u_t\right) \tag{3.9}$$

where **W** is a positive-definite weighting matrix with the same number of rows and columns as the number of instruments in $\mathbf{z}_i$. In other words, the GMM estimator chooses the parameter estimates $\widetilde{\boldsymbol{\beta}}$ to make the squared deviations of the moment conditions from zero as small as possible for a given weighting matrix **W**.

The value of the GMM objective function in (3.9) will always be nonnegative and will reach zero only in the case of an exactly identified model. If the model is overidentified, the value of the function may be used to evaluate the validity of the moment conditions used in the estimation. An overall test of the adequacy of the specified model is given by the Hansen–Sargan J test based on scaling the GMM objective function evaluated at $\widetilde{\boldsymbol{\beta}}$ by the sample size

$$J = T\,Q\left(\widetilde{\boldsymbol{\beta}}\right)$$

If the model is not properly specified, there will be a mismatch between the sample and the population moments resulting in a large value of $Q(\widetilde{\boldsymbol{\beta}})$ and hence J. Under the null hypothesis, the J statistic is distributed asymptotically as χ^2_R, where R is the number of overidentifying restrictions.

If the model is correctly specified so that the moment conditions provide a good summary of the population model, the GMM estimator $\widehat{\boldsymbol{\beta}}_{\text{GMM}}$ is a consistent estimator that satisfies the property

$$\text{plim}\left(\widetilde{\boldsymbol{\beta}}\right) = \boldsymbol{\beta}$$

where $\boldsymbol{\beta}$ is the true population parameter vector. The GMM estimator is also asymptotically normally distributed, but it is not efficient. Hansen (1982) shows that the GMM estimator has the smallest dispersion when each of the moments is weighted in terms of their relative precision, with the moments having the greatest precision given the greatest influence when computing the GMM estimates. Formally, this requires **W** to be estimated as the inverse of the covariance matrix of the moments

$$\widetilde{W} = \left(\frac{1}{T}\sum_{t=1}^{T}\widetilde{u}_t^2 \mathbf{z}_t \mathbf{z}_t'\right)^{-1}$$

where $\widetilde{u}_t$ are the residuals from a first-round estimate of the parameters of the model using an identity weighting matrix. The general form of the covariance matrix of the GMM estimators involves the derivatives of the moment conditions with respect to the parameters. For the linear regression model, however, these derivatives have a particularly straightforward form, and the covariance matrix of $\widehat{\boldsymbol{\beta}}$ is simply given by

$$\widetilde{\text{VCE}}\left(\widetilde{\boldsymbol{\beta}}\right) = T\left\{\left(\sum_{t=1}^{T}\mathbf{x}_t\mathbf{z}_t'\right)\widetilde{W}^{-1}\left(\sum \mathbf{z}_t\mathbf{x}_t'\right)\right\}^{-1}$$

The instrumental-variables or two-stage least-squares estimator and the GMM estimator are both computed using the `ivregress` command (see [R] **ivregress**). Alternatively, the `ivreg2` package written by Baum, Schaffer, and Stillman (2003, 2007)

3.5 Method-of-moments estimators and the linear model

can be used. For illustrative purposes, the official Stata command is used here. Because each of the regressors in this model are assumed to be exogenous and satisfy the zero-conditional mean assumption of (2.2), they can serve as their own instruments.

The simple instrumental-variables estimator of the model for daily PM2.5 is as follows:

```
. ivregress 2sls pm25 co temperature vv i.winddir
Instrumental variables (2SLS) regression        Number of obs   =      2,191
                                                 Wald chi2(10)   =    5410.90
                                                 Prob > chi2     =     0.0000
                                                 R-squared       =     0.7118
                                                 Root MSE        =      8.346
```

pm25	Coef.	Std. Err.	z	P>|z|	[95% Conf. Interval]	
co	22.02843	.4513482	48.81	0.000	21.14381	22.91306
temperature	.2932477	.0463087	6.33	0.000	.2024844	.3840111
vv	-1.088717	.3367156	-3.23	0.001	-1.748668	-.4287667
winddir						
ENE	-5.184654	1.453881	-3.57	0.000	-8.034208	-2.335101
ESE	-7.446572	1.265659	-5.88	0.000	-9.927217	-4.965926
SSE	-6.024663	1.015358	-5.93	0.000	-8.014728	-4.034597
SSW	-5.888658	.8517538	-6.91	0.000	-7.558065	-4.219252
WSW	-5.926742	1.006531	-5.89	0.000	-7.899507	-3.953978
WNW	-5.368964	2.781702	-1.93	0.054	-10.821	.0830722
NNW	-4.22259	3.278605	-1.29	0.198	-10.64854	2.203358
_cons	13.70695	1.364208	10.05	0.000	11.03315	16.38075

(no endogenous regressors)

The point estimates of the parameters are identical to those of ordinary least squares and maximum likelihood. The simple instrumental-variables estimator also returns the same standard errors as those estimators. This result follows from the cancellation of terms in (3.8).

The GMM estimator for the same model gives the following estimates:

```
. ivregress gmm pm25 co temperature vv i.winddir
Instrumental variables (GMM) regression        Number of obs   =      2,191
                                                Wald chi2(10)   =    4068.65
                                                Prob > chi2     =     0.0000
                                                R-squared       =     0.7118
GMM weight matrix: Robust                       Root MSE        =      8.346
```

	Coef.	Robust Std. Err.	z	P>|z|	[95% Conf. Interval]
pm25					
co	22.02843	.6066092	36.31	0.000	20.8395 23.21736
temperature	.2932477	.0457999	6.40	0.000	.2034816 .3830139
vv	-1.088717	.3269163	-3.33	0.001	-1.729461 -.447973
winddir					
ENE	-5.184654	1.889872	-2.74	0.006	-8.888734 -1.480574
ESE	-7.446572	1.518477	-4.90	0.000	-10.42273 -4.470411
SSE	-6.024663	1.228283	-4.90	0.000	-8.432053 -3.617272
SSW	-5.888658	1.016229	-5.79	0.000	-7.880431 -3.896886
WSW	-5.926742	1.1672	-5.08	0.000	-8.214412 -3.639072
WNW	-5.368964	3.562502	-1.51	0.132	-12.35134 1.613412
NNW	-4.22259	4.695904	-0.90	0.369	-13.42639 4.981213
_cons	13.70695	1.615903	8.48	0.000	10.53984 16.87406

```
(no endogenous regressors)
. display "J statistic = " %8.6f e(J)
J statistic = 0.000000
```

The GMM estimator in this case provides identical point estimates to ordinary least squares when all regressors are exogenous. Note further that when the number of moment conditions is equal to the number of unknown parameters, the GMM estimator will always satisfy each moment condition, yielding a J statistic of zero.

For any use of an instrumental-variables estimator, Stata reports the ratio of the coefficient to its standard error as a z statistic rather than the t statistic reported by **regress** (see [R] **regress**). Instrumental-variables techniques are justified by large-sample or asymptotic theory, so hypotheses on these coefficients are tested against the standard normal distribution rather than the finite-sample t distribution. Likewise, any joint tests from these models are reported as chi-squared statistics, whereas **regress** provides F statistics. For the size of the sample we are considering, inference based on the z versus t or chi-squared versus F test statistics will not differ.

3.6 Testing for exogeneity

Using method-of-moments estimation for the sake of obtaining a consistent estimator must be balanced against the inevitable loss of efficiency. In other words, there is a cost of using a moment-based estimator when the regressors, $\mathbf{x}_t$, are indeed exogenous because the asymptotic variance of the moment-based estimator is always larger than

3.6 Testing for exogeneity

the asymptotic variance of the ordinary least-squares estimator. This loss of efficiency is only worth implementing if the ordinary least-squares estimator is indeed biased and inconsistent.

The Hausman (1978) specification test provides a test of exogeneity by comparing the moment-based estimator $\widetilde{\beta}$, which is known to be consistent, with an estimator, $\widehat{\beta}$, that is known to be efficient under the null hypothesis being tested. The test is based upon a direct comparison of coefficient values in terms of the appropriately scaled, squared differences between them. The test statistic is of the form

$$H = \left(\widetilde{\beta} - \widehat{\beta}\right)' \left\{\text{VCE}\left(\widetilde{\beta} - \widehat{\beta}\right)\right\}^{-1} \left(\widetilde{\beta} - \widehat{\beta}\right)$$

and it is distributed χ_p^2 under the null hypothesis, where the degrees of freedom, p, is given by the number of potentially endogenous regressors. The key to the implementation of the test is computing $\text{VCE}(\widetilde{\beta} - \widehat{\beta})$. If $\widehat{\beta}$ is an efficient estimator under the null hypothesis, then

$$\text{VCE}\left(\widetilde{\beta} - \widehat{\beta}\right) = \text{VCE}\left(\widetilde{\beta}\right) - \text{VCE}\left(\widehat{\beta}\right) \tag{3.10}$$

The Hausman test is implemented in Stata by means of the `hausman` command (see [R] **hausman**). Stata provides two options when invoking the Hausman test. The option `sigmamore` specifies that the covariance matrix in (3.10) should be computed using the estimated residuals from the efficient estimator. The option `sigmaless` specifies that this covariance matrix should be computed using the residuals from the consistent estimator. Although the tests are asymptotically equivalent, `sigmamore` provides a proper estimate of the contrast variance for tests of exogeneity in moment-based estimation of regression models. The Hausman test requires that $\widehat{\beta}$ is indeed efficient under the null hypothesis, which will not be the case if, for example, the standard errors of the parameters require computation by robust methods (see chapter 2). An alternative, asymptotically equivalent test known as the Durbin–Wu–Hausman test, may be used even if the VCE is robust, cluster–robust, or heteroskedasticity- and autocorrelation-consistent. See section 5 of Baum, Schaffer, and Stillman (2003) for details.

The Hausman test has a similar form to a Wald test. An LM-type test of exogeneity based on an auxiliary regression approach proceeds as follows. Consider a simplified version of the classical linear regression model with only one explanatory variable, x_t, that is potentially endogenous. Suppose further that a valid instrument for x_t exists, so it is possible to formulate an auxiliary reduced-form regression for x_t as

$$x_t = \pi_0 + \pi_1 z_t + v_t, \qquad \pi_1 \neq 0$$

It follows that

$$\text{cov}(x_t, u_t) = \text{cov}(\pi_0 + \pi_1 z_t + v_t, u_t) = \text{cov}(v_t, u_t)$$

using the fact that $\text{cov}(z_t, u_t) = 0$ by virtue of z_t being a valid instrument. It therefore follows that

$$\text{cov}(x_t, u_t) = 0 \implies \text{cov}(v_t, u_t) = 0 \tag{3.11}$$

The condition in expression (3.11) may be written slightly differently by requiring that the coefficient $\boldsymbol{\theta}$ in the linear regression

$$u_t = \boldsymbol{\theta} v_t + \eta_t \tag{3.12}$$

be zero.

Of course, because u_t and v_t are disturbance terms, they are unobserved. Therefore, (3.12) cannot be estimated as it stands, and a test of $\boldsymbol{\theta} = 0$ cannot be conducted. However, substituting for u_t in the original model gives

$$y_t = \beta_0 + \beta_1 x_t + \boldsymbol{\theta} v_t + \eta_t$$

a result that suggests the following auxiliary regression-based test of endogeneity of x_t, known as the Durbin–Wu–Hausman test.

Step 1: Estimate the regression

$$x_t = \pi_0 + \pi_1 z_t + v_t$$

by ordinary least squares to obtain the estimates $\widehat{\pi}_0$ and $\widehat{\pi}_1$, and then compute the residuals $\widehat{v}_t$.

Step 2: Estimate the regression

$$y_t = \beta_0 + \beta_1 x_t + \boldsymbol{\theta} \widehat{v}_t + \eta_t$$

by ordinary least squares, and test $H_0 : \boldsymbol{\theta} = 0$ using a t test. Failure to reject the null hypothesis will indicate that condition (3.11) is not satisfied and that there is a potential endogeneity problem.

The Durbin–Wu–Hausman test of endogeneity is carried out by estimating the equation using ivregress (see [R] **ivregress**) followed by estat endogenous.[3]

In the urban air pollution model, both carbon monoxide and PM2.5 particles are caused by incomplete combustion of biomass fuels. It is possible, therefore, that using co as an exogenous explanatory variable in an equation for PM2.5 runs the risk of violating the exogeneity assumption. Instrumental-variables estimation now proceeds using lagged co as an instrument for co. Note that using lagged variables as instruments for a potentially endogenous contemporaneous regressor is common practice in a time-series setting. The lagged value of the variable is predetermined at time t and is therefore uncorrelated with the error term, while for environmental measurements, the lagged variable is usually closely related to its contemporaneous value. The results of fitting the model and performing the Durbin–Wu–Hausman test are

---

3. Alternatively, ivreg2 with the endog() option may be used.

3.6 Testing for exogeneity

```
. ivregress 2sls pm25 (co=L.co) temperature vv i.winddir
Instrumental variables (2SLS) regression     Number of obs   =    2,190
                                             Wald chi2(10)   =  3909.19
                                             Prob > chi2     =   0.0000
                                             R-squared       =   0.7113
                                             Root MSE        =   8.3546
```

pm25	Coef.	Std. Err.	z	P>|z|	[95% Conf.	Interval]
co	20.9974	.7057696	29.75	0.000	19.61412	22.38068
temperature	.2757314	.0473094	5.83	0.000	.1830067	.368456
vv	-1.438156	.3802572	-3.78	0.000	-2.183446	-.6928655
winddir						
ENE	-4.593226	1.489878	-3.08	0.002	-7.513333	-1.673119
ESE	-6.673665	1.33381	-5.00	0.000	-9.287885	-4.059446
SSE	-5.635166	1.038782	-5.42	0.000	-7.671141	-3.599191
SSW	-5.86799	.8528457	-6.88	0.000	-7.539537	-4.196443
WSW	-6.011787	1.008415	-5.96	0.000	-7.988245	-4.03533
WNW	-5.263229	2.785333	-1.89	0.059	-10.72238	.1959237
NNW	-3.788332	3.290531	-1.15	0.250	-10.23765	2.66099
_cons	15.4402	1.632574	9.46	0.000	12.24042	18.63999

```
Instrumented:  co
Instruments:   temperature vv 2.winddir 3.winddir 4.winddir 5.winddir
               6.winddir 7.winddir 8.winddir L.co

. estat endogenous co

Tests of endogeneity
Ho: variables are exogenous
Durbin (score) chi2(1)      =   3.62623  (p = 0.0569)
Wu-Hausman F(1,2178)        =   3.61234  (p = 0.0575)
```

The value of the coefficient on co is now 20.997 as opposed to the previous estimate of 22.028, and the standard error on co grows to 0.706 from 0.451. These results are marginally suggestive that co fails to satisfy the exogeneity assumption. Because the p-values of the two forms of the endogeneity test are also marginally above the 5% level, the evidence of the endogeneity of co, while not conclusive, is nevertheless worrying enough for the variable to be treated with care.

Exercises

1. `pm_daily.dta` contains daily data on various measures of air pollution in Santiago, Chile, for the period January 1, 2009 to December 31, 2014.

    a. Using the add-on package `tsspell` from the Statistical Software Components Archive written by Nick Cox (2002b), generate a time series of durations, measured as the number of successive days, for which the air quality in Santiago is greater than the World Health Organization's recommended level of $25\mu g/m^3$. Draw a histogram of the distribution of the durations, and comment on the result.

    b. Assume that the probability distribution of the durations, denoted y_t, is the exponential distribution given by

    $$f(y_t) = \alpha \exp(-\alpha y_t), \qquad \alpha > 0, \ y_t > 0$$

    Estimate α by maximum likelihood using the `mlexp` command.
    [Hint: The maximum likelihood estimator of α for this particular form of the exponential distribution is the inverse of the sample mean. This fact will enable you to check the result obtained using `mlexp`.]

    c. Now, assume that the probability distribution of the durations, y_t, is the Weibull distribution given by

    $$f(y_t) = \alpha \beta y_t^{\beta-1} \exp(-\alpha y_t^\beta), \qquad \alpha > 0, \ \beta > 0, \ y_t > 0$$

    Estimate α and β by maximum likelihood using the Stata `mlexp` command.

    d. If the restriction $\beta = 1$ is imposed, the Weibull distribution reduces to the exponential distribution. Test this restriction

    i. using a LR test,
    ii. using a Wald test, and
    iii. using a LM test.

    What do you conclude?

2. `pm_daily.dta` contains daily data on various measures of air pollution in Santiago, Chile, for the period January 1, 2009 to December 31, 2014.

    a. Assume that the distribution of the daily PM2.5 observations, denoted y_t, are drawn from a normal distribution given by

    $$f_1(y_t) = \frac{1}{\sqrt{2\pi\sigma^2}} \exp\left(-\frac{(y_t - \mu)^2}{2\sigma^2}\right)$$

    Estimate μ and σ^2 by maximum likelihood using the `mlexp` command.

3.6 Testing for exogeneity

b. Now, assume that the probability distribution of the particulate matter, y_t, is the lognormal distribution given by

$$f_2(y_t) = \frac{1}{\sqrt{2\pi\sigma^2}y_t} \exp\left(-\frac{(\log y_t - \mu)^2}{2\sigma^2}\right)$$

in which case it is the logarithm of y_t that follows a normal distribution. Estimate μ and σ^2 by maximum likelihood using the `mlexp` command.

c. Overlay a histogram of the daily PM2.5 data with the best-fitting normal and lognormal distributions. Compare the result with figure 3.1.

d. These two models of PM2.5 are not nested in that one model cannot be expressed as a subset of the other. To test that there is no significant difference in the fit of the models, the test suggested by Vuong (1989) may be used. The basic idea is to convert the likelihood functions of the two competing models into a common likelihood function using the transformation of variable technique and perform a variation of a LR test. To enable a comparison of the two models, use the transformation of variable technique to convert the distribution f_2 into a distribution of the level of y_t. Formally, the link between the two distributions is given by

$$f_1(y_t) = f_2(\log y_t)\left|\frac{d\log y_t}{dy_t}\right| = f_2(\log y_t)\left|\frac{1}{y_t}\right|$$

which now allows the log-likelihood functions of the two models to be compared.

i. Compute the difference in the log-likelihood functions of the models at each observation given by

$$d_t = \log L_{1t} - \log L_{2t}$$

ii. Compute the mean and standard deviation of d_t given by

$$\bar{d} = \frac{1}{T}\sum_{t=1}^{T} d_t, \quad s = \frac{1}{T}\sum_{t=1}^{T}(d_t - \bar{d})$$

iii. Construct the following test statistic:

$$V = \sqrt{T}\frac{\bar{d}}{s}, \quad V \xrightarrow{d} N(0,1)$$

Which model do you prefer?

4 Introducing dynamics

An important limiting feature of the basic regression model presented in chapters 2 and 3 is that it is a static model in that all the variables that appear in the model are measured at time t. To allow for environmental variables to adjust to shocks over time, it is necessary to introduce dynamics in the models so that there is a time-series dimension that reflects their adjustment processes. Dynamic models that involve only a single variable are called univariate models. In a univariate dynamic model, a single dependent variable is explained using its own past history as well as lags of other relevant variables. Univariate models will be the focus of this chapter, while multivariate extensions will be discussed in chapter 5.

As soon as a time dimension is introduced, there is a danger that the fundamental characteristics of the model will exhibit a dependence on time that is not deterministic. This property is known as nonstationarity. To abstract from this problem, the models dealt with in this chapter require that the variables involved satisfy a simplifying stationarity condition. Stationarity requires that the mean and variance of the series, as well as the covariance between adjacent observations, are not explicit functions of time. Statistical corroboration of supporting assumptions like stationarity or specific forms of nonstationarity is important in empirical work because the validity of the methods used to make inferences often depends on these assumptions. Testing for stationarity and estimation techniques for nonstationary variables are covered in chapters 6 and 7, respectively.

Dynamic time-series modeling is synonymous with the names of George Box and Gwilym Jenkins, whose seminal textbook (Box and Jenkins 1970) presented an iterative process for building time-series models. Their approach starts with the assumption that the process that generated the time series can be approximated using an autoregressive moving-average (ARMA) model if the process is stationary or an autoregressive (AR) integrated moving-average model if the stochastic process is nonstationary. By virtue of the maintained assumption of stationarity in this chapter, the discussion will focus on ARMA models.

4.1 Load-weighted electricity prices

Despite the long-term decline in costs for installed capacity of renewable energy technologies and a large demand for renewable energy in many countries, large-scale investments in these technologies remain elusive. To ensure long-term success of a renewable energy investment program, adequate attention must be paid to the preparation and financing of the project. All investments are subject to a systematic process of capital appraisal that provides a basis for selection or rejection of projects by ranking them in order of profitability or social and environmental benefits. Capital investments should not be made in projects that earn less than some minimum or hurdle rate of return. A critical component of this process is forecasting the spot price of electricity. Modeling the dynamics of the electricity price as a precursor to building a forecasting model plays an important part of the evaluation of investment in renewable energy sources.

The spot (pool) price of electricity in a competitive wholesale electricity market reflects the instantaneous balancing of demand and supply. However, this is not what electricity retailers and generators acknowledge as the wholesale price of electricity. A variable that has the potential to capture both price and volume risk for an electricity retailer is the daily load-weighted price. If price and load are available for m periods within a day (usually hourly or half-hourly), then daily load-weighted price for day t is computed from $i = 1, \ldots, m$ prices (p_{it}) and loads (q_{it}) as follows:

$$\texttt{lwp}_t = \frac{\sum\limits_{i=1}^{m} p_{it} q_{it}}{\sum\limits_{i=1}^{m} q_{it}}$$

The difference between the spot price of electricity in any half-hour interval and the load-weighted price measures the impact of that half-hour price on a retailer's revenue and is therefore a meaningful indicator of the effective wholesale value of electricity.

The analysis of this chapter focuses on historical daily load-weighted price from the Queensland region of the Australian National Electricity Market. The sample starts on January 1, 2000 and ends on August 9, 2013, providing 4970 days of data. Table 4.1 presents some descriptive statistics of the load-weighted price broken down by month. Note high average prices in January and February and June and July. The former is due to the demand for air conditioning during the hot and humid summer months in the subtropics. In June and July, high prices are not normally about demand, because the winter is quite mild, but rather reflect supply constraints. These months are when large generating plants carry out maintenance, causing reduction in capacity and making the system more vulnerable to extreme prices due to unforeseen generator outages.

4.1 Load-weighted electricity prices

Table 4.1. Summary of the Queensland load-weighted electricity price broken down by month. The data are daily for the period January 1, 2000 to August 9, 2013.

Month	Mean	Std. dev.	Min.	Max.	Skewness	Kurtosis
Jan	59.06011	124.7908	14.736	1765.796	8.425671	95.72297
Feb	55.87931	153.2503	14.48445	1759.359	8.60393	82.9134
Mar	38.9078	69.55696	9.330691	1298.408	14.27001	251.0589
Apr	32.96155	20.59206	11.76668	182.9332	2.967697	15.50462
May	37.72784	37.37961	13.28534	502.3774	7.167642	72.96917
Jun	54.35443	84.18523	13.89386	763.3647	5.156111	34.39813
Jul	44.90637	46.80732	14.49384	355.7034	4.16223	23.49422
Aug	31.38076	14.05722	1.386365	106.2175	1.655731	7.139884
Sep	29.98898	13.65651	12.4274	103.0397	1.933196	7.840898
Oct	36.96611	36.45492	6.264556	434.6444	6.272909	57.16421
Nov	40.20746	67.19658	3.788165	702.5307	7.412418	63.89256
Dec	41.66091	94.10232	6.218777	1536.139	12.59078	183.4102
Sample	42.08741	76.72463	1.386365	1765.796	12.93129	226.1483

Figure 4.1 plots the logarithm of load-weighted price. Most of the outliers are high values, but the plot in logarithms reveals extreme low values as well. The plot also reveals a great deal of structure in the load-weighted price series. In particular, there appears to be some degree of periodicity as well as periods of sustained high prices. The latter are particularly notable during 2007, a period known as the millennial drought, and again at the end of the sample, when the Australian government introduced a carbon price.[1]

---

1. For present purposes, it will simply be assumed that the series we deal with are stationary. In fact, more formal tests using the methods discussed in chapter 6 confirm that this is a reasonable assumption.

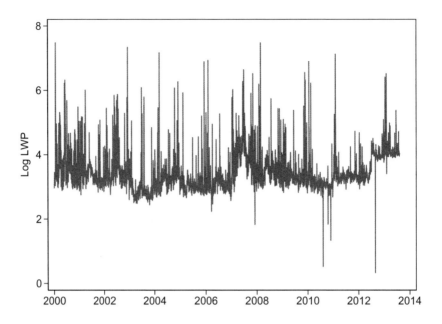

Figure 4.1. Daily load-weighted price for period January 1, 2000 to August 9, 2013 for the Queensland region of the Australian National Electricity Market.

4.2 Specifying and fitting dynamic time-series models

The Box–Jenkins approach to model building comprises the following steps:

Step 1: Identification. Use the data and all related information to help select the type of dynamic time-series model that best suits the data.

Step 2: Estimation. Use the data to estimate the parameters of the model.

Step 3: Diagnostic checking. Evaluate the fitted model and check for areas where it may be improved.

The process is iterative, so information gleaned in step 3 can then be used to start the process again at step 1 if required. Rather than starting with identification, this section will provide a brief description of dynamic time-series models and how they are fit, which will then be followed in the subsequent section by the identification process. This ordering of topics is driven by the need to provide a simple taxonomy of dynamic models to fix terminology and notation.

4.2 Specifying and fitting dynamic time-series models

AR models

The key building block in dynamic models is the AR model of order p. In this model, the information used to explain movements in y_t is contained in p of its own lags. The AR(p) model is given by

$$y_t = \phi_0 + \phi_1 y_{t-1} + \phi_2 y_{t-2} + \cdots + \phi_p y_{t-p} + v_t, \qquad v_t \sim \text{i.i.d.} \, N(0, \sigma_v^2) \qquad (4.1)$$

The fundamental reason for the importance of the AR(p) model in capturing dynamics is that all the explanatory variables are observable, and therefore ordinary least squares may be used to estimate the parameters of the model. However, because this is the first time that dependent data have been encountered, it is useful to establish the log-likelihood function for an AR model. The trick is to express the joint probability density of $\{y_1, y_2, \ldots, y_T\}$ as a sequence of conditional distributions. This sequence of distributions is

$$f(y_1) = f(y_1)$$
$$f(y_1, y_2) = f(y_2|y_1) f(y_1)$$
$$f(y_1, y_2, y_3) = f(y_3|y_2, y_1) f(y_2|y_1) f(y_1)$$
$$\vdots = \vdots \quad \vdots \quad \vdots \quad \vdots$$
$$f(y_1, y_2, \ldots, y_T) = f(y_T|y_{T-1}, \ldots y_1) f(y_{T-1}|y_{T-2}, \ldots y_1) \ldots f(y_3|y_2, y_1) f(y_2|y_1) f(y_1)$$

where y_1 is the initial value and $f(y_1)$ is its (marginal) probability density function. It follows that the joint probability density for all T observations can be written as

$$f(y_1, y_2, \ldots, y_T) = f(y_1) \prod_{t=2}^{T} f(y_t|y_{t-1}, \ldots, y_1)$$

where, for simplicity, the explicit dependence upon parameters has been suppressed. The log-likelihood function is simply a reinterpretation of the joint probability distribution that takes the parameters as arguments. Let $\boldsymbol{\theta} = \{\phi_0, \phi_1, \ldots \phi_p\}$ then

$$\log L(\boldsymbol{\theta}) = \log f(y_1; \boldsymbol{\theta}) + \sum_{t=2}^{T} \log f(y_t|y_{t-1}, y_{t-2}, \ldots, y_1; \boldsymbol{\theta})$$

Note that as $T \to \infty$, the contribution from the first term becomes less and less important as the number of terms in the summation increases. Consequently, it is often simply ignored in the estimation. To fix ideas, the log-likelihood function of the AR(p) model in (4.1) based on the distributional assumptions of v_t is given by

$$\log L(\boldsymbol{\theta}) = -\frac{T}{2} \log 2\pi - \frac{T}{2} \log \sigma_v^2 - \frac{1}{2} \sum_{t=p+1}^{T} \frac{v_t^2}{\sigma_v^2}$$
$$v_t^2 = \{y_t - (\phi_0 + \phi y_{t-1} + \phi_2 y_{t-2} + \cdots + \phi_p y_{t-p})\}^2$$

Moving-average models

An alternative way to introduce dynamics into univariate models is to allow the lags in the dependent variable y_t to be implicitly determined by current and q lagged values of the disturbance term. The model is

$$y_t = \phi_0 + v_t + \psi_1 v_{t-1} + \psi_2 v_{t-2} + \cdots + \psi_q v_{t-q}, \qquad v_t \sim \text{i.i.d. } N(0, \sigma_v^2)$$

in which v_t is a disturbance term with zero mean and constant variance σ_v^2 and $\psi_0, \psi_1, \ldots, \psi_q$ are unknown parameters. Unlike a linear regression or AR model, the moving average model is no longer linear in the parameters. The log-likelihood function of an MA(q) model with $\boldsymbol{\theta} = \{\phi_0, \psi_1, \ldots \psi_q\}$ is

$$\log L(\boldsymbol{\theta}) = -\frac{T}{2} \log 2\pi - \frac{T}{2} \log \sigma_v^2 - \frac{1}{2} \sum_{t=1}^{T} \frac{v_t^2}{\sigma_v^2}$$

$$v_t^2 = \{y_t - (\phi_0 + \psi v_{t-1} + \psi_2 v_{t-2} + \cdots + \psi_p v_{t-q})\}^2$$

The q starting values for $v_1, \ldots, v_q$ may be set to zero. The summation runs from 1 to T unlike the AR case.

If a regression-based approach to fitting models that incorporate MA terms is needed, Durbin (1959) suggests the following two-step procedure:

Step 1: Fit an AR(p) model where p is chosen generously, and compute the residuals

$$\widehat{v}_t = y_t - \left(\phi_0 + \widehat{\phi}_1 y_{t-1} + \widehat{\phi}_2 y_{t-2} + \cdots + \widehat{\phi}_p y_{t-p} \right)$$

Step 2: Fit the model with the lagged residuals from step 1 replacing the unobservable MA error terms. In the case of the MA(p,q) model

$$y_t = \phi_0 + \psi_1 \widehat{v}_{t-1} + \cdots + \psi_q \widehat{v}_{t-q} + v_t^*$$

where v_t^* is the regression disturbance term.

Although the Durbin procedure is not recommended in place of the full maximum-likelihood method, it is useful to obtain starting values and sometimes provides a valuable perspective on the estimation problem.

ARMA models

The ARMA(p,q) model combines the AR(p) and MA(q) models and is given by

$$y_t = \phi_0 + \phi_1 y_{t-1} + \cdots + \phi_p y_{t-p} + v_t + \psi_1 v_{t-1} + \cdots + \psi_q v_{t-q} \qquad v_t \sim \text{i.i.d. } N(0, \sigma_v^2)$$

Once again, because of the presence of the MA(q) component in the model, parameter estimation is by maximum likelihood. The log-likelihood function is

$$\log L(\boldsymbol{\theta}) = -\frac{T}{2}\log 2\pi - \frac{T}{2}\log \sigma_v^2 - \frac{1}{2}\sum_{t=q+1}^{T}\frac{v_t^2}{\sigma_v^2}$$

$$v_t^2 = \{y_t - (\phi_0 + \phi_1 y_{t-1} + \cdots + \phi_p y_{t-p} + v_t + \psi_1 v_{t-1} + \cdots + \psi_q v_{t-q})\}^2$$

with $\boldsymbol{\theta} = \{\phi_0, \phi_1, \phi_2, \ldots, \phi_p, \psi_1, \psi_2, \ldots, \psi_q\}$.

The extension of these dynamic models to include explanatory variables is straightforward. The ARMA-X(p, q) model is given, which is an ARMA model that includes additional explanatory variables

$$y_t = \phi_0 + \phi_1 y_{t-1} + \phi_2 y_{t-2} + \cdots + \phi_p y_{t-p} + \beta_1 x_{1t} + \beta_2 x_{2t} + \cdots + \beta_K x_{Kt}$$
$$+ v_t + \psi_1 v_{t-1} + \cdots + \psi_q v_{t-q} \qquad v_t \sim \text{i.i.d. } N(0, \sigma_v^2)$$

The only differences in the log-likelihood function are that the computation of $\widehat{v}_t$ must be adjusted to reflect the additional explanatory variables and $\boldsymbol{\theta}$ must be expanded to accommodate the extra parameters.

All of these single dependent variable models can be fit by Stata's `arima` command (see [TS] **arima**). By default, Stata will report standard errors based on the outer product of the gradient matrix estimate of the covariance matrix of the parameters, known as the BHHH method (Berndt et al. 1974). This choice is driven by the experience that in time-series models when the log-likelihood and its derivative depend on recursive computations, the estimates based on the outer product of gradients are numerically more stable. A covariance estimate based on the Hessian matrix is obtained by explicitly using the `vce(oim)` option.

4.3 Exploring the properties of dynamic models

While the mechanics of estimating the log-likelihood function for time-series models are relatively straightforward, it is more difficult to choose the correct model with optimal lag orders p and q. An important tool for exploring the dynamics of time-series data and for identifying the appropriate dynamic model is the autocorrelation function (ACF). The kth-order autocovariance of the series y_t is given by

$$\gamma_k = E\left[\{y_t - E(y_t)\}\{y_{t-k} - E(y_{t-k})\}\right] \qquad k = 0, 1, 2 \ldots$$

where γ_0 is the unconditional variance of y_t. The ACF of y_t is then defined as

$$\rho_k = \frac{\gamma_k}{\gamma_0}$$

To compute the ACF, the following sequence of AR models is fit by ordinary least squares, one equation at a time:

$$y_t = \phi_{10} + \rho_1 y_{t-1} + v_{1t}$$
$$y_t = \phi_{20} + \rho_2 y_{t-2} + v_{2t}$$
$$\vdots = \quad \vdots \quad \vdots \quad \vdots$$
$$y_t = \phi_{k0} + \rho_k y_{t-k} + v_{kt}$$

The estimated ACF is given by $\{\widehat{\rho}_1, \widehat{\rho}_2, \ldots, \widehat{\rho}_k\}$. The notation adopted for the constant term emphasizes that this term will be different for each equation.

Figure 4.2 illustrates the ACF using 5,000 observations simulated from the following three models:

$$\begin{aligned} \text{AR}(2) &: y_t = 0.5 y_{t-1} + 0.3 y_{t-2} + v_t \\ \text{MA}(2) &: y_t = -0.4 v_{t-1} + 0.2 v_{t-2} + v_t \\ \text{ARMA}(2,2) &: y_t = 0.5 y_{t-1} + 0.3 y_{t-2} - 0.4 v_{t-1} + 0.2 v_{t-2} + v_t \end{aligned} \quad (4.2)$$

with $v_t \sim N(0, 0.5^2)$. Starting values are generated from a standard normal distribution, and the first 500 observations are discarded prior to estimation.

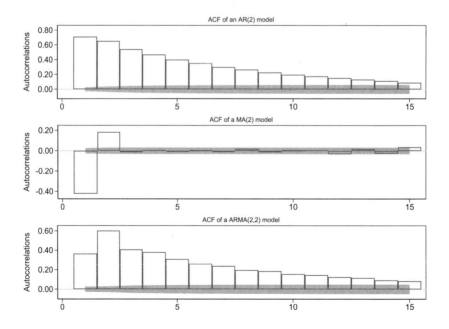

Figure 4.2. ACFs of data simulated from the models in (4.2)

4.3 Exploring the properties of dynamic models

The following general observations may be made about the patterns of ACFs on the basis of the examples in figure 4.2.

1. The ACF of an AR(2) model declines for increasing k, so the effects of previous values on y_t quickly diminish. In general, the autocorrelations of an AR(p) model decay exponentially. For higher-order AR models, the properties of the ACF are generally more complicated but always exhibit exponential decay.
2. The ACF of an MA(q) model cuts off at the MA order, in this case 2.
3. The autocorrelations of the ARMA(p,q) model are more difficult to interpret. In this case, there is some cyclical behavior evident in the first few autocorrelations.

Another measure of the dynamic properties of a time series is the partial autocorrelation function (PACF), which measures the relationship between y_t and y_{t-k} after the intermediate lags have been partialed out of the series. The PACF at lag k is denoted as ϕ_{kk}. By implication, the PACF for an AR(p) model is zero for lags greater than p. For example, in the AR(1) model, the PACF has a spike at lag 1 and thereafter is $\phi_{kk} = 0$, $\forall\, k > 1$. This is in contrast with the ACF, which in general has nonzero values for higher lags. Note that, by construction, the ACF and PACF at lag 1 are equal.

To compute the PACF, the following sequence of models is fit by ordinary least squares, one equation at a time:

$$y_t = \phi_{10} + \phi_{11} y_{t-1} + v_{1t}$$
$$y_t = \phi_{20} + \phi_{21} y_{t-1} + \phi_{22} y_{t-2} + v_{2t}$$
$$y_t = \phi_{30} + \phi_{31} y_{t-1} + \phi_{32} y_{t-2} + \phi_{33} y_{t-3} + v_{3t}$$
$$\vdots = \quad \vdots \quad\quad \vdots \quad\quad \vdots \quad\quad \vdots$$
$$y_t = \phi_{k0} + \phi_{k1} y_{t-1} + \phi_{k2} y_{t-2} + \cdots + \phi_{kk} y_{t-k} + v_{kt}$$

The estimated PACF is therefore given by $\{\widehat{\phi}_{11}, \widehat{\phi}_{22}, \ldots, \widehat{\phi}_{kk}\}$.

Figure 4.3 illustrates the ACFs using the same 5,000 observations simulated from the three models in (4.2) used in figure 4.2. The following general observations may be made about the patterns of PACFs:

1. The PACF of an AR(2) model cuts off at lag 2. More generally, the partial autocorrelations of an AR(p) model are zero for lags higher than p.
2. The PACF of the MA(2) model does not cut off at the MA order. In general, the PACF of an MA process behaves more like the autocorrelation of an AR model.
3. The partial autocorrelations of the ARMA(2,2) model, like the autocorrelations, exhibit cyclical behavior that is quite different from the forms of the AR and MA models. It is difficult, however, to exactly identify the order of the ARMA model from the plot.

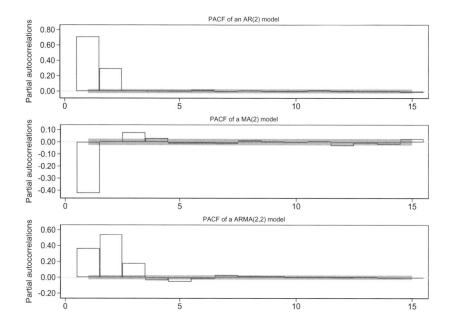

Figure 4.3. PACFs of data simulated from the models in (4.2)

Once the basic type of model (AR, MA, or ARMA) has been decided upon, the next important question is the choice of lag length. As is apparent in the discussion of figures 4.3 and 4.2, the lag structure is not always immediately apparent. Even if it is, it is necessary to verify this choice statistically. If p and q are too small, important dynamics are excluded from the model. If p and q are too large, there are redundant lags that reduce the precision of the parameter estimates. The most common data-driven way of selecting the optimum lag order is to use information criteria. An information criterion is a scalar that is a simple but effective way of balancing the improvement in the fit of the equations with the loss of degrees of freedom that results from increasing the lag order of a time-series model. The three most commonly used information criteria for selecting a parsimonious time-series model are the Akaike information criterion (AIC) (Akaike 1974, 1976), the Hannan–Quinn information criterion (HQIC) (Hannan and Quinn 1979, Hannan 1980), and the Schwarz (1978) or Bayesian information criterion (SBIC) (Schwarz 1978). If $\log L$ is the value of the log-likelihood function evaluated at the model parameters, Stata computes these information criteria as follows:

$$\text{AIC} = -2\frac{\log L}{T} + 2\frac{k}{T}$$
$$\text{HQIC} = -2\frac{\log L}{T} + 2\log\{\log(T)\}\frac{k}{T}$$
$$\text{SBIC} = -2\frac{\log L}{T} + \log(T)\frac{k}{T}$$

in which k is the total number of parameters. The optimal model, and hence the appropriate choice of lag order, is given when the information criteria are minimized. Of course, as will quickly become apparent, in empirical work the information criteria often provide conflicting evidence on the optimal choice of p and q. Both judgment and the context of the problem must be used in formulating the final choice.

4.4 ARMA models for load-weighted electricity price

Figure 4.4 plots the ACF and the PACF for the logarithm of load-weighted price, `llwp`, out to 21 days. There is clearly a strong AR element in the data, but the fact that the PACF remains significant out to the seventh lag suggests that an AR model of at least order 7 is required.

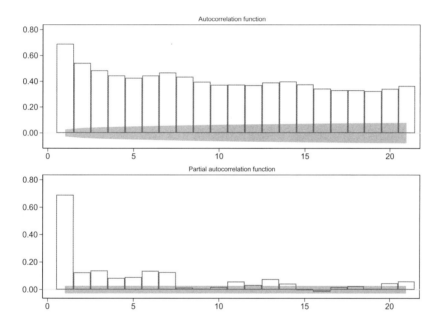

Figure 4.4. ACF of the log of Queensland load-weighted electricity price. Data are daily for the period January 1, 2000 to August 9, 2013.

To select the optimum lag order of the AR model, the Stata command `varsoc` is used, and the maximum lag of 28 is specified. This command is usually used in the context of multiple-equation models (see chapter 5) but can be used here with a single variable as the argument. The procedure fits an AR model up to the order specified in the option `maxlag()` and provides a table of all the computed information criteria. The results are as follows:

```
. use http://www.stata-press.com/data/eeus/llwp
. varsoc llwp, maxlag(28)
   Selection-order criteria
   Sample:  29jan2000 - 09aug2013              Number of obs     =     4942
```

lag	LL	LR	df	p	FPE	AIC	HQIC	SBIC
0	-4363.24				.342428	1.76618	1.76665	1.7675
1	-2782.79	3160.9	1	0.000	.180705	1.12699	1.12791	1.12962
2	-2741.98	81.622	1	0.000	.177817	1.11088	1.11226	1.11483
3	-2689.34	105.29	1	0.000	.174139	1.08998	1.09182	1.09524
4	-2668.03	42.614	1	0.000	.172714	1.08176	1.08407	1.08834
5	-2646.81	42.431	1	0.000	.171307	1.07358	1.07635	1.08148
6	-2597.17	99.284	1	0.000	.167968	1.05389	1.05713	1.06311
7	-2555.75	82.848	1	0.000	.165242	1.03753	1.04123	1.04807
8	-2555.12	1.2615	1	0.261	.165267	1.03768	1.04184	1.04953
9	-2555.01	.21722	1	0.641	.165327	1.03804	1.04266	1.05121
10	-2554.22	1.5811	1	0.209	.165341	1.03813	1.04321	1.05261
11	-2545.55	17.326	1	0.000	.164829	1.03503	1.04057	1.05082
12	-2542.65	5.8008	1	0.016	.164702	1.03426	1.04026	1.05137
13	-2527.57	30.168	1	0.000	.163766	1.02856	1.03502	1.04699
14	-2522.68	9.7822	1	0.002	.163508	1.02698	1.03391	1.04673*
15	-2522.66	.03079	1	0.861	.163573	1.02738	1.03477	1.04844
16	-2522.07	1.1916	1	0.275	.1636	1.02755	1.03539	1.04992
17	-2521.24	1.6515	1	0.199	.163612	1.02762	1.03593	1.05131
18	-2519.41	3.6706	1	0.055	.163556	1.02728	1.03605	1.05229
19	-2519.28	.262	1	0.609	.163614	1.02763	1.03686	1.05396
20	-2513.53	11.494	1	0.001	.1633	1.02571	1.0354	1.05335
21	-2504.01	19.028	1	0.000	.162738	1.02226	1.03242	1.05122
22	-2502.57	2.8808	1	0.090	.162709	1.02209	1.0327	1.05236
23	-2497.12	10.913	1	0.001	.162416	1.02028	1.03136	1.05188
24	-2494.05	6.141	1	0.013	.16228	1.01944	1.03099	1.05235
25	-2493.64	.81148	1	0.368	.162319	1.01968	1.03169	1.05391
26	-2493.63	.0303	1	0.862	.162384	1.02008	1.03255	1.05563
27	-2484.1	19.057*	1	0.000	.161824*	1.01663*	1.02956*	1.05349
28	-2483.41	1.3775	1	0.241	.161845	1.01676	1.03015	1.05493

```
   Endogenous:  llwp
   Exogenous:   _cons
```

The information criteria give different results for the optimal lag structure of the preferred AR model. In fact, these results reflect a commonly observed pattern, namely, that the SBIC chooses a more parsimonious lag order (14 lags) than do the other criteria (27 lags). This is why applications where parsimonious models are required to preserve degrees of freedom, the SBIC choice is often used. Fitting an AR(14) model given by

$$y_t = \phi_0 + \phi_1 y_{t-1} + \phi_2 y_{t-2} + \cdots + \phi_{14} y_{t-14} + v_t$$

4.4 ARMA models for load-weighted electricity price

yields the following results:

```
. regress llwp L(1/14).llwp
```

Source	SS	df	MS		Number of obs	=	4,956
					F(14, 4941)	=	393.03
Model	913.147549	14	65.224825		Prob > F	=	0.0000
Residual	819.98539	4,941	.165955351		R-squared	=	0.5269
					Adj R-squared	=	0.5255
Total	1733.13294	4,955	.349774559		Root MSE	=	.40738

llwp	Coef.	Std. Err.	t	P>|t|	[95% Conf. Interval]	
llwp						
L1.	.5262947	.0142126	37.03	0.000	.4984316	.5541578
L2.	.0195612	.0160463	1.22	0.223	-.0118967	.051019
L3.	.0666695	.0160484	4.15	0.000	.0352076	.0981314
L4.	.0116227	.0160682	0.72	0.470	-.0198782	.0431236
L5.	.0010979	.0160667	0.07	0.946	-.0304	.0325957
L6.	.053052	.0160657	3.30	0.001	.0215562	.0845479
L7.	.1069958	.0160833	6.65	0.000	.0754654	.1385262
L8.	.0038521	.0160833	0.24	0.811	-.0276784	.0353826
L9.	-.0083028	.0160657	-0.52	0.605	-.0397988	.0231931
L10.	-.0186135	.0160659	-1.16	0.247	-.0501099	.0128829
L11.	.0359889	.0160673	2.24	0.025	.0044899	.067488
L12.	-.0072851	.0160475	-0.45	0.650	-.0387453	.0241751
L13.	.0533227	.0160455	3.32	0.001	.0218664	.084779
L14.	.0438922	.0142136	3.09	0.002	.0160273	.071757
_cons	.3864188	.0472455	8.18	0.000	.2937965	.479041

It is clear that the model picks up a lot of AR structure and the coefficients on the higher-order lags are significant, so much so that the choice of 14 seems to be somewhat conservative. On the other hand, there are several lag coefficients of quite low-order that are insignificant. A more parsimonious approach may be to try a low-order ARMA model. There is evidence of some cyclicality in the autocorrelations and partial autocorrelations of load-weighted price. Figures 4.3 and 4.4 suggest that a low-order ARMA model is capable of generating some cyclicality. Consequently, we adopt an ARMA(2,2) model given by

$$y_t = \phi_0 + \phi_1 y_{t-1} + \phi_2 y_{t-2} + v_t + \psi_1 v_{t-1} + \psi_2 v_{t-2}$$

The results obtained from fitting the model are as follows:

```
. arima llwp, ar(1 2) ma(1 2) nolog vsquish
ARIMA regression
Sample: 01jan2000 - 09aug2013                Number of obs    =      4970
                                             Wald chi2(4)     = 503648.75
Log likelihood = -2602.784                   Prob > chi2      =    0.0000
```

	llwp	Coef.	OPG Std. Err.	z	P>|z|	[95% Conf. Interval]	
llwp							
	_cons	3.46054	.1055089	32.80	0.000	3.253746	3.667334
ARMA							
	ar						
	L1.	1.410631	.0240213	58.72	0.000	1.36355	1.457712
	L2.	-.4167726	.0231171	-18.03	0.000	-.4620812	-.3714639
	ma						
	L1.	-.8853634	.025374	-34.89	0.000	-.9350955	-.8356313
	L2.	-.0324997	.0195238	-1.66	0.096	-.0707656	.0057663
	/sigma	.4084591	.0016524	247.19	0.000	.4052204	.4116978

Note: The test of the variance against zero is one sided, and the two-sided
      confidence interval is truncated at zero.

Apart from the second MA lag, the estimated coefficients are statistically significant. What is troubling, however, is that the large coefficient estimate on the first-order AR component is greater than 1 and may indicate instability. Although this result anticipates the formal discussion of the detection of nonstationarity in chapter 6, it may be noted in passing that the stability of the equation may be checked using the `estat aroots` command after an `arima` estimation.

```
. estat aroots, nograph
  Eigenvalue stability condition
```

Eigenvalue	Modulus
.9893885	.989389
.4212426	.421243

```
  All the eigenvalues lie inside the unit circle.
  AR parameters satisfy stability condition.
  Eigenvalue stability condition
```

Eigenvalue	Modulus
.9206636	.920664
-.03530026	.0353

```
  All the eigenvalues lie inside the unit circle.
  MA parameters satisfy invertibility condition.
```

4.4 ARMA models for load-weighted electricity price

These diagnostics show that there is no cause for concern about the stability of the estimated equation.

A final model to try in the category of simple ARMA models is the ARMA-X model. In this case, the ARMA(2,2) model is augmented by day-of-the-week dummy variables. Figure 4.4 reveals that there are noticeable spikes in the PACF of the data, indicating that a periodic weekly effect may be present. Note that the arima command does not accept factor variables, so the prefix command xi is used to create the dummy variables on the fly. The model to be fit is

$$y_t = \phi_0 + \phi_1 y_{t-1} + \phi_2 y_{t-2} + \sum_{i=1}^{6} \gamma_i D_{it} + v_t + \psi_1 v_{t-1} + \psi_2 v_{t-2}$$

where D_{it} are the day of the week dummies for six of the seven days.

```
. xi: arima llwp i.dd, ar(1 2) ma(1 2) nolog vsquish
i.dd              _Idd_0-6            (naturally coded; _Idd_0 omitted)

ARIMA regression

Sample:  01jan2000 - 09aug2013                  Number of obs     =       4970
                                                Wald chi2(10)     =  709838.50
Log likelihood = -2481.294                      Prob > chi2       =     0.0000
```

	llwp	Coef.	OPG Std. Err.	z	P>|z|	[95% Conf. Interval]	
llwp							
	_Idd_1	.2315756	.0166885	13.88	0.000	.1988668	.2642843
	_Idd_2	.2448261	.0203283	12.04	0.000	.2049833	.2846688
	_Idd_3	.2183937	.0230436	9.48	0.000	.173229	.2635584
	_Idd_4	.2397728	.0232204	10.33	0.000	.1942618	.2852839
	_Idd_5	.1714898	.0229736	7.46	0.000	.1264624	.2165171
	_Idd_6	.0533986	.0193903	2.75	0.006	.0153943	.091403
	_cons	3.296375	.1145318	28.78	0.000	3.071897	3.520853
ARMA							
ar							
	L1.	1.479747	.0231556	63.90	0.000	1.434363	1.525131
	L2.	-.4847357	.0223479	-21.69	0.000	-.5285368	-.4409345
ma							
	L1.	-.9630535	.0247175	-38.96	0.000	-1.011499	-.9146081
	L2.	.0337486	.0192302	1.75	0.079	-.003942	.0714392
/sigma		.3985951	.0016005	249.04	0.000	.3954582	.401732

Note: The test of the variance against zero is one sided, and the two-sided confidence interval is truncated at zero.

The results are positive with the day of the week dummies, illustrating contrasts to the omitted day (Sunday), proving to be very significant. Only the second lag on the MA part of the equation is statistically insignificant, confirming the result from the previous model.

The third stage of the Box–Jenkins procedure is diagnostic checking. Figure 4.5 plots the ACF of the residuals of the AR(14), ARMA(2,2), and ARMA-X(2,2) models for the log of the Queensland load-weighted electricity price. What is immediately apparent is that there is a strong weekly effect in these data that is particularly evident in the ARMA(2,2) residuals. Worryingly, the ACF of the AR(14) model is flat out to 14 lags, but then the seasonal effect kicks in again and appears to become worse. This pattern is also evident in the ARMA-X(2,2) model, which otherwise does a good job of modeling the data. These observations based on the ACFs in figure 4.5 suggest that a different approach to the periodic component in the data is required to obtain a satisfactory result.

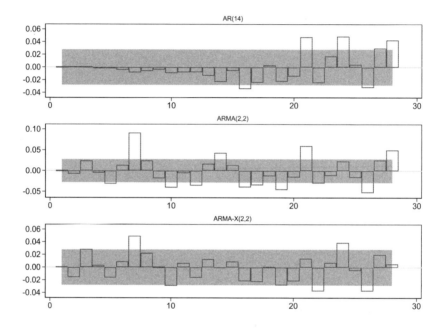

Figure 4.5. ACF of the residuals of the AR(14), ARMA(2,2), and ARMA-X(2,2) models of the log of Queensland load-weighted electricity price. Data are daily for the period January 1, 2000 to August 9, 2013.

4.5 Seasonal ARMA models

What is apparent in the simple ARMA models for load-weighted price is that the strong day of the week effect is not being adequately modeled. This effect can be thought of as a weekly seasonal effect and should be modeled accordingly. The most straightforward way to deal with seasonal effects in an ARMA framework is simply to add AR and MA terms at the seasonal lag. In this instance, a basic ARMA(2,1) model is augmented to include, in a simple additive way, AR and moving-average terms of order 7. The fitted model is

$$y_t = \phi_0 + \phi_1 y_{t-1} + \phi_2 y_{t-2} + \phi_7 y_{t-7} v_t + \psi_1 v_{t-1} + \psi_2 v_{t-2} + \psi_7 v_{t-7}$$

```
. arima llwp, ar(1/2 7) ma(1 7) nolog vsquish
ARIMA regression
Sample:  01jan2000 - 09aug2013              Number of obs   =      4970
                                            Wald chi2(5)    =  2.64e+07
Log likelihood = -2584.431                  Prob > chi2     =    0.0000
```

	llwp	Coef.	OPG Std. Err.	z	P>|z|	[95% Conf. Interval]	
llwp	_cons	3.493684	.1643145	21.26	0.000	3.171634	3.815735
ARMA	ar						
	L1.	1.586146	.0156264	101.50	0.000	1.555519	1.616773
	L2.	-.5339893	.0119218	-44.79	0.000	-.5573556	-.5106231
	L7.	-.052607	.0051657	-10.18	0.000	-.0627316	-.0424824
	ma						
	L1.	-1.064403	.0141458	-75.25	0.000	-1.092128	-1.036678
	L7.	.0742921	.0103132	7.20	0.000	.0540787	.0945056
	/sigma	.4069462	.0016318	249.39	0.000	.403748	.4101444

Note: The test of the variance against zero is one sided, and the two-sided
      confidence interval is truncated at zero.

All the estimated parameters are statistically significant, and the stability of the fitted model is verified using `estat aroots`.

The strength of the periodicity of the seasonal component of the load-weighted price data makes it unlikely that a simple additive model will be an adequate representation of the data. A multiplicative approach to dealing with the seasonality is adopted instead. Stationary seasonal ARMA models require that the seasonal AR and moving-average terms enter the equation multiplicatively, and this gives rise to a new notation that can be a little tricky to decipher. The basic notation for a seasonal ARMA model is $(p,q) \times (P,Q)_s$ where p and q are interpreted as usual, while P_s, Q_s refer to multiplicative seasonals. There are two points to bear in mind.

1. The subscript s is the seasonal period, and P and Q represent the AR and MA seasonal lags. For the current example, $s = 7$, so $P = Q = 1$ will mean y_{t-7} is the

appropriate AR term, and v_{t-7} is the MA term. On the other hand, if $P = Q = 2$ is specified, the required lag orders are y_{t-7}, y_{t-14}, v_{t-7}, and v_{t-14}.

2. The seasonal lags enter multiplicatively and not additively. This multiplication results in higher-order terms appearing in the equation to be estimated whose coefficients are subject to nonlinear constraints.

The pertinent question now is how to multiply lag polynomials. To do so, it is useful to use lag operator notation. Define the lag operator L such that

$$y_t = L^0 y_t, \quad y_{t-1} = L y_t \quad y_{t-2} = L^2 y_t \quad \cdots \quad y_{t-p} = L^P y_t$$

Using this construction, it becomes clear that

$$y_{t-1} \times y_{t-2} = L y_t \times L^2 y_t = L^3 y_t = y_{t-3}$$

As an example, consider the $(2,1) \times (1,2)_4$ seasonal ARMA model, and examine the AR and MA components of the model separately.

1. AR terms:
   The AR part of the estimated equation will be obtained by multiplying out the following expression in the lag polynomial

   $$\left(1 - \phi_1 L - \phi_2 L^2\right)\left(1 - \phi_4 L^4\right) y_t$$

   which has two ordinary lag terms and one seasonal lag. As a consequence, the terms y_{t-1} and y_{t-2} appear, as well as y_{t-4} from the seasonal polynomial. In addition, both y_{t-1} and y_{t-2} must be multiplied by y_{t-4}, which means that y_{t-5} and y_{t-6} will also appear in the equation.

2. MA terms:
   The MA part of the estimated equation will be obtained by multiplying out the following expression in the lag polynomial

   $$\left(1 + \psi_1 L\right)\left(1 + \psi_4 L^4 + \psi_8 L^8\right) v_t$$

   which has one ordinary lag and two seasonal lags. The terms v_t, v_{t-1}, v_{t-4}, and v_{t-8} will appear, together with v_{t-4} and v_{t-8} each multiplied by v_{t-1}.

The entire model to be fit is then

$$y_t = \phi_0 + \phi_1 y_{t-1} + \phi_2 y_{t-2} + \phi_4 y_{t-4} - \phi_1 \phi_4 y_{t-5} - \phi_2 \phi_4 y_{t-6}$$
$$+ v_t + \psi_1 v_{t-1} + \psi_4 v_{t-4} + \psi_8 v_{t-8} + \psi_1 \psi_4 v_{t-5} + \psi_1 \psi_8 v_{t-9}$$

Note that the additional lagged terms in both the AR part and the MA part do not require the estimation of additional parameters because their coefficients are combinations of the coefficients on earlier lags.

4.5 Seasonal ARMA models

Seasonal ARMA models are fit in Stata using the sarima() option after the arima command. The model is fit by maximum likelihood, with the nonlinear constraints on the coefficients imposed. The fact that the model is fit subject to the nonlinear coefficient constraints means that only the primitive parameter estimates are reported by Stata and not the derived coefficients on the higher-order multiplicative terms. The model to be fit has the form $(2,1) \times (1,1)_7$ and is given by

$$y_t = \phi_0 + \phi_1 y_{t-1} + \phi_2 y_{t-2} + \phi_7 y_{t-7} - \phi_1 \phi_7 y_{t-8} - \phi_2 \phi_7 y_{t-9} + v_t$$
$$+ \psi_1 v_{t-1} + \psi_7 v_{t-7} + \psi_1 \psi_8 v_{t-8}$$

Note that estimates will be reported only for ϕ_0, ϕ_1, ϕ_2, ϕ_7, ψ_1, ψ_7, and σ_v—a total of seven parameters.

```
. arima llwp, arima(2,0,1) sarima(1,0,1,7) nolog vsquish
ARIMA regression
Sample:  01jan2000 - 09aug2013                  Number of obs     =      4970
                                                Wald chi2(5)      =  1.06e+06
Log likelihood = -2485.839                      Prob > chi2       =    0.0000
```

	llwp	Coef.	OPG Std. Err.	z	P>|z|	[95% Conf. Interval]	
llwp							
	_cons	3.600051	.2962483	12.15	0.000	3.019415	4.180687
ARMA							
ar							
	L1.	1.41778	.0127322	111.35	0.000	1.392825	1.442735
	L2.	-.4300421	.0107018	-40.18	0.000	-.4510173	-.4090669
ma							
	L1.	-.899965	.0102908	-87.45	0.000	-.9201346	-.8797955
ARMA7							
ar							
	L1.	.9968419	.0011892	838.24	0.000	.9945111	.9991727
ma							
	L1.	-.9774831	.0034569	-282.76	0.000	-.9842585	-.9707077
/sigma		.3987281	.0014573	273.61	0.000	.3958719	.4015843

Note: The test of the variance against zero is one sided, and the two-sided confidence interval is truncated at zero.

Returning once more to the third stage of the Box–Jenkins procedure, diagnostic testing, figure 4.6 plots the ACF of the additive and multiplicative seasonal models. The additive seasonal model performs particularly poorly, but the signs are extremely positive for the multiplicative seasonal model. The $(2,1) \times (1,1)_7$ specification of the seasonal ARMA model comes close to capturing all the structure in the model residuals.

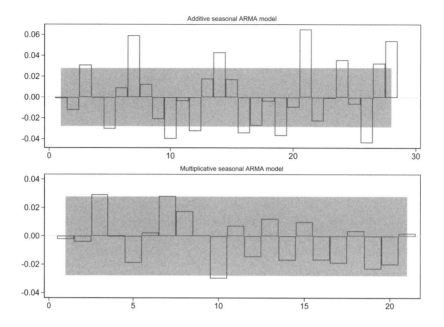

Figure 4.6. ACF of the residuals of the additive seasonal model and the multiplicative seasonal model of the log of Queensland load-weighted electricity price. Data are daily for the period January 1, 2000 to August 9, 2013.

In conclusion, there are many flavors of dynamic models that have been introduced in this chapter. The presence of seasonality can often complicate the process. One thing that is absolutely certain is the presence of persistence in many environmental time series. Models that ignore this persistence and do not account for dynamics in some form or another are bound to be misspecified. The power and ease of use of the Stata `arima` command greatly facilitate dynamic model building along the lines originally suggested by Box and Jenkins.

Exercises

1. `llwp.dta` contains daily data on load-weighted price, `lwp`, for the Queensland region of the Australian National Electricity Market for the period January 1, 2000 to August 9, 2013.

    a. Plot the time series of `llwp`, the logarithm of the load-weighted price.
    b. Fit an AR(7) regression model for `llwp`, and interpret the results.
    c. Generate indicator variables that capture the effect on load-weighted prices
        - when the millennium drought most affected the electricity market (June 1, 2007 to July 31, 2007), and

4.5 Seasonal ARMA models

- during the period of the carbon tax (July 1, 2012 to July 31, 2014).

Include these dummy variables in the model, and test for their joint significance.

d. The month of the year and the day of the week are potentially important in explaining the dynamics of electricity prices. Use Stata's factor-variable notation to include these effects in the model. Test for the significance of the monthly effects and the daily effects separately, and then provide a joint test of all of these deterministic effects. What do you conclude about the relative strengths of these effects?

e. Use Stata's factorial interaction operator to estimate and test whether the first-order autocorrelation coefficient on `llwp` depends on the day of the week.

f. Based on your modeling results thus far, fit a preferred model, and provide diagnostic tests on the residuals of the model. Interpret the results.

2. `pm_hourly.dta` contains hourly data on various measures of air pollution in Santiago, Chile, for the period January 1, 2009 00:00:00 to December 31, 2014 23:00:00.

   a. Plot the autocorrelation and PACFs for `pm25`, using a horizon of 96 hours. Comment on the results.

   b. Generate a daily dataset for `pm25`, and plot the autocorrelation and PACFs using a horizon of 32 days. Comment on the results.

   c. Based on your results in exercise 2b, build a dynamic model for `tpm25` using only its own lags and deterministic terms. Test the resultant equation for residual autocorrelation and heteroskedasticity.

   d. Now, return to the hourly data for `pm25` and build a dynamic model for the hourly data using only lags of `pm25` and deterministic terms. Test the residuals of your dynamic equation for heteroskedasticity and autocorrelation.

   e. Use the Stata factorial interaction operator to estimate a dynamic regression in which the first lag of `pm25` in a dynamic regression varies for every hour. Interpret the results.

5 Multivariate time-series models

In chapter 4, we introduced a general model of the dynamics of a single variable, known as the autoregressive moving-average (ARMA) model. Its multivariate extension is the vector autoregressive moving-average (VARMA) model, which, although conceptually straightforward, is usually fit using the techniques that will be introduced in chapter 7. This chapter will focus primarily on a restricted version of the general VARMA model, in which the dynamics of the dependent variables are driven solely by their own lags. This special case is known as a vector autoregression (VAR) model (Mann and Wald 1943; Sims 1980).

The popularity of VAR models is partly due to the fact that there is often no relevant theoretical foundation for determining the empirical relationship between variables that appear to be jointly determined. An attractive feature of VAR models is that one does not have to specify a specific structure, such as delimiting some variables as dependent variables and others as explanatory variables. Rather, the model is driven solely by the stochastic nature of variables. One of the problems of VAR modeling is that the number of parameters to be estimated grows very quickly, so interpretation of the model is complicated. This chapter will discuss the basic tools for simplifying and interpreting the dynamics of VAR models.

In addition to the statistical approaches to aid interpretation, there are many instances in environmental model building where physical laws help with specifying model structure. For example, one might argue with some validity that CO_2 emissions are largely due to economic activity, but you would be less confident in specifying the nature of this interaction in detail. Hybrid models known as structural vector autoregressions (SVARs), in which some minimal structure is imposed on the dynamics of the VAR, are becoming increasingly popular in environmental econometrics. This chapter deals with two varieties of restrictions that are commonly used in the area of SVAR modeling. Note that the assumption of stationarity of the variables invoked for the previous chapter will be maintained here.

5.1 CO_2 emissions and growth

There appears to be a strong link between per capita CO_2 emissions and the business cycle, a link that has captured the attention of policymakers, theorists, and empirical researchers alike since its discovery in the early 1990s (Grossman and Krueger 1993, 1995). Figure 5.1 demonstrates this link by showing that the CO_2 to output ratio declines in a linear fashion from the start of the sample period in 1973.

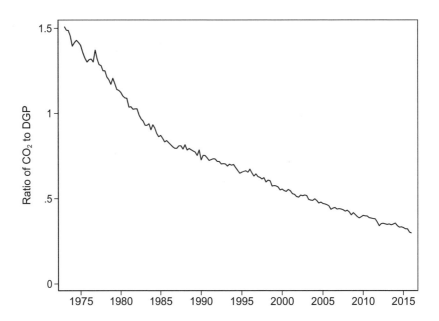

Figure 5.1. Plot of the scaled ratio of per capita CO_2 emissions to real per capita gross domestic product (GDP) in the United States. The data are quarterly for the period 1973Q1 to 2016Q1.

The theoretical and empirical literature both have mostly focused on developing models that replicate the inverted U-shaped relationship between emissions and gross domestic product (GDP), a phenomenon that has become known as the environmental Kuznets curve (EKC). The EKC is formulated in terms of the levels of CO_2 emissions and real GDP, a specification that runs into problems in terms of satisfying the stationarity condition. Accordingly, testing the EKC is best left until after the techniques for dealing with nonstationary variables are introduced.

Casual empirical evidence of another form of the link between CO_2 emissions and real GDP is given by figure 5.2, which plots the cyclical components of per capita CO_2 emissions and the logarithm of real GDP using quarterly U.S. data.[1] The degree of correlation of the two cyclical components is quite remarkable. Carbon emissions are highly procyclical, suggesting that they are linked with the shocks that drive fluctuations in the economy. These stylized facts raise two immediate questions. The first is whether such fluctuations in carbon emissions should be accounted for in improving environmental policy. A better understanding of carbon emissions' response to business-cycle shocks can help evaluate the relevance of optimal environmental policy.

---

[1]. The Hodrick–Prescott filter is used to compute the cyclical components shown in figure 5.2. See Hodrick and Prescott (1997) for a discussion of the filter. To compute the Hodrick–Prescott filter in Stata, use `tsfilter hp` (see [TS] **tsfilter hp**) with `smooth(1600)`.

5.2 The VARMA model

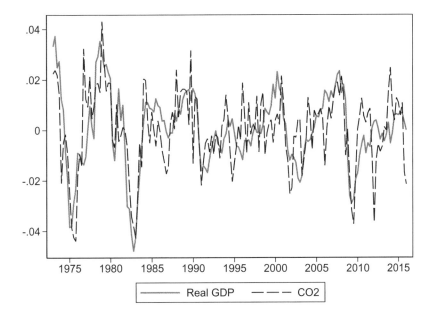

Figure 5.2. Plot of the cyclical components of real GDP and per capita CO_2 emissions in the United States. The cyclical components are extracted using a Hodrick–Prescott filter with smoothing parameter set to 1600. The data are quarterly for the period 1973Q1 to 2016Q1.

These questions will be addressed in this chapter, first in a completely data-driven way and then by introducing limited information in terms of the identification of the effect of structural technology shocks on carbon emissions.

5.2 The VARMA model

The natural extension to the univariate ARMA class of models discussed in chapter 4 is one in which $\mathbf{y}_t$ represents a vector of N time series. The VARMA(p,q) model is given by

$$\mathbf{y}_t = \mathbf{\Phi}_0 + \mathbf{\Phi}_1 \mathbf{y}_{t-1} + \cdots + \mathbf{\Phi}_p \mathbf{y}_{t-p} + v_t + \mathbf{\Psi}_1 v_{t-1} + \cdots + \mathbf{\Psi}_q v_{t-q}, \quad \mathbf{v}_t \sim N(0, V) \quad (5.1)$$

in which $\mathbf{y}_t$ and $\mathbf{v}_t$ are the N dimensional vectors

$$\mathbf{y}_t = \begin{bmatrix} y_{1t} \\ y_{2t} \\ \vdots \\ y_{Nt} \end{bmatrix} \quad \mathbf{v}_t = \begin{bmatrix} v_{1t} \\ v_{2t} \\ \vdots \\ v_{Nt} \end{bmatrix}$$

where $\boldsymbol{\Phi}_0$ is an $(N \times 1)$ vector of constants and $\boldsymbol{\Phi}_i$ with $i = 1,\ldots,p$ and $\boldsymbol{\Psi}_j$ with $j = 1,\ldots,q$ are both $(N \times N)$ parameter matrices. $\boldsymbol{\Phi}_0$, $\boldsymbol{\Phi}_i$, and $\boldsymbol{\Psi}_j$ are given, respectively, by

$$\boldsymbol{\Phi}_0 = \begin{bmatrix} \phi_{10} \\ \vdots \\ \phi_{N0} \end{bmatrix}, \quad \boldsymbol{\Phi}_i = \begin{bmatrix} \phi_{i11} & \cdots & \phi_{i1N} \\ \vdots & \ddots & \vdots \\ \phi_{iN1} & \cdots & \phi_{iNN} \end{bmatrix}, \quad \boldsymbol{\Psi}_j = \begin{bmatrix} \psi_{j11} & \cdots & \psi_{j1N} \\ \vdots & \ddots & \vdots \\ \psi_{jN1} & \cdots & \psi_{jNN} \end{bmatrix}$$

and $\mathbf{V}$ is the $(N \times N)$ covariance matrix of v_t. Note that the constant terms in the model have been suppressed for notational convenience.

The parameters of the model in (5.1) may be estimated by maximum likelihood. Let the parameter vector to be $\Theta = \{\boldsymbol{\Phi}_0, \boldsymbol{\Phi}_1, \boldsymbol{\Phi}_2, \ldots, \boldsymbol{\Phi}_p, \boldsymbol{\Psi}_1, \boldsymbol{\Psi}_2, \ldots, \boldsymbol{\Psi}_q\}$ and let $s = \max(p, q)$ be the maximum lag in the model. The log-likelihood function to be maximized is given by

$$\log L(\Theta) = \sum_{t=s+1}^{T} \log f(\mathbf{y}_t | \mathbf{y}_{t-1}, \mathbf{y}_{t-2}, \ldots, y_1; \Theta)$$

in which $\log f(\mathbf{y}_t | \mathbf{y}_{t-1}, \mathbf{y}_{t-2}, \ldots, y_1; \Theta)$ is the joint probability density function of the sample of T observations of $\mathbf{y}_t$ conditional on its own lags. Based on the assumption that the distribution of v_t is multivariate normal as in (5.1), and that for simplicity $p = q = s = 1$, the log-likelihood function for the VARMA$(1,1)$ model given a sample of T observations on the N components of the vector $\mathbf{y}_t$ is

$$\log L(\Theta) = -\frac{TN}{2}\log 2\pi - \frac{T}{2}\log |\mathbf{V}| - \frac{1}{2}\sum_{t=2}^{T} v_t' \mathbf{V}^{-1} v_t \qquad (5.2)$$

with[2]

$$\mathbf{V} = \frac{1}{T}\sum_{t=2}^{T} v_t v_t' \qquad (5.3)$$

In principle, the estimation of the VARMA model is straightforward, but there is no official Stata command to implement it. However, it is possible to fit VARMA models using state-space modeling, as presented in chapter 9.

5.3 The VAR model

If the restriction $q = 0$ is imposed, then the VARMA model has only an autoregressive structure with no moving-average part. The resulting model is an autoregressive model of order p given by

$$\mathbf{y}_t = \boldsymbol{\Phi}_0 + \boldsymbol{\Phi}_1 \mathbf{y}_{t-1} + \cdots + \boldsymbol{\Phi}_p \mathbf{y}_{t-p} + v_t, \quad \mathbf{v}_t \sim N(0, \mathbf{V}) \qquad (5.4)$$

---

[2]. The denominator $T - s$ could be used in computing the covariance matrix, but the maximum likelihood estimator uses T.

5.3 The VAR model

The important distinguishing features of the VAR model are that there is an equation for each of the N variables in the system and that each equation has the same set of explanatory variables, namely, the lagged values of all the variable in the system. For $N = 2$, the VAR model is

$$y_{1t} = \phi_{10} + \sum_{i=1}^{p} \phi_{i11} y_{1t-i} + \sum_{i=1}^{p} \phi_{i12} y_{2t-i} + v_{1t}$$

$$y_{2t} = \phi_{20} + \sum_{i=1}^{p} \phi_{i21} y_{1t-i} + \sum_{i=1}^{p} \phi_{i22} y_{2t-i} + v_{2t} \quad (5.5)$$

where y_{1t} and y_{2t} are the dependent variables; p is the lag length, which is the same for all equations; and v_{1t} and v_{2t} are disturbance terms.

Higher-dimensional VARs containing N variables $\{y_{1t}, y_{2t}, \ldots, y_{Nt}\}$ are specified in exactly the same way, although it is convenient to use matrix notation as in (5.4). The disturbances $\mathbf{v}_t = \{v_{1t}, v_{2t}, \ldots, v_{Nt}\}$ have zero mean with covariance matrix

$$\mathbf{V} = \begin{bmatrix} \sigma_1^2 & \sigma_{12} & \cdots & \sigma_{1N} \\ \sigma_{21} & \sigma_2^2 & \cdots & \sigma_{2N} \\ \vdots & \vdots & \ddots & \vdots \\ \sigma_{N1} & \sigma_{N2} & \cdots & \sigma_N^2 \end{bmatrix} \quad (5.6)$$

This matrix is a symmetric matrix with nonzero off-diagonal elements, $\sigma_{ij} = \sigma_{ji} \neq 0$, for all $i \neq j$. The maximum likelihood estimates of the model parameters, $\Theta = \{\mathbf{\Phi}_1, \mathbf{\Phi}_2, \ldots, \mathbf{\Phi}_p\}$, based on a sample of T observations on the N components of the vector $\mathbf{y}_t$, are obtained by maximizing the log-likelihood function given in (5.2) and constructing an estimate of $\mathbf{V}$ using (5.3).

The usual situation in VAR modeling is that N, the number of equations in the system, is much smaller than T, the number of observations. The system may then be regarded as a seemingly unrelated regression (SUR) model as proposed by Zellner (1962), in which each equation is a valid model in its own right and can be estimated equation by equation. Although the latter approach yields consistent estimates, full maximum-likelihood estimation provides an estimator that is also efficient because the estimation accounts for the additional information in the covariance matrix, $\mathbf{V}$. There is one instance where the ordinary least squares, SUR, and maximum likelihood approaches are identical, namely, when the regressors on the right-hand side of each equation are exactly the same. This is exactly the situation in the VAR model.

Consider the simple VAR(1) linking the growth rate of CO_2 emissions and the growth rate of real GDP:

$$\Delta \text{gdp}_t = \phi_{10} + \phi_{11} \Delta \text{gdp}_t + \phi_{12} \Delta \text{co2}_{t-1} + v_{1t}$$
$$\Delta \text{co2}_t = \phi_{20} + \phi_{21} \Delta \text{gdp}_t + \phi_{22} \Delta \text{co2}_{t-1} + v_{2t}$$

Lowercase variable names denote logarithms, and v_{1t} and v_{2t} are disturbance terms. Equation-by-equation ordinary least squares produces the following results:

```
. use http://www.stata-press.com/data/eeus/vars
. regress  Dgdp L.Dco2 L.Dgdp if tin(1973q3,2016q1)
```

Source	SS	df	MS
Model	.001560271	2	.000780136
Residual	.009136189	168	.000054382
Total	.010696461	170	.00006292

Number of obs =	171
F(2, 168) =	14.35
Prob > F =	0.0000
R-squared =	0.1459
Adj R-squared =	0.1357
Root MSE =	.00737

Dgdp	Coef.	Std. Err.	t	P>|t|	[95% Conf. Interval]	
Dco2						
L1.	.0503804	.0274662	1.83	0.068	-.0038429	.1046038
Dgdp						
L1.	.3332546	.0726398	4.59	0.000	.1898502	.4766591
_cons	.0044558	.0007514	5.93	0.000	.0029724	.0059392

```
. regress  Dco2 L.Dco2 L.Dgdp if tin(1973q3,2016q1)
```

Source	SS	df	MS
Model	.008958785	2	.004479393
Residual	.066035211	168	.000393067
Total	.074993996	170	.000441141

Number of obs =	171
F(2, 168) =	11.40
Prob > F =	0.0000
R-squared =	0.1195
Adj R-squared =	0.1090
Root MSE =	.01983

Dco2	Coef.	Std. Err.	t	P>|t|	[95% Conf. Interval]	
Dco2						
L1.	-.2124942	.0738421	-2.88	0.005	-.3582721	-.0667163
Dgdp						
L1.	.8378277	.1952899	4.29	0.000	.4522891	1.223366
_cons	-.0089928	.0020201	-4.45	0.000	-.0129809	-.0050048

The results indicate that GDP growth is an important predictor of the growth rate of CO_2 emissions but that the reverse relationship is much weaker. The growth rate of CO_2 emissions has a t statistic of 1.85 with an associated p-value of 0.066 in the equation for the growth rate of real GDP.

5.3 The VAR model

In Stata, the `sureg` command (see [R] **sureg**) will estimate the system by feasible generalized least squares. The `sureg` results are

```
. sureg (Dgdp Dco2 = L.Dgdp L.Dco2) if tin(1973q3,2016q1), small dfk
Seemingly unrelated regression
```

Equation	Obs	Parms	RMSE	"R-sq"	F-Stat	P
Dgdp	171	2	.0073744	0.1459	14.35	0.0000
Dco2	171	2	.0198259	0.1195	11.40	0.0000

	Coef.	Std. Err.	t	P>|t|	[95% Conf. Interval]	
Dgdp						
Dgdp L1.	.3332546	.0726398	4.59	0.000	.1903686	.4761407
Dco2 L1.	.0503804	.0274662	1.83	0.067	-.003647	.1044078
_cons	.0044558	.0007514	5.93	0.000	.0029777	.0059338
Dco2						
Dgdp L1.	.8378277	.1952899	4.29	0.000	.4536827	1.221973
Dco2 L1.	-.2124942	.0738421	-2.88	0.004	-.3577452	-.0672432
_cons	-.0089928	.0020201	-4.45	0.000	-.0129665	-.0050192

The `sureg` options `dfk` and `small` cause the commands to use small-sample degrees-of-freedom adjustments when computing the estimated covariance matrix, $\widehat{\mathbf{V}}$, and to report small-sample t and F statistics instead of the large-sample normal and chi-squared statistics. These options make the results equivalent to the ordinary least-squares results.

Of course, if there are restrictions on the parameters of the VAR, then ordinary least squares is no longer appropriate. The Stata command `var` (see [TS] **var**) is the correct command to use when fitting a VAR. If there are no restrictions and the explanatory variables are identical, then `var` will use ordinary least squares applied to each equation. If restrictions are imposed, `var` will use an iterative SUR estimator.

```
. var Dgdp Dco2, lags(1) small dfk

Vector autoregression
Sample:  1973q3 - 2016q1                Number of obs   =         171
Log likelihood =   1029.465             AIC             =   -11.97035
FPE            =   2.17e-08             HQIC            =   -11.92562
Det(Sigma_ml)  =   2.02e-08             SBIC            =   -11.86012
```

Equation	Parms	RMSE	R-sq	F	P > F
Dgdp	3	.007374	0.1459	14.34546	0.0000
Dco2	3	.019826	0.1195	11.39601	0.0000

		Coef.	Std. Err.	t	P>|t|	[95% Conf. Interval]	
Dgdp							
	Dgdp L1.	.3332546	.0726398	4.59	0.000	.1898502	.4766591
	Dco2 L1.	.0503804	.0274662	1.83	0.068	-.0038429	.1046038
	_cons	.0044558	.0007514	5.93	0.000	.0029724	.0059392
Dco2							
	Dgdp L1.	.8378277	.1952899	4.29	0.000	.4522891	1.223366
	Dco2 L1.	-.2124942	.0738421	-2.88	0.005	-.3582721	-.0667163
	_cons	-.0089928	.0020201	-4.45	0.000	-.0129809	-.0050048

An important part of the specification of a VAR is the choice of the optimal lag length, p. If p is too small, important parts of the dynamics are excluded from the model. If p is too large, then there are redundant lags, which can reduce the precision of the parameter estimates, inflating the standard errors and yielding t statistics that are relatively too small. Moreover, in choosing a lag structure in a VAR, care must be exercised because degrees of freedom can quickly diminish for even moderate lag lengths.

Just as in the univariate case outlined in chapter 4, the Akaike information criterion (AIC) (Akaike 1974, 1976), the Hannan–Quinn information criterion (HIC) (Hannan and Quinn 1979; Hannan 1980), and the Schwarz (1978) or Bayesian information criterion (SBIC) (Schwarz 1978) are often used to select the optimum lag order. In the case of a VAR, the Stata command varsoc will automatically search for the optimal lag order based on the information criteria, up to the user-specified option maxlag().

5.4.1 Granger causality testing

```
. varsoc Dgdp Dco2, maxlag(8)
   Selection-order criteria
   Sample:  1975q2 - 2016q1                    Number of obs     =       164
```

lag	LL	LR	df	p	FPE	AIC	HQIC	SBIC
0	970.712				2.5e-08	-11.8136	-11.7982	-11.7758
1	993.747	46.07	4	0.000	2.0e-08	-12.0457	-11.9997*	-11.9323*
2	998.693	9.8926	4	0.042	2.0e-08*	-12.0572*	-11.9805	-11.8682
3	1001.49	5.6011	4	0.231	2.0e-08	-12.0426	-11.9352	-11.778
4	1003.44	3.8852	4	0.422	2.1e-08	-12.0175	-11.8794	-11.6773
5	1007.12	7.3727	4	0.117	2.1e-08	-12.0137	-11.8449	-11.5979
6	1008.99	3.7333	4	0.443	2.1e-08	-11.9877	-11.7882	-11.4962
7	1011.08	4.1743	4	0.383	2.2e-08	-11.9643	-11.7341	-11.3973
8	1020.17	18.189*	4	0.001	2.1e-08	-12.0265	-11.7656	-11.3838

```
   Endogenous:  Dgdp Dco2
   Exogenous:   _cons
```

The results obtained from `varsoc` show that the AIC chooses a lag order of 2, while the HIC and SBIC select 1 as the optimal lag length. Generally speaking, these results are fairly typical in that the AIC is known to favor longer lag lengths, while the HIC and SBIC favor shorter lag lengths. When there is a conflict between the optimal lag lengths selected by the information criteria, individual judgment must be exercised. In this instance, parsimony is an important consideration because the sample size is relatively small. The choice of a lag length of 1 in this instance seems justified.

5.4 Analyzing the dynamics of a VAR

The fact that VARs produce many parameters creates difficulties in understanding the dynamic interrelationships among the variables in the system. There are three common methods used in empirical work to help with the interpretation of the dynamics of a VAR model, namely, Granger causality testing, impulse–response functions (IRF), and variance decompositions.

5.4.1 Granger causality testing

In a VAR model, all lags are assumed to contribute to the prediction of each dependent variable. However, in most empirical applications, many of the estimated coefficients are statistically insignificant. It is then a question of crucial importance to determine if at least one of the parameters on the lagged values of the explanatory variables in any equation is not zero. In the bivariate VAR case, this suggests that a test of the information content of y_{2t} on y_{1t} in (5.5) is given by testing the joint restrictions

$$\phi_{112} = \phi_{212} = \phi_{312} = \cdots = \phi_{p12} = 0$$

If y_{2t} is important in predicting future values of y_{1t} over and above lags of y_{1t} alone, then y_{2t} is said to cause y_{1t} in the sense of Granger (1969). It is, of course, also possible to test for Granger causality in the reverse direction by performing a joint test of the lags of y_{1t} in the y_{2t} equation. The restrictions needed to establish these causal patterns are tested using Wald tests, which have χ_r^2 distributions under the null hypothesis, where r is the number of zero restrictions being tested. It is important to remember that the formulation of the tests in terms of zero restrictions on the coefficients of the VAR mean that in the case of, say, testing for Granger causality from y_{2t} to y_{1t}, the null and alternative hypotheses are

$$H_0 : \quad y_{2t} \text{ does not cause } y_{1t}$$
$$H_1 : \quad y_{2t} \text{ does cause } y_{1t}$$

The tests are routinely provided by Stata using the vargranger (see [TS] **vargranger**) command after estimation of a VAR.

Consider the VAR linking the growth rate of CO_2 emissions and the growth rate of real GDP, but now set the lag length to 1 as chosen by the HIC and SBIC criteria. The results of Granger causality tests after estimation of the VAR are as follows:

```
. quietly var Dgdp Dco2, lags(1) small dfk
. vargranger
   Granger causality Wald tests
```

Equation	Excluded	F	df	df_r	Prob > F
Dgdp	Dco2	3.3645	1	168	0.0684
Dgdp	ALL	3.3645	1	168	0.0684
Dco2	Dgdp	18.406	1	168	0.0000
Dco2	ALL	18.406	1	168	0.0000

These results indicate that excluding real GDP growth from the equation for the growth rate of CO_2 emissions is strongly rejected. The conclusion is that real GDP growth Granger causes the growth of CO_2 emissions. The reverse null hypothesis—that the growth rate of CO_2 emissions can be excluded from the real GDP equation—cannot be rejected at the 5% level, because the p-value of the test is 0.068. In other words, the growth of CO_2 emissions does not Granger cause real GDP growth at the 5% level.

Note, however, that Granger causality tests depend strongly on the choice of lag length and the sample over which the VAR is estimated. For example, assume that the optimal lag length is chosen as 2, the value returned by the AIC statistic. In this instance, the results of the Granger causality test from CO_2 to real GDP growth is more decisive. The p-value of the Wald test is 0.35, and Granger causality is strongly rejected.

5.4.2 Impulse–responses

```
. quietly var Dgdp Dco2, lags(1/2) small dfk
. vargranger
  Granger causality Wald tests
```

Equation	Excluded	F	df	df_r	Prob > F
Dgdp	Dco2	1.0447	2	165	0.3541
Dgdp	ALL	1.0447	2	165	0.3541
Dco2	Dgdp	14.721	2	165	0.0000
Dco2	ALL	14.721	2	165	0.0000

Various methods have been proposed to try and account for this observed instability in Granger causality test results with respect to the estimation period. Original contributions by Thoma (1994) on the use of a forward expanding window for Granger causality testing and by Swanson (1998) on a rolling window version prompted interest in the problem of dealing with the time-varying nature of causal relationships in economics. Recent work by Shi, Phillips, and Hurn (2018) and Shi, Hurn, and Phillips (2020) suggests an exhaustive search over all possible subsamples to determine the dynamic evolution of causal relationships over time.

5.4.2 Impulse–responses

Although Granger causality is a helpful tool for understanding the dynamics of a VAR, the relations between the variables, namely, CO_2 and GDP in the example used in this chapter, are difficult to see directly from the parameter matrices. Consequently, another tool is often used called impulse–response analysis, which is based on the counterfactual experiment of tracing the marginal effect of a shock to only one variable in the system and evaluating the responses of the variables this impulse. For example, the counterfactual question might be what happens to CO_2 emissions over time if there is a positive demand shock that raises GDP in the current period. Tracing the response of both variables displays the full dynamics of the system and how the variables interact with each other over time. Exactly how only one variable can be shocked at a time requires a little manipulation, so there are several steps involved in constructing impulse–responses, each of which will be considered in turn.

Vector moving-average form

The first step is to express the VAR(p) model given in (5.4) in an equivalent form. In the case of stationary variables, the Wold representation theorem states that, subject to some technical conditions—in particular, invertibility of the y polynomial—$\mathbf{y}_t$ can be expressed as an infinite vector moving-average model given by

$$\mathbf{y}_t = \boldsymbol{\mu} + v_t + \boldsymbol{\Psi}_1 v_{t-i} + \boldsymbol{\Psi}_2 v_{t-2} + \boldsymbol{\Psi}_3 v_{t-3} + \cdots + \boldsymbol{\Phi}_h v_{t-h} + \cdots \qquad (5.7)$$

where $\boldsymbol{\mu}$ is an $(N \times 1)$ vector of constants. Furthermore, the construction of the $\boldsymbol{\Psi}$ matrices requires no further estimation but can be constructed from the estimated VAR matrices, $\boldsymbol{\Phi}$, using the recursion

$$\begin{aligned}
\boldsymbol{\Psi}_1 &= \boldsymbol{\Phi}_1 \\
\boldsymbol{\Psi}_2 &= \boldsymbol{\Phi}_1\boldsymbol{\Psi}_1 + \boldsymbol{\Phi}_2 \\
\boldsymbol{\Psi}_3 &= \boldsymbol{\Phi}_1\boldsymbol{\Psi}_2 + \boldsymbol{\Phi}_2\boldsymbol{\Psi}_1 + \boldsymbol{\Phi}_3 \\
\boldsymbol{\Psi}_4 &= \boldsymbol{\Phi}_1\boldsymbol{\Psi}_3 + \boldsymbol{\Phi}_2\boldsymbol{\Psi}_2 + \boldsymbol{\Phi}_3\boldsymbol{\Psi}_1 + \boldsymbol{\Phi}_4 \\
&\vdots \\
\boldsymbol{\Psi}_i &= \sum_{j=1}^{i} \boldsymbol{\Phi}_j \boldsymbol{\Psi}_{i-j}
\end{aligned}$$

with $\boldsymbol{\Phi}_j = 0, j > p$.

The Wold form is useful for tracing the effects of shocks. From (5.7), it follows that

$$\begin{aligned}
\mathbf{y}_t &= \boldsymbol{\mu} + v_t + \boldsymbol{\Psi}_1 v_{t-1} + \boldsymbol{\Psi}_2 v_{t-2} + \boldsymbol{\Psi}_3 v_{t-3} + \cdots \\
\mathbf{y}_{t+1} &= \boldsymbol{\mu} + v_{t+1} + \boldsymbol{\Psi}_1 \mathbf{v}_t + \boldsymbol{\Psi}_2 v_{t-1} + \boldsymbol{\Psi}_3 v_{t-2} + \cdots \\
\mathbf{y}_{t+2} &= \boldsymbol{\mu} + v_{t+2} + \boldsymbol{\Psi}_1 v_{t+1} + \boldsymbol{\Psi}_2 \mathbf{v}_t + \boldsymbol{\Psi}_3 v_{t-1} + \cdots \\
&\vdots \qquad \vdots \qquad \vdots \qquad \vdots \\
\mathbf{y}_{t+h} &= \boldsymbol{\mu} + v_{t+h} + \boldsymbol{\Psi}_1 v_{t+h-1} + \boldsymbol{\Psi}_2 v_{t+h-2} + \cdots + \boldsymbol{\Psi}_h \mathbf{v}_t + \cdots
\end{aligned} \qquad (5.8)$$

This form emphasizes that a shock to $\mathbf{v}_t$ affects $\mathbf{y}_{t+1}$ through $\boldsymbol{\Psi}_1$, $\mathbf{y}_{t+2}$ through $\boldsymbol{\Psi}_2$, and hence also $\mathbf{y}_{t+h}$ through $\boldsymbol{\Psi}_h$. So we may be tempted to claim that the response of $\mathbf{y}_{t+h}$ to a change in v_t is

$$\frac{\partial \mathbf{y}_{t+h}}{\partial v_t} = \boldsymbol{\Psi}_h$$

This claim is, however, premature.

Orthogonalized impulses

Interpreting the dynamics of the VAR in terms of $\mathbf{v}_t = \{v_{1t}, v_{2t}, \ldots, v_{Nt}\}$ is complicated by the fact that $\mathbf{V}$, as shown in (5.6), is a nondiagonal matrix. Therefore, the disturbances are correlated with each other. Tracing the effect of a shock to v_{it} on the system by means of the $\boldsymbol{\Psi}$ matrices as claimed previously is impossible because v_{it} is correlated with all the other elements of v_t. To make interpretation easier, we must define additional structure on the system.

5.4.2 Impulse–responses

The way in which this structure is imposed in this simple VAR framework is to define a recursive order on the disturbances v_t. Impulse–response analysis transforms v_{it} into another disturbance term z_{it}, which satisfies the recursive ordering while being uncorrelated by construction with other elements of z. Consider a matrix $\mathbf{S}$ that is lower triangular, and define

$$\begin{bmatrix} S_{11} & 0 & \cdots & 0 \\ S_{21} & S_{22} & \cdots & 0 \\ \vdots & \vdots & \ddots & \vdots \\ S_{N1} & S_{N2} & \cdots & S_{NN} \end{bmatrix} \begin{bmatrix} z_{1t} \\ z_{2t} \\ \vdots \\ z_{Nt} \end{bmatrix} = \begin{bmatrix} v_{1t} \\ v_{2t} \\ \vdots \\ v_{Nt} \end{bmatrix} \qquad (5.9)$$

This form of $\mathbf{S}$ implies the following recursive structure:

$$S_{11} z_{1t} = v_{1t}$$
$$S_{21} z_{1t} + S_{22} z_{2t} = v_{2t}$$
$$S_{31} z_{1t} + S_{32} z_{2t} + S_{33} z_{3t} = v_{3t}$$
$$\vdots \qquad \vdots \qquad \vdots$$
$$S_{N1} z_{1t} + S_{N2} z_{2t} + \cdots + S_{NN} z_{Nt} = v_{Nt}$$

It can then be deduced that a shock to the structural disturbance z_{1t}, which is orthogonal in that all other elements of $\mathbf{z}_t$ are zero, will have immediate impact on all the elements of $\mathbf{y}_t$. A shock to z_{2t} will affect all the variables in $\mathbf{y}_t$ with the exception of y_{1t}. Similarly, a shock to z_{3t} will affect all the variables in $\mathbf{y}_t$ but not y_{1t} and y_{2t}. So a shock to z_{ht} will generally have a contemporaneous effect on all y_{jt}, where $j \geq h$.

Sims (1980) suggested that the matrix $\mathbf{S}$ be obtained by a technique known as a Choleski decomposition, in which the covariance matrix $\mathbf{V}$ in (5.6) is decomposed into the product of a lower triangular matrix $\mathbf{S}$ and its transpose, $\mathbf{V} = \mathbf{S}\mathbf{S}'$. Using the relationship in (5.9), the Choleski approach implies that

$$\mathbf{S}\, E(\mathbf{z}_t \mathbf{z}_t')\, \mathbf{S}' = E(v_t v_t') = \mathbf{S}\mathbf{S}' = \mathbf{V}$$

It is now clear that $\mathbf{S}$ is the matrix equivalent of the square root of the variance, $\mathbf{V}$. Because $\mathbf{z}_t = \mathbf{S}^{-1} v_t$, these transformed errors are standardized orthogonal shocks. The impulse–responses can now be obtained from

$$\mathbf{y}_t = \boldsymbol{\mu} + \mathbf{S}\mathbf{z}_t + \boldsymbol{\Psi}_1 \mathbf{S} z_{t-1} + \boldsymbol{\Psi}_2 \mathbf{S} z_{t-2} + \boldsymbol{\Psi}_3 \mathbf{S} z_{t-3} + \cdots$$

and, by inspection of the expressions in (5.8), the IRF are

$$\mathrm{IRF}_h = \frac{\partial \mathbf{y}_{t+h}}{\partial \mathbf{z}_t} = \boldsymbol{\Psi}_h \mathbf{S}$$

Impulse–response computation in Stata is part of the `var postestimation` (see [TS] **var postestimation**) suite, where the command `irf` is used to create IRFs using a three-stage process:

Step 1: Fit a VAR model.

Step 2: Create an IRF and a related IRF file using the `irf create` command.

Step 3: Graph or summarize the impulse–responses in a table using `irf graph` or `irf table`.

The IRF for the bivariate VAR for the growth rates of real GDP and CO_2 may be implemented using the following commands:

```
. quietly var Dgdp Dco2, lags(1) small dfk
. local endash = ustrunescape("\u2013")
. irf create co2, step(8) set(co2) replace
. irf graph oirf, irf(co2) impulse(Dgdp) response(Dco2) individual scheme(s1mono)
> xtitle("Forecast horizon (quarters)") title("") legend(off)
> title("Impulse`endash´response: GDP &rarr COsub:2",
> size(medsmall)) iname(g1, replace) xlab(#5) subtitle("")
. irf graph coirf, irf(co2) impulse(Dgdp) response(Dco2) individual  scheme(s1mono)
> xtitle("Forecast horizon (quarters)") title("") legend(off)
> title("Cumulative impulse`endash´response: GDP &rarr COsub:2",
> size(medsmall)) iname(g2, replace) xlab(#5) subtitle("")
. graph combine g11 g21, ycommon rows(1)
```

After fitting the VAR, the `irf create` command sets up the current IRF and names it co2. From the results stored in co2.irf, the command `irf graph oirf` graphs the effect of an orthogonal one-standard-deviation shock to the growth rate of real GDP on the growth rate of CO_2. The use of the `individual` option requires that only the requested IRF is graphed. Because the data are in growth rates, it may be that the cumulative impact of the shock is the main point of interest. The command `irf graph coirf` will plot the required cumulative responses to a one-standard-deviation orthogonalized shock. The results obtained from these commands are shown in figure 5.3.

5.4.2 Impulse–responses

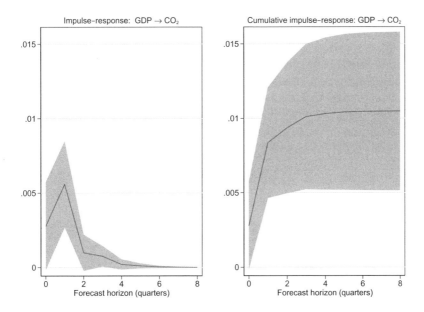

Figure 5.3. Impulse–response functions illustrating the dynamic interaction of real GDP and CO_2 emissions. The data are quarterly for the period 1972q1 to 2012q1.

The response of the growth rate of CO_2 to the GDP shock is positive and also contemporaneous because the effect is seen at lag 0. The effect of the impulse intensifies at lag 1; this result follows from the fact that the VAR is specified with one lag. Thereafter, the impulse rapidly falls away to zero. Note, however, that the cumulative impulse–response, defined in terms of the level of CO_2, is not zero but settles down to a higher equilibrium level following the initial impulse. The gray bands around the two IRFs represent the 95% confidence intervals. Arguably, only the impulse at lag 1 is significantly different from zero in the left-hand panel, but in the right-hand panel the new equilibrium is established at a level that is significantly different from zero.

The same information can be obtained in tabular form using the `irf table` command.

```
. irf table oirf, irf(co2)  noci
                Results from co2

         |    (1)       (2)       (3)       (4)
   step  |   oirf      oirf      oirf      oirf
---------+------------------------------------------
   0     |  .007374   .002781     0        .01963
   1     |  .002598   .005588   .000989   -.004171
   2     |  .001147   .000989   .000119    .001715
   3     |  .000432   .000751   .000126   -.000264
   4     |  .000182   .000202   .000029    .000162
   5     |  .000071   .000109   .000018   -.00001
   6     |  .000029   .000036   5.4e-06    .000017
   7     |  .000012   .000017   2.7e-06    8.9e-07
   8     |  4.7e-06   6.1e-06   9.3e-07    2.0e-06

(1) irfname = co2, impulse = Dgdp, and response = Dgdp
(2) irfname = co2, impulse = Dgdp, and response = Dco2
(3) irfname = co2, impulse = Dco2, and response = Dgdp
(4) irfname = co2, impulse = Dco2, and response = Dco2
```

Examination of the table of impulse–responses highlights the importance of variable ordering to the propagation of shocks. The ordering imposed by the estimation of the VAR is that a shock to real GDP will affect all variables in the current period, but a shock to CO_2 emissions will only have an effect on real GDP in the next quarter. The use of the Choleski decomposition to identify the orthogonal structural disturbances is responsible for 0 appearing as the first element of the third column.

5.4.3 Forecast-error variance decomposition

Impulse–response analysis provides information on the dynamics of the VAR system of equations and how each variable responds and interacts to shocks in the other variables in the system. Forecast-error variance decomposition analysis, on the other hand, provides insight into the *relative importance* of shocks on the movements in the variables in the system. The analysis decomposes movements in each variable over the horizon of the impulse–response analysis into the separate relative effects of each shock with the results expressed as a percentage of the overall movement.

From (5.8), the actual and expected values of $\mathbf{y}_{t+1}$ conditional on information at time t are, respectively,

$$\mathbf{y}_{t+1} = \boldsymbol{\mu} + v_{t+1} + \boldsymbol{\Psi}_1 \mathbf{v}_t + \boldsymbol{\Psi}_2 v_{t-1} + \boldsymbol{\Psi}_3 v_{t-2} + \cdots$$
$$E_t(\mathbf{y}_{t+1}) = \boldsymbol{\mu} + \boldsymbol{\Psi}_1 \mathbf{v}_t + \boldsymbol{\Psi}_2 v_{t-1} + \boldsymbol{\Psi}_3 v_{t-2} + \cdots$$

where the notation $E_t(\cdot)$ emphasizes that conditioning is on information at time t. It follows that

$$\mathbf{y}_{t+1} - E_t(\mathbf{y}_{t+1}) = v_{t+1}$$

5.4.3 Forecast-error variance decomposition

At time $t+2$ and conditional still on information at time t, we have

$$\mathbf{y}_{t+2} = \boldsymbol{\mu} + v_{t+2} + \boldsymbol{\Psi}_1 v_{t+1} + \boldsymbol{\Psi}_2 \mathbf{v}_t + \boldsymbol{\Psi}_3 v_{t-1} + \cdots$$
$$E_t(\mathbf{y}_{t+2}) = \boldsymbol{\mu} + \boldsymbol{\Psi}_2 \mathbf{v}_t + \boldsymbol{\Psi}_3 v_{t-1} + \cdots$$

so that

$$\mathbf{y}_{t+2} - E_t(\mathbf{y}_{t+2}) = v_{t+2} + \boldsymbol{\Psi}_1 v_{t+1}$$

Similarly, at time $t+3$

$$y_{t+3} - E_t(y_{t+3}) = v_{t+3} + \boldsymbol{\Psi}_1 v_{t+2} + \boldsymbol{\Psi}_2 v_{t+1}$$

and in general

$$\mathbf{y}_{t+h} - E_t(\mathbf{y}_{t+h}) = v_{t+h} + \boldsymbol{\Psi}_1 v_{t+h-1} + \cdots \boldsymbol{\Psi}_{h-1} v_{t+1}$$

This result may be written as

$$\mathbf{y}_{t+h} - E_t(\mathbf{y}_{t+h}) = \sum_{i=0}^{h-1} \boldsymbol{\Psi}_i v_{t+h-i} = \sum_{i=0}^{h-1} \boldsymbol{\Psi}_i \mathbf{S} z_{t+h-i}$$

with $\boldsymbol{\Psi}_0 = I_N$ and where the second equality on the right-hand side follows from the decomposition of the shocks, v_t, given in (5.9). In other words, the error in forecasting $\mathbf{y}_{t+h}$ depends on all the shocks up to $h-1$. Recognizing that $E(\mathbf{z}_t \mathbf{z}_t') = I_N$, the variance of this forecast error can be expressed as

$$\mathbf{V}D_h = \sum_{i=0}^{h-1} \boldsymbol{\Psi}_i \mathbf{S}\mathbf{S}' \boldsymbol{\Psi}_i'$$

This summation can now be separated into the contributions of each of the shocks as

$$\mathbf{V}D_h = \sum_{i=0}^{h-1} \boldsymbol{\Psi}_i S_1 S_1' \boldsymbol{\Psi}_i' + \sum_{i=0}^{h-1} \boldsymbol{\Psi}_i S_2 S_2' \boldsymbol{\Psi}_i' + \cdots + \sum_{i=0}^{h-1} \boldsymbol{\Psi}_i S_N S_N' \boldsymbol{\Psi}_i'$$

in which the variance is written in terms of the separate contributions of each of the N shocks in the system to the h-step ahead variances and where S_i is the ith column of $\mathbf{S}$.

For the bivariate VAR(4) for the growth rate of GDP and CO_2 emissions, the variance decomposition for eight quarters ahead is computed as follows:

```
. irf table fevd, irf(co2) noci
                Results from co2
```

step	(1) fevd	(2) fevd	(3) fevd	(4) fevd
0	0	0	0	0
1	1	.019679	0	.980321
2	.984253	.088197	.015747	.911803
3	.984358	.089616	.015642	.910384
4	.984157	.090753	.015843	.909247
5	.984153	.090831	.015847	.909169
6	.984149	.090855	.015851	.909145
7	.984149	.090858	.015851	.909142
8	.984149	.090859	.015851	.909141

```
(1) irfname = co2, impulse = Dgdp, and response = Dgdp
(2) irfname = co2, impulse = Dgdp, and response = Dco2
(3) irfname = co2, impulse = Dco2, and response = Dgdp
(4) irfname = co2, impulse = Dco2, and response = Dco2
```

Columns 1 and 3 contain the variance decomposition of real GDP, and columns 2 and 4 contain the variance decomposition of CO_2 emissions. Taking real GDP first, at time $t+1$ the contribution to variance is entirely due to the shock to GDP itself. This follows directly from the causal ordering imposed. By contrast, for CO_2 emissions, while the vast majority of 98% is made up of the shock to CO_2, there is a 2% contribution from the GDP shock. Thereafter, the variance decomposition is much as expected. The contribution of CO_2 to the forecast variance of the GDP hardly changes and reaches 1.6% by $t+8$, while the influence of the GDP shock on CO_2 is much stronger, contributing 9.1% of the variance by $t+8$.

5.5 SVARs

Recall that the VAR(p) is given by

$$\mathbf{y}_t = \mathbf{\Phi}_0 + \mathbf{\Phi}_1 \mathbf{y}_{t-1} + \mathbf{\Phi}_2 \mathbf{y}_{t-2} + \cdots \mathbf{\Phi}_p \mathbf{y}_{t-p} + \mathbf{v}_t \qquad v_t \sim N(0, \mathbf{V}) \qquad (5.10)$$

In examining the dynamics of a VAR model, it was necessary to impose a specific recursive ordering on the variables to derive a set of orthogonal structural disturbances that give the impulse–responses and variance decompositions structural interpretations. This purely statistical approach uses a Choleski decomposition to estimate the lower triangular matrix $\mathbf{S}$, which satisfies the condition $\mathbf{V} = \mathbf{SS}'$. This approach imposes dynamics that may not necessarily be consistent with the true structure of the underlying processes. Moreover, different recursive orderings give different results, and the number of alternative orderings increases dramatically as the number of variables in the model increases.

5.5.1 Short-run restrictions

An alternative approach is to specify a VAR model with additional structure imposed to ensure that the model disturbances themselves are orthogonal to each other and therefore have the interpretation of being structural disturbances. SVAR models do not require the imposition of a strict recursive structure on the model but instead specify restrictions that generally are motivated by economic theory. For a comprehensive review of SVAR modeling, see Ouliaris, Pagan, and Restrepo (2016).

5.5.1 Short-run restrictions

Short-run restrictions involve explicitly defining the contemporaneous relationships between the variables $\mathbf{y}_t$. Consider writing the VAR(p) as

$$\mathbf{A}\mathbf{y}_t = A_0 + A_1 \mathbf{y}_{t-1} + A_2 \mathbf{y}_{t-2} + \cdots + A_p \mathbf{y}_{t-p} + \mathbf{B}\mathbf{z}_t \qquad (5.11)$$

in which $\mathbf{B}$ is a diagonal matrix. The fact that $\mathbf{B}$ is a diagonal matrix means that the disturbances, $\mathbf{z}_t$, are uncorrelated (a condition implied by the assumption that the structural model is correctly specified) and therefore have the interpretation of being structural disturbances. The matrix $\mathbf{A}$ expresses the contemporaneous relationships between the variables, $\mathbf{y}_t$, and is defined to be consistent with economic theory. The relationship between the parameter matrices of the VAR and those of the SVAR are given by $\mathbf{\Phi}_i = \mathbf{A}^{-1} \mathbf{A}_i$. In addition, comparing (5.10) and (5.11) reveals that $\mathbf{v}_t = \mathbf{A}^{-1} \mathbf{B} \mathbf{z}_t$, so $\mathbf{A}^{-1}\mathbf{B}$ plays the same role as the Choleski decomposition matrix $\mathbf{S}$ encountered previously.

Estimation of the structural parameters of the model in Stata is performed by maximum likelihood using the svar (see [TS] **var svar**) command and specifying the restrictions on the two matrices $\mathbf{A}$ and $\mathbf{B}$. An intuitive approach to obtaining structural parameter estimates from the reduced-form VAR is useful to fix ideas. The process is as follows:

Step 1: Estimate each equation in the reduced-form VAR by ordinary least squares to obtain the estimates $\widehat{\mathbf{\Phi}}_1, \widehat{\mathbf{\Phi}}_2, \ldots, \widehat{\mathbf{\Phi}}_p$ and compute the VAR residuals $\widehat{v}_t$.

Step 2: Compute the estimated covariance matrix of the VAR residuals

$$\widehat{\mathbf{V}} = \frac{1}{T-p} \sum_{t=p+1}^{T} \widehat{v}_t \widehat{v}_t'$$

Step 3: The Choleski decomposition of $\widehat{\mathbf{V}}$, given by

$$\widehat{\mathbf{V}} = \widehat{\mathbf{S}}\widehat{\mathbf{S}}'$$

will provide an estimate of the matrix $\mathbf{S}$. The matrices $\widehat{\mathbf{A}}^{-1}$ and $\widehat{\mathbf{B}}$ are then obtained by solving

$$\widehat{\mathbf{S}} = \widehat{\mathbf{A}}^{-1}\widehat{\mathbf{B}}$$

In general, the matrix multiplication on the right-hand side results in a nonlinear combination of the elements of **A** and **B**, which are set equal to the matching elements of the matrix **S**. The result is a system of nonlinear equations that must be solved.

Step 4: Once $\widehat{\mathbf{A}}^{-1}$ is available, the $\widehat{\mathbf{A}}_i$s are then computed from $\widehat{\mathbf{A}}_i = \widehat{\mathbf{A}}^{-1}\widehat{\boldsymbol{\Phi}}_i$ where the $\widehat{\boldsymbol{\Phi}}_i$ are estimated in step 1.

Consider a bivariate SVAR(1) model of the growth rates of real labor productivity (Dlp) and CO_2 emissions (Dco2). The main interest here is that a shock to labor productivity may be interpreted as an unanticipated technology shock. This scenario is consistent with the theory of real business cycles, and this shock has an immediate short-run effect on CO_2 emissions. To fit this model in Stata, the short-run restrictions must be imposed. The matrix **A** in (5.11) and the matrix **B** are defined as

$$\mathbf{A} = \begin{bmatrix} 1 & 0 \\ a_{21} & 1 \end{bmatrix}, \quad \mathbf{B} = \begin{bmatrix} b_{11} & 0 \\ 0 & b_{22} \end{bmatrix}$$

in which there is only one contemporaneous short-run effect, a_{12}, to estimate, together with two elements on the main diagonal of **B**. The elements of matrices **A** and **B** to be estimated are designated as missing values (.). Estimation of the SVAR proceeds as follows:

```
. use http://www.stata-press.com/data/eeus/vars, clear
. matrix A = (1,0\.,1)
. matrix B = (.,0\0,.)
. svar Dlp Dco2 if tin(1973q1,2012q1), aeq(A) beq(B) lags(1) var dfk small nolog
> vsquish
Vector autoregression
Sample: 1973q3 - 2012q1                    Number of obs   =         155
Log likelihood =   956.0014                AIC             =   -12.25808
FPE            =    1.63e-08               HQIC            =   -12.21023
Det(Sigma_ml)  =    1.51e-08               SBIC            =   -12.14027
```

Equation	Parms	RMSE	R-sq	F	P > F
Dlp	3	.006044	0.0426	3.377982	0.0367
Dco2	3	.020763	0.0216	1.676415	0.1905

	Coef.	Std. Err.	t	P>|t|	[95% Conf. Interval]
Dlp					
Dlp L1.	.1939277	.0797213	2.43	0.016	.0364229 .3514325
Dco2 L1.	-.0174376	.0239197	-0.73	0.467	-.0646957 .0298204
_cons	.0031809	.0005822	5.46	0.000	.0020306 .0043312

5.5.1 Short-run restrictions

```
Dco2
    Dlp
    L1.       .044577   .2738648    0.16   0.871    -.4964959    .58565

    Dco2
    L1.     -.1484234   .0821709   -1.81   0.073    -.310768    .0139213

    _cons   -.0033957      .002    -1.70   0.092    -.0073472   .0005558
```

Estimating short-run parameters
Structural vector autoregression
(1) [/A]1_1 = 1
(2) [/A]1_2 = 0
(3) [/A]2_2 = 1
(4) [/B]1_2 = 0
(5) [/B]2_1 = 0

Sample: 1973q3 − 2012q1   Number of obs   =   155
Exactly identified model   Log likelihood =   952.972

		Coef.	Std. Err.	t	P>|t|	[95% Conf. Interval]
/A						
	1_1	1	(constrained)			
	2_1	.2588672	.2751437	0.94	0.348	-.2847325 .8024669
	1_2	0	(constrained)			
	2_2	1	(constrained)			
/B						
	1_1	.0060441	.0003433	17.61	0.000	.0053659 .0067224
	2_1	0	(constrained)			
	1_2	0	(constrained)			
	2_2	.0207042	.0011759	17.61	0.000	.018381 .0230275

The results indicate that the contemporaneous effect of the technology shock on CO_2 emissions is not statistically significant, calling the model into question. Khan et al. (2015) provide a fairly extensive discussion of the counterintuitive result that a positive technology shock has an immediate negative effect on emissions (recall that the matrix $\mathbf{A}$ contains the negatives of the contemporaneous effects because it appears on the left-hand side in the SVAR specification), but they do not consider the statistical significance of their result.

These SVAR short-run restrictions are exactly those imposed by the Choleski decomposition approach to the identification of the VAR impulse–responses used previously. This may be easily demonstrated by showing that $\mathbf{A}^{-1}\mathbf{B} = \mathbf{S}$. The matrices $\widehat{\mathbf{A}}$ and $\widehat{\mathbf{B}}$ are stored by svar as e(A) and e(B), respectively. Furthermore, the Choleski decomposition matrix can be obtained from the estimated covariance matrix of the unrestricted VAR residuals estimated in step 1 and stored as e(Sigma).

```
. matrix Aest = e(A)
. matrix Best = e(B)
. matrix chol_est = inv(Aest)*Best
```

```
. matrix list chol_est
chol_est[2,2]
            Dlp        Dco2
 Dlp    .00604413          0
 Dco2  -.00156463   .02070425

. matrix sig_var = e(Sigma)

. matrix chol_var = cholesky(sig_var)

. matrix list chol_var
chol_var[2,2]
            Dlp        Dco2
 Dlp    .00604413          0
 Dco2  -.00156463   .02070425
```

It is evident that the matrix `chol_est` obtained from the SVAR is identical to the matrix `chol_var`, which is obtained from the Choleski decomposition of the unrestricted VAR residuals.

5.5.2 Long-run restrictions

Another way to impose restrictions is to constrain the impulse–responses in the long run to have values that are motivated by economic theory. This does not impose restrictions in the short run but ensures that the restrictions are binding in the limit as $t \to \infty$. Consider again (5.11), and assume that $\mathbf{A}$ is the identity matrix. In the long-run steady state, where all change in $\mathbf{y}_t$ ceases, it follows that

$$(I_n - \boldsymbol{\Phi}_1 - \boldsymbol{\Phi}_2 - \cdots - \boldsymbol{\Phi}_p)\mathbf{y}_t = \mathbf{B}\mathbf{z}_t$$
$$\Rightarrow \quad \mathbf{y}_t = (I_n - \boldsymbol{\Phi}_1 - \boldsymbol{\Phi}_2 - \cdots - \boldsymbol{\Phi}_p)^{-1}\mathbf{B}\mathbf{z}_t$$

The sum of the coefficient matrices is commonly written in short-hand as $\boldsymbol{\Phi}(1)$, so

$$\mathbf{y}_t = \mathbf{C}\mathbf{z}_t, \qquad \mathbf{C} = \boldsymbol{\Phi}(1)^{-1}\mathbf{B}$$

where $\mathbf{C}$ is a matrix that summarizes the long-run responses to the orthogonalized structural shocks. Consequently, the long-run identifying restrictions are specified in terms of the elements of the matrix $\mathbf{C}$. For example, constraining $\mathbf{C}_{12} = 0$ forces the long-run response of variable y_{1t} to a shock in y_{2t} to be zero.

Stata estimates of the model parameters when there are long-run restrictions are obtained using `svar` and specifying the $\mathbf{C}$ matrix. The intuitive approach using the reduced form is similar to the short-run restrictions case, with the exception that step 3 is slightly different.

Step 1: Estimate each equation in the reduced-form VAR by ordinary least squares to obtain the estimates $\widehat{\boldsymbol{\Phi}}_1, \widehat{\boldsymbol{\Phi}}_2, \ldots, \widehat{\boldsymbol{\Phi}}_p$, and compute the VAR residuals $\widehat{v}_t$.

Step 2: Compute the estimated covariance matrix of the VAR residuals

$$\widehat{\mathbf{V}} = \frac{1}{T-p} \sum_{t=p+1}^{T} \widehat{v}_t \widehat{v}_t'$$

5.5.2 Long-run restrictions

Step 3: The Choleski decomposition of $\widehat{\mathbf{V}}$ is given by

$$\widehat{\mathbf{V}} = \widehat{\mathbf{S}}\widehat{\mathbf{S}}'$$

and in this instance, estimates of $\mathbf{C}$ can be obtained by solving

$$\widehat{\mathbf{S}} = \widehat{\mathbf{\Phi}}(1)\widehat{\mathbf{C}}$$

Once again, the matrix multiplication on the right-hand side will generally be a nonlinear combination of the terms in $\mathbf{\Phi}(1)$ and $\mathbf{C}$. Solving this problem amounts to solving a system of nonlinear equations.

In the real business cycle literature, variation in productivity is viewed as the main driver of business cycles, while fluctuations in the price of investment goods relative to the price of consumption goods are also regarded as important drivers of U.S. business cycles. The relationship between these shocks and CO_2 emissions can be expressed in a long-run SVAR specified as

$$\mathbf{y}_t = \begin{bmatrix} \mathtt{Dp}_t \\ \mathtt{Dlp}_t \\ \mathtt{Dco2}_t \end{bmatrix}$$

in which $\mathtt{Dp}_t$ is the change in logarithm of the relative price of investment, $\mathtt{Dlp}_t$ is the growth rate of labor productivity, and $\mathtt{Dco2}_t$ is the growth rate of per capita CO_2 emissions.

The first structural shock to the system, z_{1t}, is an investment shock and relies on the assumption that it has a long-run effect on the relative price of investment. The second shock, z_{2t}, is a technology shock that is assumed to affect labor productivity in the long run. The third shock, z_{3t}, has no explicit structural interpretation. The matrix of long-run restrictions, $\mathbf{C}$, is given by

$$\mathbf{C} = \begin{bmatrix} c_{11} & 0 & 0 \\ c_{21} & c_{22} & 0 \\ c_{31} & c_{32} & c_{33} \end{bmatrix}$$

which encapsulates the following restrictions:

1. An investment shock affects only the long-run relative price of investment: $c_{12} = c_{13} = 0$.
2. Investment and technology shocks affect labor productivity in the long-run, but CO_2 emissions do not: $c_{23} = 0$.

Intuitively, the effect of these positive business cycle shocks on CO_2 emissions could be either positive or negative depending on the relative strength of two effects. On one hand, the income effect predicts that an increase in wealth due to the improved relative price of investment or labor productivity will result in a higher demand for a clean environment and hence lower per capita CO_2 emissions. On the other hand, the substitution effect predicts that as investment is relatively cheaper, the opportunity cost of investing in abatement instead of capital is higher. This argument suggests lower demand for a clean environment and hence higher per capita CO_2 emissions.

The cyclical pattern illustrated in figure 5.2 suggests strongly that the income effect should dominate the substitution effect, and positive shocks to the business cycle should be accompanied by increased emissions. The SVAR is specified in growth rates so that the cumulative impulse–responses represent the long-run effects of shocks.

Estimation of the unrestricted VAR reveals an optimal lag order of at most $p = 1$. This result is slightly at odds with the choice of $p = 4$ proposed by Khan et al. (2015), but the more generous choice of lag length will be maintained here. Fitting the SVAR(4) with long-run restrictions gives the following results:

```
. matrix C = (.,0,0\.,.,0\.,.,.)
. matrix list C

C[3,3]
    c1  c2  c3
r1   .   0   0
r2   .   .   0
r3   .   .   .

. svar Dp Dlp Dco2 if tin(1973q1,2012,q1), lreq(C) lags(1/4) nolog
Estimating long-run parameters

Structural vector autoregression

 ( 1)  [/C]1_2 = 0
 ( 2)  [/C]1_3 = 0
 ( 3)  [/C]2_3 = 0

Sample:  1974q2 - 2012q1                    Number of obs   =       152
Exactly identified model                    Log likelihood  =   1429.71
```

	Coef.	Std. Err.	z	P>|z|	[95% Conf. Interval]	
/C						
1_1	.0162933	.0009345	17.44	0.000	.0144617	.0181248
2_1	.002474	.0006153	4.02	0.000	.0012679	.00368
3_1	-.0014069	.0012843	-1.10	0.273	-.0039241	.0011102
1_2	0	(constrained)				
2_2	.007382	.0004234	17.44	0.000	.0065522	.0082118
3_2	-.0021252	.0012759	-1.67	0.096	-.004626	.0003756
1_3	0	(constrained)				
2_3	0	(constrained)				
3_3	.015659	.0008981	17.44	0.000	.0138987	.0174192

The results of the estimation are somewhat disappointing. The long-run effects on CO_2 emissions of the investment shock, c_{31}, are statistically insignificant with a p-value of 0.273. The effect of the technology shock, c_{32}, shows more promise with a p-value of 0.096, indicating significance at the 10% level.

5.5.2 Long-run restrictions

The cumulative impulse–responses from the SVAR tracing the impact of investment and productivity shocks on per capita CO_2 emissions are shown in figure 5.4.

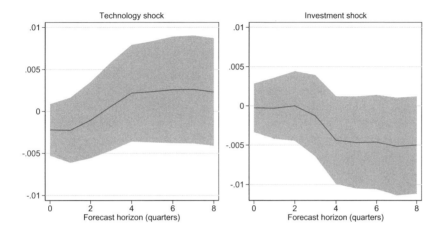

Figure 5.4. Plot of the cumulative IRFs from the SVAR tracing the impact of investment and technology shocks on per capita CO_2 emissions

Emissions are unchanged following a positive unanticipated investment shock before falling to the new equilibrium level. By contrast, after a positive unanticipated technology shock, they increase sharply before settling down to the steady-state level after about a year.

The immediate negative response of emissions following the technology shock means that it is extremely difficult to find support for the strong procyclicality of per capita CO_2 emissions being driven by real business cycle theory, so the task is to find alternative shocks that are consistent with the procyclicality observed in the data.

Exercises

1. `stern&kaufmann14.dta` contains annual data on several climate change variables used in Kaufmann and Stern (1997) and Stern and Kaufmann (2014), including temperature anomalies in the northern and southern hemispheres and variables linked to radiative forcing from both natural and anthropogenic sources.

a. Plot the temperature anomalies for the northern and southern hemispheres. The term *temperature anomaly* means a departure from the reference value or long-term average, in this case the 1951–1980 mean temperature. A positive anomaly indicates that the observed temperature was warmer than the reference value, while a negative anomaly indicates that the observed temperature was cooler than the reference value. Comment on the plots.

b. Fit a simple bivariate VAR(2) model for the temperature anomalies in the southern and northern hemispheres. Include a linear deterministic time trend in the VAR as an exogenous variable. Interpret the results.

c. A useful rule of thumb for setting the maximum possible lag length when trying to estimate the optimal lag length of a VAR is that to take `maxlag` to $(T^{1/3})$ rounded to the nearest integer, a maximum possible lag length must be chosen. Using this rule of thumb, estimate the optimal lag length for the VAR estimated in exercise 1b. Discuss your results.

d. Using the lag length suggested by the AIC, reestimate the bivariate VAR and check the stability of the system.

e. Perform Granger causality tests using the `vargranger` command. Do your results support the conclusion that the southern hemisphere temperature anomaly causes the northern hemisphere temperature but not vice versa?

f. Global temperature is determined by the balance between solar radiation (incoming) and infrared radiation (outgoing). This balance is disturbed by both natural (`rfnat`) and anthropogenic (`rfanth`) emissions of gases in a process known as *radiative forcing*, measured in Watts per square meter (W/m2). Now, expand the VAR system to four variables by including `rfnat` and `rfanth`. Fit a VAR(4). Is there any support for the hypothesis that anthropogenic radiative forcing influences temperatures in the northern hemisphere? Similarly, is there continued support for the hypothesis that southern temperatures cause northern temperatures?

g. Now, fit a VAR(2). Do your conclusions change? What do these results illustrate about Granger causality tests?

2. A VAR model (in which the T dimension—the number of observations—is much larger than the N dimension—the number of equations) is a restricted version of the more general case of a SUR model (Zellner 1962), the restriction being that all the variables on the right-hand side of the regression equations are identical across all equations. In a SUR model, each equation is allowed to have its own set of coefficients and regressors. Each equation can be estimated by ordinary least squares, and each of these equations satisfies the zero-conditional mean assumption by containing only exogenous (or predetermined) regressors. If each of the equations can be estimated over the same time period, the underlying data can be considered as a wide-form panel dataset; see chapter 12. The Stata `sureg` (see [R] **sureg**) command implements a SUR estimator, which accounts for the likely correlation of the error terms across equations. These correlations represent common shocks, and incorporating them in estimation process can provide gains

5.5.2 Long-run restrictions

in efficiency. The SUR estimator allows each equation to have its own error variance, and the estimator is constructed by taking the covariance of the errors into account.

phh.dta contains data on pollution abatement costs and foreign direct investment as well as several control variables for five states of the United States, namely, New Hampshire (NH), Iowa (IA), North Carolina (NC), Washington (WS), and Michigan (MI).

  a. Use the Stata notes facility to become familiar with the data.
  b. The pollution haven hypothesis posits that production within polluting industries will shift to locations with lax environmental regulation. As such, there should be a negative relationship between state foreign direct investment and pollution abatement costs, the proxy for environmental regulation. Compute the correlation matrix and Spearman's rank correlation between the log of foreign direct investment and the log of pollution abatement costs for each state. Comment on the results.
  c. Fit a SUR model for the five states in which foreign direct investment is regressed on abatement costs and all the control variables. Interpret the impact of abatement costs on foreign direct investment.
  d. Look at the residual covariance matrix, and comment on the suitability of the SUR estimator. Refit the model using the corr option to obtain the correlation matrix of the residuals. Interpret the results, including the result of the Breusch–Pagan test of independence.
  e. Test if the estimated coefficients on lrac are statistically distinguishable across states. Interpret the results.

6 Testing for nonstationarity

An important property of many environmental time series is that they exhibit strong trends. Series that are characterized by trending behavior are referred to as being nonstationary. The presence of nonstationarity in the time-series representation of a variable has important implications for both the econometric method used and the economic interpretation of the model in which that variable appears. This chapter focuses on identifying and testing for nonstationarity in environmental time series, while chapter 7 deals with modeling with nonstationary variables.

A variable y_t is said to be stationary if its distribution, or some important aspects of its distribution, is constant over time. There are two commonly used definitions of stationarity, known as weak (or covariance) and strong (or strict) stationarity. It is the former that will be of primary interest here. A process is weakly stationary if both the population mean and the population variance are constant over time and if the covariance between two observations is a function only of the distance between them and not of time.[1] Although the concept of nonstationarity is often mentioned in terms of a unit-root process, it is important to note that any time series with a time-varying mean or time-varying variance is a nonstationary process.

6.1 Per capita CO_2 emissions

The Earth absorbs energy from the sun and emits energy into space with the difference between incoming and outgoing radiation being known as radiative forcing. When incoming energy is greater than outgoing energy, positive radiative forcing will cause the planet to warm. Natural phenomena that contribute to radiative forcing include changes in the sun's energy output, changes in Earth's orbit, and volcanic activity. Anthropogenic (or human-caused) radiative forcing includes emissions of heat-trapping greenhouse gases.

It is well known that per capita CO_2 emissions, which are an important component of anthropogenic radiative forcing, have grown over time. Figure 6.1 plots annual per capita CO_2 emissions for the United States for the period 1870 to 2000. The data are provided by the United States Department of Energy through its Carbon Dioxide Information Analysis Center at Oak Ridge National Laboratory. An observation is an annual number giving national emissions in metric tons of carbon from fossil fuel burning, cement manufacturing, and gas flaring.

---

1. Strict stationarity is a stronger requirement than that of weak stationarity and pertains to all the moments of the distribution, not just the first two.

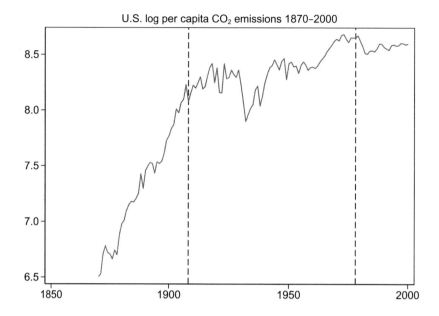

Figure 6.1. Annual per capita CO_2 emissions for the United States for the period 1870 to 2000, with three phases of emission development demarcated by the vertical lines

Per capita CO_2 emissions in industrialized countries are often characterized in terms of three phases as in Lanne and Liski (2004). The first phase (1870–circa 1908) was that of fast growth of per capita emissions because early industrialization and development in general was primarily dependent on coal. The second phase (circa 1908–circa 1978) was characterized by slower growth due to the shift from solid to nonsolid fuels (from coal to oil and gas). This diversification in national fuel compositions was partly induced by local but CO_2-related pollution problems and technological progress associated with the demand for higher energy density fuels. The third phase followed the oil price shocks of the 1970s, which may have permanently changed the structure of emissions from fossil fuels and possibly led to downward sloping per capita emission trends.

One of the important questions of environmental econometrics is to assess if this growth is consistent with deterministic behavior with respect to time or if the dependence on time is stochastic. A major challenge in time-series modeling is determining the source of nonstationary behavior, which may be due to either a deterministic trend or a stochastic trend. A deterministic trend is a nonrandom function of time

$$y_t = \alpha + \delta t + v_t \qquad (6.1)$$

6.1 Per capita CO_2 emissions

in which t in this instance is a simple time trend taking integer values from 1 to T and v_t is an independently and identically distributed (i.i.d.) random variable. In this model, shocks to the system have a transitory effect in that the process always reverts to its mean of $\alpha + \delta t$. This implies that removing the deterministic trend from y_t will produce a stationary series. A series that is stationary once its deterministic trend has been removed is therefore called a *trend-stationary process*.

By contrast, a time series that exhibits a stochastic trend is random and varies over time, for example,

$$y_t = \alpha + y_{t-1} + v_t \qquad (6.2)$$

which is known as a random walk with drift model. In this model, the best guess for the next value of series is the current value plus some constant rather than a deterministic mean value. This is a generalization of the pure random-walk process, in which α would not appear. As a result, this kind of model is also called a "local trend" or "local level" model. The appropriate course of action here is to difference the data to obtain a stationary series as follows:

$$\Delta y_t = \alpha + v_t$$

A process that has a stochastic trend that is removed by differencing the series is known as a *difference-stationary process*.[2]

If the behavior of per capita CO_2 is deterministic with respect to time, then the emissions series may be appropriately detrended. If this is not the case, then a new set of econometric tools dealing with fundamentally nonstationary variables must be used. However, distinguishing trend stationarity from difference stationarity is a difficult task that is made more so by the practical modeling problems encountered in the data. There are at least two problems evident in CO_2 emissions data plotted in figure 6.1.

1. It is obvious from the plot that a simple linear deterministic trend is unlikely to be an adequate representation of any deterministic behavior with respect to time.

2. Related to this point is the idea that the series may be characterized by structural breaks consistent with the phases of development of an industrialized country.

---

2. Of course, first differencing, a series that has a linear deterministic trend, such as (6.1) will automatically remove this trend. Why, then, is the solution not simply to difference all series, irrespective of whether they are trend or difference stationary? In first difference form, (6.1) becomes

$$\Delta y_t = \delta + v_t - v_{t-1}$$

so the process of taking the first difference introduces a moving average error term that has a coefficient of 1 on the lagged error. The nonstationarity has been transferred from the linear trend to a more complex form of nonstationarity in the error structure. This is known as overdifferencing, which introduces several econometric problems and is best avoided.

6.2 Unit roots

Consider again the random walk with drift model in (6.2). Recursively substituting for the lagged value of y_t on the right-hand side yields

$$y_t = 2\alpha + y_{t-2} + v_t + v_{t-1}$$
$$y_t = 3\alpha + y_{t-3} + v_t + v_{t-1} + v_{t-2}$$
$$\vdots = \vdots \quad \vdots$$
$$y_t = t\alpha + y_0 + v_t + v_{t-1} + v_{t-2} + \cdots + v_1$$

This demonstrates that y_t is the summation of all past disturbances. The element of summation of the disturbances in a nonstationary process yields an important concept— the order of integration of a series. A process is said to be integrated of order d, denoted by $I(d)$, if it can be rendered stationary by differencing d times. That is, y_t is nonstationary but $\Delta^d y_t$ is stationary. Accordingly, a process is said to be integrated of order one, denoted by $I(1)$, if it can be rendered stationary by differencing once; y_t is nonstationary but $\Delta y_t = y_t - y_{t-1}$ is stationary. If $d = 2$, then y_t is $I(2)$ and must be differenced twice to achieve stationarity.

A series that is $I(1)$ is also said to have a unit root, and tests for nonstationarity are commonly known as tests for unit roots. Consider the general nth order autoregressive (AR) process

$$y_t = \phi_1 y_{t-1} + \phi_2 y_{t-2} + \cdots + \phi_n y_{t-n} + v_t$$

This may be rewritten using the lag operator, L, defined in chapter 4 so that

$$y_t = \phi_1 L y_t + \phi_2 L^2 y_t + \cdots + \phi_n L^n y_t + v_t$$

or

$$\Phi(L) y_t = v_t$$

where

$$\Phi(L) = 1 - \phi_1 L - \phi_2 L^2 - \cdots - \phi_n L^n$$

is a polynomial in the lag operator of order n. The roots of this polynomial are the values of L, which satisfy the equation

$$1 - \phi_1 L - \phi_2 L^2 - \cdots - \phi_n L^n = 0$$

If all the roots of this equation are greater than 1 in absolute value, then y_t is stationary. If, on the other hand, any of the roots lie within the unit circle, then y_t is nonstationary.

For the AR(1) model

$$(1 - \phi_1 L) y_t = v_t$$

the roots of the equation

$$1 - \phi_1 L = 0$$

6.2 Unit roots

are of interest. The single root of this equation is given by $L^* = 1/\phi_1$, and the root is greater than 1 if and only if $|\phi_1| < 1$. If this is the case, then the AR(1) process is stationary. If, on the other hand, the root of the equation is unity, then $|\phi_1| \geq 1$ and the AR(1) process is nonstationary.

In the AR(2) model
$$\left(1 - \phi_1 L - \phi_2 L^2\right) y_t = v_t$$
there are two unit roots, corresponding to the roots of the equation
$$1 - \phi_1 L - \phi_2 L^2 = 0$$
and y_t will have a unit root if either of the roots is unity. In a AR(2) model, the solution could also consist of a complex conjugate pair of roots. In that case, the modulus of the complex pair must lie inside the unit circle if the series is to be stationary. The presence of complex roots introduces cyclical behavior in the series.

To illustrate these alternatives, the following series are simulated for three AR(2) processes:

1. $y_t = 0.9 y_{t-1} - 0.2 y_{t-2} + v_t$
2. $y_t = 0.8 y_{t-1} - 0.5 y_{t-2} + v_t$
3. $y_t = 1.8 y_{t-1} - 0.8 y_{t-2} + v_t$

The first process has two real roots outside the unit circle, implying that both (φ_1, φ_2) are less than 1 in absolute value. The series is stationary as shown in figure 6.2.

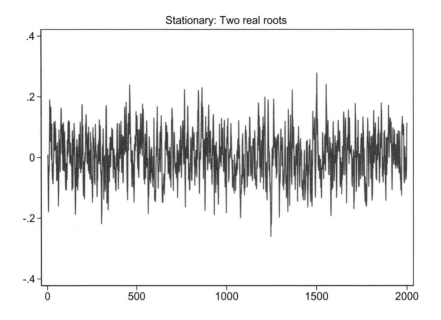

Figure 6.2. AR(2) model with two real roots

6.2 Unit roots

The second process has a pair of complex roots, with its modulus inside the unit circle, and also exhibits stationary behavior, as shown in figure 6.3.

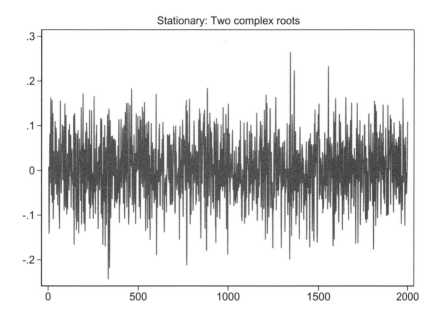

Figure 6.3. AR(2) model with complex roots

The third process has one unit root because the sum of its lag coefficients is 1. Its second root of 1.44 is outside the unit circle. The presence of a unit root defines an $I(1)$ process, drifting arbitrarily far from its starting values, as shown in figure 6.4.

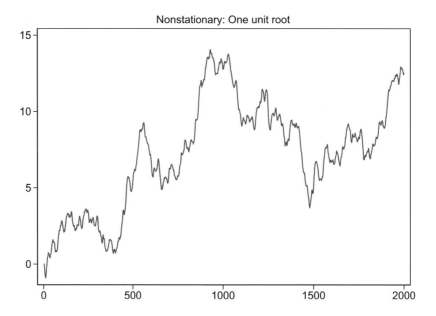

Figure 6.4. AR(2) model with one unit root

In the event of $\phi_1 = 2$ and $\phi_2 = -1$, then both absolute roots of the equation are one. The y_t series has two unit roots and is therefore $I(2)$. We do not illustrate a process with two unit roots such as that described above with coefficients of $(2, -1)$, because its behavior is monotonic, increasing without bound.

6.3 First-generation unit-root tests

6.3.1 Dickey–Fuller tests

The original testing procedures for unit roots were developed by Dickey and Fuller (1979, 1981). Consider again the simple AR(1) model

$$y_t = \alpha + \rho y_{t-1} + v_t \qquad (6.3)$$

in which v_t is a disturbance term with zero mean and constant variance σ^2. The null and alternative hypotheses are, respectively,

$$H_0: \quad \rho = 1 \quad \text{(Variable is nonstationary)}$$
$$H_1: \quad \rho < 1 \quad \text{(Variable is stationary)}$$

To carry out the test, (6.3) is fit by ordinary least squares, and a t statistic is constructed to test that $\rho = 1$

$$t_\rho = \frac{\widehat{\rho} - 1}{\operatorname{se}(\widehat{\rho})} \qquad (6.4)$$

6.3.1 Dickey–Fuller tests

The problem is that the distribution of the statistic in (6.4) is not a Student t distribution. In fact, the distribution of this statistic under the null hypothesis of nonstationarity is nonstandard and is known as the Dickey–Fuller distribution. Consequently, the t statistic given in (6.4) is commonly known as the Dickey–Fuller unit-root test to recognize that even though it is a t statistic by construction, its distribution is not the standard t distribution.

In practice, (6.3) is transformed as follows. Subtract y_{t-1} from both sides and collect terms to give
$$y_t - y_{t-1} = \alpha + (\rho - 1)y_{t-1} + v_t$$
or by defining $\beta = \rho - 1$ and $\Delta y_t = y_t - y_{t-1}$, we have
$$\Delta y_t = \alpha + \beta y_{t-1} + v_t \tag{6.5}$$

There is a real possibility that the disturbance term in (6.5) will exhibit autocorrelation. One common way to correct for autocorrelation is to include lags of the dependent variable Δy_t in regression (6.5). The equation then becomes
$$\Delta y_t = \beta y_{t-1} + \sum_{i=1}^{p} \phi_i \Delta y_{t-i} + v_t$$
and the test is known as the augmented Dickey–Fuller (ADF) test.

There are in fact three forms of the ADF test, namely,

Model 1: $\quad \Delta y_t = \beta y_{t-1} + \sum_{i=1}^{p} \phi_i \Delta y_{t-i} + v_t$

Model 2: $\quad \Delta y_t = \alpha + \beta y_{t-1} + \sum_{i=1}^{p} \phi_i \Delta y_{t-i} + v_t$

Model 3: $\quad \Delta y_t = \alpha + \delta t + \beta y_{t-1} + \sum_{i=1}^{p} \phi_i \Delta y_{t-i} + v_t$

For each of these three models, the null hypothesis of the Dickey–Fuller test remains the same, namely, $H_0 : \beta = 0$. The choice of which deterministic components to include in the test depends primarily on the nature of the process, y_t, under the alternative hypothesis.

1. Model 1, in which there is no constant or time trend in the test regression, assumes that the process has zero mean under the alternative hypothesis.
2. Model 2 is used in situations where the process has nonzero mean under the alternative, but the data do not appear to be trending.
3. Model 3 is appropriate when the data are visibly trending and may in fact be stationary around a deterministic trend.

Dickey–Fuller tests are carried out in Stata using the `dfuller` (see [TS] **dfuller**) command. Model 2 is the default, and model 3 is implemented using the `trend` option. The default is to fit a model without any included lags of the dependent variable. One of the options for the command is lags(), which specifies the number of lagged dependent variables to include in the ADF version of the test. If the chosen lag length is too small, autocorrelation may remain in the error term, v_t, which will affect the statistical size of the test. Including an excessive number of lags will, however, reduce the power of the test. Stata does not have an automatic lag selection capability for the dfuller command. Methods to deal with this problem include the following:

1. Perform the test for different choices of lag length, and look for a robust pattern to emerge.
2. Include lags until the t statistic on the lagged changes of the dependent variable is statistically insignificant. This test statistic follows the standard t distribution.
3. Use information criteria, as outlined in chapter 4, to inform the choice.[3]

Applying the Dickey–Fuller testing framework to the United States CO_2 emission data gives the following results:

```
. use http://www.stata-press.com/data/eeus/logpercapitaco2
. tsset year, yearly
        time variable:  year, 1870 to 2000
                delta:  1 year
. // Model 2
. dfuller co2_usa
Dickey-Fuller test for unit root                   Number of obs   =       130

                               ---------- Interpolated Dickey-Fuller ---------
                  Test         1% Critical       5% Critical      10% Critical
              Statistic            Value             Value             Value
------------------------------------------------------------------------------
 Z(t)            -3.054            -3.500            -2.888            -2.578
------------------------------------------------------------------------------
MacKinnon approximate p-value for Z(t) = 0.0302
. // Model 3
. dfuller co2_usa, trend
Dickey-Fuller test for unit root                   Number of obs   =       130

                               ---------- Interpolated Dickey-Fuller ---------
                  Test         1% Critical       5% Critical      10% Critical
              Statistic            Value             Value             Value
------------------------------------------------------------------------------
 Z(t)            -2.350            -4.030            -3.446            -3.146
------------------------------------------------------------------------------
MacKinnon approximate p-value for Z(t) = 0.4065
```

The test statistic from model 2 indicates rejection of the null hypothesis of a unit root (implying CO_2 emissions are stationary) at the 5% level, whereas the test statistic from model 3, where a trend is included in the test regression, does not reject the null hypothesis of a unit root (CO_2 emissions are nonstationary). This kind of ambiguity is often found in unit-root testing. The difficulty here stems from the fact that model 2

---

3. A lag-length selection procedure that has good properties in unit-root testing is the modified Akaike information criterion method proposed by Ng and Perron (2001).

6.3.1 Dickey–Fuller tests

may not be the appropriate choice of model given the strong trending behavior in the series and that model 3 is the correct test regression to use. Alternatively, it may be that the use of the Dickey–Fuller test without any lag augmentation causes residual autocorrelation which is affecting the results.

Because the data are annual observations, a lag length of two lags in this case seems sufficient to capture any adjustment in emissions. Accordingly, ADF tests are first performed using two lags of the dependent variable.

```
. dfuller co2_usa, lags(2)
Augmented Dickey-Fuller test for unit root        Number of obs  =       128
                              ——————— Interpolated Dickey-Fuller ———————
                 Test         1% Critical      5% Critical     10% Critical
              Statistic          Value            Value            Value

    Z(t)        -3.191          -3.501           -2.888           -2.578

MacKinnon approximate p-value for Z(t) = 0.0205
. dfuller co2_usa, trend lags(2)
Augmented Dickey-Fuller test for unit root        Number of obs  =       128
                              ——————— Interpolated Dickey-Fuller ———————
                 Test         1% Critical      5% Critical     10% Critical
              Statistic          Value            Value            Value

    Z(t)        -1.961          -4.031           -3.446           -3.146

MacKinnon approximate p-value for Z(t) = 0.6224
```

The ambiguity in the results is not resolved by the use of the ADF test. Model 2 still rejects the null hypothesis of a unit root at the 5% level, while model 3 fails to reject the null at the 1% level. On balance, given the strong trending behavior shown by the CO_2 emission data for the United States in figure 6.1, a model that does not allow a deterministic trend under the alternative hypothesis is likely to be misspecified. The results from model 3 therefore may be more persuasive than those obtained from model 2.

The broad thrust of the results appears to be robust to variations in lag length because increasing the lag length to 4 produces similar results.

```
. dfuller co2_usa, lags(4)
Augmented Dickey-Fuller test for unit root        Number of obs  =       126
                              ——————— Interpolated Dickey-Fuller ———————
                 Test         1% Critical      5% Critical     10% Critical
              Statistic          Value            Value            Value

    Z(t)        -3.078          -3.501           -2.888           -2.578

MacKinnon approximate p-value for Z(t) = 0.0282
```

```
. dfuller co2_usa, trend lags(4)
Augmented Dickey-Fuller test for unit root         Number of obs   =      126
                              ———————— Interpolated Dickey-Fuller ————————
                 Test           1% Critical      5% Critical     10% Critical
              Statistic            Value            Value            Value

    Z(t)       -2.035             -4.031           -3.447           -3.147

MacKinnon approximate p-value for Z(t) = 0.5823
```

6.3.2 Phillips–Perron tests

Phillips and Perron (1988) propose an alternative method for adjusting the Dickey–Fuller test for autocorrelated errors. Their test is based on the test regression in (6.3), where the constant can be excluded or a trend included. Estimation is by ordinary least squares, but a nonparametric approach to correct for the autocorrelation is used. The Phillips–Perron test statistic is

$$\tilde{t}_\beta = t_\beta \left(\frac{\widehat{\gamma}_0}{\widehat{f}_0}\right)^{1/2} - \frac{T(\widehat{f}_0 - \widehat{\gamma}_0)\text{se}(\widehat{\beta})}{2\widehat{f}_0^{1/2} s}$$

where t_β is the ADF statistic, s is the standard error of the test regression, and $\widehat{f}_0$ is known as the long-run variance. The long-run variance is computed as

$$\widehat{f}_0 = \widehat{\gamma}_0 + 2\sum_{j=1}^{p}\left(1 - \frac{j}{p}\right)\widehat{\gamma}_j$$

where p is the length of the lag and $\widehat{\gamma}_j$ is the jth estimated autocovariance of the ordinary least-squares residuals obtained from estimating the test regression.

$$\widehat{\gamma}_j = \frac{1}{T}\sum_{t=j+1}^{T}\widehat{v}_t\widehat{v}_{t-j}$$

The critical values are the same as the Dickey–Fuller critical values when the sample size is large.

The Stata command for invoking Phillips–Perron tests for unit roots is pperron (see [TS] **pperron**). The default is to include a constant but no trend in the test regression. The option noconstant suppresses the constant term while the option trend dictates that a trend term be included in the associated regression. This option may not be used if noconstant is specified. Whereas the default in the dfuller command is to set the lag length to zero, for pperron the default lag length is

$$p = \text{int}\left\{4\left(\frac{T}{100}\right)^{2/9}\right\}$$

6.4.1 KPSS test

For the current CO_2 emissions data, the results returned by `pperron` are similar to those obtained by `dfuller`. The null hypothesis of a unit root is rejected if the test regression does not include a deterministic trend, but the null hypothesis cannot be rejected if a trend is included in the regression.

```
. pperron co2_usa, lags(2)
Phillips-Perron test for unit root              Number of obs   =        130
                                                Newey-West lags =          2
                                   Interpolated Dickey-Fuller
                    Test          1% Critical      5% Critical    10% Critical
                 Statistic           Value            Value           Value

Z(rho)            -4.223           -19.900          -13.760         -11.040
Z(t)              -3.392            -3.500           -2.888          -2.578

MacKinnon approximate p-value for Z(t) = 0.0112

. pperron co2_usa, trend  lags(2)
Phillips-Perron test for unit root              Number of obs   =        130
                                                Newey-West lags =          2
                                   Interpolated Dickey-Fuller
                    Test          1% Critical      5% Critical    10% Critical
                 Statistic           Value            Value           Value

Z(rho)            -5.853           -27.600          -20.820         -17.600
Z(t)              -2.278            -4.030           -3.446          -3.146

MacKinnon approximate p-value for Z(t) = 0.4461
```

An assessment of the results produced by these first-generation tests is that CO_2 emissions contain a unit root.

6.4 Second-generation unit-root tests

Several extensions and alternatives to the Dickey–Fuller, ADF, and Phillips–Perron unit-root tests have been proposed and are available in Stata or as downloadable packages from the Statistical Software Components (SSC) Archive.

6.4.1 KPSS test

The Dickey–Fuller testing framework for unit-root testing is based on the null hypothesis that a time series y_t is nonstationary or $I(1)$. There is, however, a test commonly known as the KPSS test, after Kwiatkowski, Phillips, Schmidt, and Shin (1992) that is often reported in the empirical literature and which has a null hypothesis of stationarity or $I(0)$. Consider the regression model

$$y_t = \delta t + r_t + \epsilon_t$$

where ϵ_t is a disturbance term and r_t is a random walk:

$$r_t = r_{t-1} + u_t, \quad u_t \sim \text{i.i.d.}(0, \sigma_u^2)$$

Under the null hypothesis that y_t is a trend-stationary series, $\sigma_u^2 = 0$, in which case r_t is simply a constant.[4] The Lagrange multiplier (LM) test of this hypothesis requires estimation of the model under the null hypothesis. Define $\widehat{e}_t$ as the ordinary least-squares residuals from regression of y_t on a constant and a deterministic trend, and define s_t as the partial sum process of $\widehat{e}_t$. The standardized test statistic based on the LM principle is given by

$$\mathrm{LM} = \frac{1}{T^2 \widehat{f}_0} \sum_{t=1}^{T} S_t^2$$

in which $\widehat{f}_0$ is a consistent estimator of the long-run variance of e_t. A routine for implementing this test is available as the command `kpss` (Baum 2000), which can be installed from the SSC Archive.[5]

Applying the KPSS test to the emissions data yields

```
. use http://www.stata-press.com/data/eeus/logpercapitaco2
. tsset year, yearly
        time variable:  year, 1870 to 2000
                delta:  1 year
. kpss co2_usa

KPSS test for co2_usa

Maxlag = 4 chosen by Schwert criterion
Autocovariances weighted by Bartlett kernel

Critical values for H0: co2_usa is trend stationary

10%: 0.119  5% : 0.146  2.5%: 0.176  1% : 0.216

Lag order     Test statistic
    0              2.28
    1              1.18
    2              .806
    3              .617
    4              .503
```

The results of the test indicate rejection of the null hypothesis of stationarity at the 1% level, a conclusion that is robust over all the lag lengths for which the test is computed. These results tend to confirm those obtained using the `dfuller` and `pperron` commands.

---

4. If $\delta = 0$, the null hypothesis is that y_t is stationary around a level r_0.
5. To install the package, type `ssc describe kpss`.

6.4.2 Elliott–Rothenberg–Stock DFGLS test

Dickey–Fuller tests struggle to reject the null hypothesis of nonstationarity, or $I(1)$, when it is in fact false. For this reason, Elliott, Rothenberg, and Stock (1996) developed a second-generation test based on the premise that the test is more likely to reject the null hypothesis of a unit root if the process is close to being nonstationary under the alternative hypothesis. Consider the model

$$y_t = \alpha + \delta t + u_t \qquad (6.6)$$
$$u_t = \phi u_{t-1} + v_t$$

in which u_t is a disturbance term with zero mean and constant variance σ^2. This is the fundamental equation from which model 3 of the Dickey–Fuller test is derived. Testing for a unit root can proceed by fitting (6.6) by ordinary least squares to obtain $\widehat{\alpha}$ and $\widehat{\delta}$ and then testing for a unit root using the detrended data

$$y_t^* = y_t - \left(\widehat{\alpha} + \widehat{\delta t}\right)$$

using an ADF test. This procedure is asymptotically equivalent to the single-step ADF test based on model 3.

There is, however, the potential for the ordinary least-squares estimates of the detrending coefficients, $\widehat{\alpha}$ and $\widehat{\delta}$, to be inefficient because of the serial correlation in the residuals in (6.6). Essentially, the value $\phi = 0$ is being imposed when the coefficients α and δ are estimated in this simple approach. Elliott, Rothenberg, and Stock (1996) suggest a generalized least-squares approach to estimating the detrending parameters by choosing a value of ϕ defined as

$$\widetilde{\phi} = 1 + \overline{c}/T \qquad (6.7)$$

so that $\overline{c}$ is chosen to ensure that $\widetilde{\phi}$ is in the vicinity of $\phi = 1$, the null hypothesis.[6] The hope is that the increased efficiency in the estimates $\widehat{\alpha}$ and $\widehat{\delta}$ will be reflected by increased power in the region where the alternative is close to the null, which is exactly the area where the test is likely to suffer problems. The actual proposed values of $\overline{c}$ depend upon whether the detrending equation has only a constant or both a constant and a time trend, given, respectively, by

$$\begin{cases} \overline{c} = -7 & \to \widetilde{\phi} = 0.9650 \quad \{\text{Constant } (\alpha \neq 0, \delta = 0)\} \\ \overline{c} = -13.5 & \to \widetilde{\phi} = 0.9325 \quad \{\text{Trend}(\alpha \neq 0, \delta \neq 0)\} \end{cases} \qquad (6.8)$$

While the ADF test critical values depend on the assumption of simple detrending, the critical values of this test depend on the value of $\overline{c}$.

The so-called generalized least-squares version of the augmented Dickey–Fuller test (DF-GLS) is implemented in Stata using the `dfgls` (see [TS] **dfgls**) command. This

---

6. Of course, choosing $\overline{c} = -T$ means that $\widetilde{\phi} = \phi$, and the test procedure reverts to the simple Dickey–Fuller test.

command will automatically select the best lag length for the test. Although there is no one uniformly most powerful test for a unit root, the DF-GLS test is approximately the most powerful for this problem.[7]

```
. dfgls co2_usa

DF-GLS for co2_usa                                      Number of obs =     118
Maxlag = 12 chosen by Schwert criterion

              DF-GLS tau      1% Critical      5% Critical     10% Critical
   [lags]   Test Statistic       Value            Value            Value

     12         -0.866          -3.543           -2.789           -2.514
     11         -0.977          -3.543           -2.812           -2.535
     10         -0.627          -3.543           -2.833           -2.556
      9         -0.711          -3.543           -2.854           -2.575
      8         -0.947          -3.543           -2.875           -2.594
      7         -0.770          -3.543           -2.894           -2.612
      6         -0.580          -3.543           -2.913           -2.629
      5         -0.517          -3.543           -2.931           -2.646
      4         -0.538          -3.543           -2.947           -2.661
      3         -0.548          -3.543           -2.963           -2.675
      2         -0.358          -3.543           -2.977           -2.688
      1         -0.478          -3.543           -2.990           -2.699

Opt Lag (Ng-Perron seq t) = 11 with RMSE   .0691484
Min SIC  = -5.070476 at lag  1 with RMSE   .076103
Min MAIC = -5.135413 at lag 11 with RMSE   .0691484
```

The results of the unit-root tests when using the DF-GLS test indicate an unambiguous failure to reject the null hypothesis of nonstationarity. There is thus agreement between the second-generation approaches that CO_2 emissions in the United States are in fact nonstationary.

As a final word on second-generation unit-root tests, it should be noted that another approach to improve the power of unit-root tests has been to include additional stationary covariates in the test regressions. The intuition of this approach is that the correlation between y_t and a stationary covariate, x_t, may help reduce the regression error variance resulting in more precise parameter estimates. The more the covariate is correlated with y_t, the higher the power of the test. As R^2 increases, there is a larger gain in using the information contained in the stationary covariate over a univariate test, and the power increases; see Hansen (1995). Elliott and Jansson (2003) extend the procedure suggested by Hansen (1995) by combining generalized least-squares detrending combined with the likelihood-ratio principle. The idea is to estimate the system (6.8) using value $\phi = 1$ and then reestimate the equation with $\phi = \widetilde{\phi}$ as in (6.7) with the appropriate value of $\bar{c}$, given the choice of deterministic terms, as given in (6.8). The likelihood function returned in each case is then used to compute a likelihood-ratio

---

7. The assumption of using y_1 as the starting value for the detrended data has important implications for the power of the test. Specifically, power declines as the initial value grows. Harvey, Leybourne, and Taylor (2009) suggest a simple practical approach to dealing with this problem: compute the tests using both generalized and ordinary least squares detrending, and reject the null hypothesis if either test or both tests reject it. This is known as the union of rejections approach.

6.5.1 Known breakpoint

test of $\phi = 1$. An implementation of the Elliott–Jansson test is available from the SSC Archive as `urcovar` (Baum 2007).

6.5 Structural breaks

Unit-root tests are particularly sensitive to breaks in the structure of a relationship. Neglecting a structural break in unit-root tests makes it likely that you will be unable to reject the null hypothesis of a unit root. The extent to which this occurs varies with the type of test and the timing of the break, but the overall finding is very clear; instability in the underlying data generating process is likely to be considered as nonstationary behavior.

6.5.1 Known breakpoint

Consider the general model of trending data with a possible break in the intercept and the trend given by

$$y_t = \alpha + \beta t + \gamma \mathrm{DU}_{bt} + \delta \mathrm{DT}_{bt} + u_t$$

where DU_{bt} is an intercept-break dummy variable and DT_{bt} is a trend break-dummy variable given, respectively, by

$$\mathrm{DU}_{bt} = \begin{cases} 0, & t \leq T_b \\ 1, & t > T_b \end{cases} \qquad \mathrm{DT}_{bt} = \begin{cases} 0, & t \leq T_b \\ t - T_b, & t > T_b \end{cases}$$

The breakpoint, should it occur, is denoted by T_b with $1 < T_b < T$. Additionally, let τ be the fraction of the sample at which the break occurs.

The simplest approach is to chose an appropriate detrending regression from this general form and test the detrended data, y_t^*, for a unit root using model 1 of the ADF test—the test regression without intercept or trend. Note, however, that the critical values for these tests will not be the usual Dickey–Fuller critical values but those reported by Perron and Vogelsang (1993). The fundamental problem is that critical values of the test depend upon where exactly in the sample the break occurs. Perron and Vogelsang (1993) extend the analysis of structural breaks to include generalized least squares detrending, as in the DF-GLS test. The situation is further complicated because not only are the critical values dependent upon the position of the structural break but also the appropriate value for the crucial detrending parameter $\bar{c}$, defined in (6.8), varies with the position of the break.

In this example, the hypothesized phases of CO_2 emissions are taken to be 1912 and 1978. The approach is to detrend the series using ordinary least squares with a constant and linear trend, where the latter has two breakpoints. An efficient way of coding this in Stata is to use the `mkspline` command. The relevant command to use `mkspline` to place knots at 1912 and 1978, obtain a broken trend over the sample period, and generate the detrended series is as follows:

```
. mkspline t1 1912 t2 1978 t3=year
. regress co2_us t1 t2 t3
  (output omitted)
. predict double that, xb
. predict double uhat, residual
```

Figure 6.5 plots the CO_2 emissions data with the broken trend superimposed. It is interesting that the slope of the trend line after the second break point does not appear to be much different from that prior to the break.

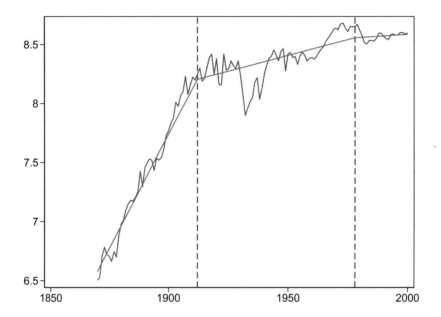

Figure 6.5. Ordinary least squares detrending of the annual per capita CO_2 emissions for the United States for the period 1870 to 2000 using `mkspline` with breakpoints at 1912 and 1978

The residuals from the linear detrending process are now tested for a unit root using the `dfgls` testing procedure.

```
. dfgls uhat, maxlag(1)
```

DF-GLS for uhat   Number of obs =   129

[lags]	DF-GLS tau Test Statistic	1% Critical Value	5% Critical Value	10% Critical Value
1	-3.584	-3.543	-2.978	-2.688

6.5.2 Single-break unit-root tests

The results indicate rejection of a unit root, so CO_2 emissions can be regarded as stationary around the broken linear trend. This result is in sharp contrast with the results of both first- and second-generation unit-root tests and emphasizes the potential importance of allowing for structural breaks when testing for unit roots.

6.5.2 Single-break unit-root tests

Perron and Vogelsang (1992) propose a class of test statistics that allow for two alternative forms of change: the additive outlier (AO) model for capturing a sudden change and the innovational outlier (IO) model which is appropriate for modeling a gradual shift in the mean of the series. The test statistics do not require a priori knowledge of the breakpoint, because their computation involves searching over the sample for a single break date. The AO model assumes that under the null hypothesis of a unit root, the dynamics of y_t are given by the model

$$y_t = y_{t-1} + \delta \text{DTB}_t + u_t$$

where $\text{DTB}_t = 1$ if $t = T_b + 1$ and 0 otherwise for the break date T_b. The mean of the series changes by δ at $t = T_b + 1$.

Under the alternative hypothesis, the series does not contain a unit root. The data generating process for y_t is the intercept shift model

$$y_t = \alpha + \gamma \text{DU}_t + v_t \tag{6.9}$$

where $\text{DU}_t = 1$ if $t > T_b$ and 0 otherwise. The mean of the series is α up to $t = T_b$ and $\alpha + \gamma$ afterward.

The test strategy is then to compute the detrended series y_t^* from (6.9) and regress it on lagged values, lagged differences, and a set of dummy variables; the latter is required to ensure that the distribution of the test statistic on $\widehat{\beta}$ will be manageable. The test regression is

$$y_t^* = \sum_{i=0}^{k} \delta_i \text{DTB}_{t-i} + \beta y_{t-1}^* + \sum_{i=1}^{k} \theta_i \Delta y_{t-i}^* + u_t$$

This regression, similar in nature to the ADF model, yields an estimate of β, which will be significantly less than 1 in the presence of stationarity. Perron and Vogelsang (1992) provide critical values for the AO and IO tests in their Tables 3 and 4.

Alternatively, it may be that the change in the mean is not instantaneous but gradual. This is known as the IO model, and the test regression becomes

$$y_t = \alpha + \gamma \text{DU}_{bt} + \delta \text{DTB}_t + \beta y_{t-1} + \sum_{i=1}^{k} \theta_i \Delta y_{t-i} + e_t$$

which again yields a test of β differing from one in the presence of stationarity.

In both the AO and the IO models, the appropriate values of T_b (the breakpoint) and k (the AR order) are unknown. This is resolved for T_b by fitting the model for each feasible breakpoint and following one of several proposed rules to identify the optimal single breakpoint. In this case, the minimum t statistic on $\widehat{\beta}$ is used to identify the breakpoint. Conditional on the value of T_b, (Perron 1990) shows how to select the AR order k, starting from an appropriately chosen maximum order. The unit-root test statistics returned by the AO and IO models will account for a one-time level shift that might otherwise be identified as a departure from stationarity.

The commands for unit-root tests in the presence of structural breaks are available as clemao1 and clemio1 (Clemente, Montañés, and Reyes 1998), which can be installed from the SSC Archive.[8]

```
. use http://www.stata-press.com/data/eeus/logpercapitaco2, clear
. tsset year
        time variable:  year, 1870 to 2000
                delta:  1 unit
. clemao1 co2_usa, maxlag(1)  trim(0.10)
Clemente-Montas-Reyes unit-root test with single mean shift, AO model
co2_usa      T =  105       optimal breakpoint :      1900
AR( 0)                  du1            (rho - 1)       const

Coefficient:          1.22305           -0.15215       7.15680
t-statistic:           23.117            -3.514
P-value:                0.000            -3.560    (5% crit. value)
. clemio1 co2_usa, maxlag(1)  trim(0.10)
Clemente-Montas-Reyes unit-root test with single mean shift, IO model
co2_usa      T =  105       optimal breakpoint :      1896
AR( 1)                  du1            (rho - 1)       const

Coefficient:          0.07572           -0.08742       0.66625
t-statistic:            2.113            -3.486
P-value:                0.037            -4.270    (5% crit. value)
```

Both models indicate that CO_2 emissions are stationary once the single break in trend is accounted for and reaffirm the results obtained using unit-root tests based on known breakpoints. Interestingly, both models estimate a significant break point that is slightly earlier than the previous discussion of the three phases of emissions suggested.

6.5.3 Double-break unit-root tests

Clemente, Montañés, and Reyes (1998) extend the Perron–Vogelsang methodology to allow for double mean shifts. They demonstrate that a two-dimensional grid search for breakpoints, T_{b1} and T_{b2}, may be used for either the AO or IO models and provide critical values for the tests. The AO test involves searching for the minimal t ratio of the hypothesis $\beta = 1$ in the test regression

---

8. To install the package type search clemao.

6.5.3 Double-break unit-root tests

$$y_t^* = \sum_{i=0}^{k} \delta_{1i} \text{DTB1}_{t-i} + \sum_{i=0}^{k} \delta_{2i} \text{DTB2}_{t-i} + \beta y_{t-1}^* + \sum_{i=1}^{k} \theta_i \Delta y_{t-i}^* + v_t$$

For the IO model, the modified equation to be estimated becomes

$$y_t = \alpha + \gamma_1 \text{DU1}_t + \gamma_2 \text{DU2}_t + \delta_1 \text{DTB1}_t + \delta_2 \text{DTB2}_t + \beta y_{t-1} + \sum_{i=1}^{k} \theta_i \Delta y_{t-i} + v_t$$

Once more, the search is for the minimal t ratio for the hypothesis $\beta = 1$. When searching for the breakpoints, these tests are customarily applied to a trimmed sample with a common choice being 5% of the sample trimmed from each end. The double structural break unit-root tests are available as commands `clemao2` and `clemio2` from the same package as the single structural break tests.

The results of the AO tests for a double shift in the time series find the optimal breakpoints to be 1900 and 1960; see figure 6.6. The latter is much earlier than expected and well before the hypothesized change point after the oil shocks of the 1970s. When the IO test is applied, the optimal breakpoints appear in 1897 and 1937, during the interwar era. Conditional on these two breakpoints, United States per capita CO_2 emissions are stationary.

```
. clemao2 co2_usa, maxlag(1) trim(0.10) graph
Clemente-Montas-Reyes unit-root test with double mean shifts, AO model
co2_usa     T =   105      optimal breakpoints :     1900 ,     1960
AR( 0)                du1              du2        (rho - 1)       const

Coefficients:      1.09428          0.32192         -0.20312       7.15680
t-statistics:       22.742            7.250          -4.186
P-values:            0.000            0.000          -5.490 (5% crit. value)
. clemio2 co2_usa, maxlag(1) trim(0.10)
Clemente-Montas-Reyes unit-root test with double mean shifts, IO model
co2_usa     T =   105      optimal breakpoints :     1897 ,     1937
AR( 0)                du1              du2        (rho - 1)       const

Coefficients:      0.10683          0.03596         -0.12188       0.90150
t-statistics:        2.892            1.979          -4.022
P-values:            0.005            0.050          -5.490 (5% crit. value)
```

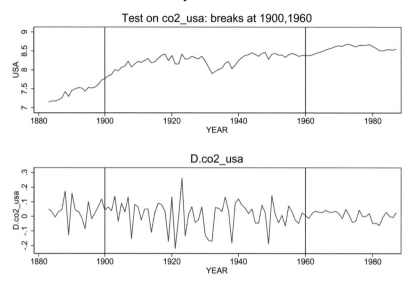

Figure 6.6. AO testing for a unit root in the presence of a double shift. Annual per capita CO_2 emissions for the United States for the period 1870 to 2000 has optimal break points at 1900 and 1960.

The results of the exhaustive unit-root testing of U.S. per capita CO_2 emissions over the period 1870–2000 reveals an interesting dichotomy. When a simple deterministic trend is included under the alternative hypothesis, the null hypothesis of a unit root cannot be rejected. Emissions are therefore not found to be stationary around the deterministic trend. As soon as a broken deterministic trend is allowed, the unit-root tests suggest that emissions are stationary around this broken trend. Given the time span of the data and the substantial changes in the patterns of production during that period, allowing for structural breaks seems like a reasonable strategy.

6.5.3 Double-break unit-root tests

Exercises

1. twobalance.dta contains annual data from 1955 to 2011 on temperature for three separate layers of the earth: the shallow ocean and atmosphere layer (gissv3$_t$) with low heat capacity; a deep ocean layer (ohc$_t$) with higher heat capacity; and radiative forcing (forcing$_t$) from factors like greenhouse gases, solar irradiance, and reflective aerosols from volcanic eruptions.

    a. Plot the level and differences of the series, and comment on the results.
    b. Test the three series for a unit root using the ADF and Phillips–Perron tests. Do the tests both with and without a linear trend in the test regression. What do you conclude?
    c. Test the three series for a unit root using the KPSS and DF-GLS tests. How do the results compare with those obtained in exercise 1b?

    item The radiative forcing variable, forcing$_t$, shows several negative outliers. Identify the dates of these outliers and create impulse dummy variables for the largest two outliers. Following the method of Frances and Haldrup (1994), test the series for a unit root using the Dickey–Fuller model 3 test regression augmented with the dummy variables. Does the result change your conclusion?

2. commodity.dta contains monthly observations on U.S. traded commodity price indices for copper, cotton, maize, sugar, wheat, wool, and zinc for the period January 1957 to November 2003.

    a. Compute the natural logarithm of the commodity price indices, and test for stationarity
        i. using the Dickey–Fuller and Phillips–Perron testing methods; and
        ii. using the DF-GLS test.

    Interpret the results of the tests.

    b. Test for unit roots allowing for a single structural break in the series. Does this approach change your conclusions?
    c. The Clemente, Montañés, and Reyes (1998) tests choose the break point fraction τ (equivalently the breakpoint T_b) where the Dickey–Fuller t statistic is minimized. Harris et al. (2009) suggest estimating τ by using the residuals from a first-differenced version of the detrending regression in (6.9) given by

    $$\Delta y_t = \alpha + \gamma \text{DI}_t + v_t$$

    where DI$_t$ is now an impulse dummy that takes the value of 1 at the current breakpoint and zero elsewhere. The choice of τ is now given by

    $$\hat{\tau} = \arg\min_{\tau} \sum_{t=1}^{T} \hat{v}_t(\tau)^2$$

    Implement this method for the commodity data, and see how the results compare with the results from exercise 2b.

7 Modeling nonstationary variables

As discussed in chapter 6, many important environmental variables exhibit trending behavior and are therefore nonstationary. An important implication is that nonstationary time series can be rendered stationary through differencing. An alternative method of achieving stationarity is to form linear combinations of the nonstationary series that are themselves stationary. This phenomenon is known as cointegration, introduced by Engle and Granger (1987). Cointegration between two or more nonstationary time series implies the existence of a long-run equilibrium relationship between them in which the weights of the linear combination used to achieve stationarity are the parameters of the equilibrium relationship. For early surveys of cointegration, see Dolado, Jenkinson, and Sosvilla-Riverso (1990) and Muscatelli and Hurn (1992).

Having uncovered the long-run relationships between two or more variables by establishing evidence of cointegration, the short-run dynamic properties of these variables are modeled by combining the information from the lags of the variables with the long-run relationships obtained from the cointegrating relationship. This model is known as an error-correction model. In the multivariate case, the model is a vector error-correction model (VECM), which may be shown to be a restricted form of the vector autoregression (VAR) model discussed in chapter 5. Although establishing cointegration is logically prior to fitting a model based on its existence, the discussion here will first focus on estimation. This aids the exposition because the test statistics require some prior understanding of the structure of the model and its estimation.

7.1 The crush spread

Figure 7.1 provides a time-series plot of the monthly prices of soybeans, soybean oil, and soybean meal over the period 1980 to 2017. Application of a simple augmented Dickey–Fuller test (with one lag) to each of the series using Stata's `dfuller` (see [TS] **dfuller**) command provides prima facie evidence that they are $I(1)$. The test statistics for the prices of soybeans, soybean oil, and soybean meal are, respectively, -2.546, -2.609, and -2.605. The 5% and 10% critical values are -2.872 and -2.570, indicating that the null hypothesis of a random walk cannot be rejected at the 5% level in all cases.

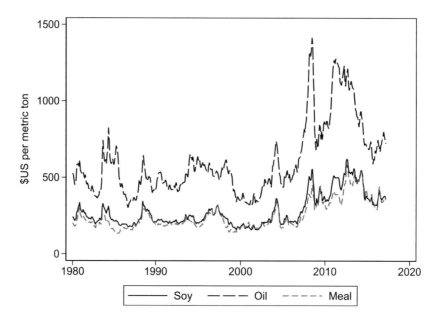

Figure 7.1. Time-series plots of the price of soybeans, soybean oil, and soybean meal. The data are monthly, 1980:1 to 2017:3.

Despite the fact that there is evidence to support the conclusion that each of these series is nonstationary, figure 7.1 also provides casual empirical evidence to suggest, as expected, that they exhibit long-run equilibrium behavior in that they all track each other quite closely. The central concern of this chapter is to formalize a statistical notion of the phenomenon of equilibrium relationships between nonstationary variables in which linear combinations of these $I(1)$ variables may in fact be stationary or $I(0)$, in which case these variables are cointegrated.

A key component in the soybean market is the crush spread. Soybeans are processed into two products—soybean meal and soybean oil—by a process known as crushing. The crush spread is then defined as the difference between the combined value of these two byproducts, and the value of the soybeans and may loosely be interpreted as the profit margin for the soybean processor. The rule of thumb is that one unit of soybeans produces 80% soybean meal and 18% soybean oil, with the remainder going to waste (CME Group 2015). To determine the crush margin on a unit of soybeans, therefore, the calculation is[1]

$$\text{Crush margin} = 0.80 \times \text{Price soybean meal} + 0.18 \times \text{Price soybean oil} - \text{Price of soybeans}$$

---

1. The crush spread is usually specified in terms of prices of futures contracts, but the data used here refer to the spot prices of the commodities.

The crush spread is simply a linear combination of the prices in the soybean market. Therefore, it exactly corresponds with the notion of cointegration.

Figure 7.1 also suggests that variances of the prices series are larger in the latter part of the sample. This feature is often observed in applied econometric analysis, and the usual strategy adopted to deal with this is to model the logarithms of the series. The analysis for this chapter will, however, be conducted on the levels of the series rather than the logarithms mainly because the linear interpretation of the crush spread is thereby maintained.

7.2 Illustrating equilibrium relationships

Assume that we have two variables, y_t and x_t, that share a long-run equilibrium relationship given by

$$y_t = \beta_0 + \gamma_0 t + \beta_1 x_t \tag{7.1}$$

in which t is a deterministic time trend. This long-run relationship is simply a linear combination of y_t and x_t, with the weights of the linear combination given by $(1, -\beta_1)$. There is no particular need to normalize with respect to y_t, but this assumption is simply maintained here for simplicity.

If this long-run relationship is an accurate description of the relationship between the two variables, a shock to the system would imply that both y_t and x_t adjust to restore equilibrium. The system dynamics can be described by the following bivariate system:

$$\Delta y_t = \delta_1 + \alpha_1(y_{t-1} - \beta_0 - \gamma_0 t - \beta_1 x_{t-1}) \tag{7.2}$$
$$\Delta x_t = \delta_2 + \alpha_2(y_{t-1} - \beta_0 - \gamma_0 t - \beta_1 x_{t-1}) \tag{7.3}$$

In this model, the parameters α_1 and α_2 capture the speed of adjustment to the equilibrium specified in (7.1). The term $(y_{t-1} - \beta_0 - \gamma_0 t - \beta_1 x_{t-1})$ represents the previous disequilibrium that the changes Δy_t and Δx_t seek to redress. This representation is therefore known as an error-correction model (Sargan 1964; Hendry and Anderson 1977; Davidson et al. 1978). For a simple derivation of a similar system of equations, see [TS] **vec intro**, and for a survey of error-correction models, see Alogoskoufis and Smith (1991).

Both the nature of the long-run equilibrium relationship and the adjustment to this equilibrium following a shock can be demonstrated by simulating these equations. Figure 7.2 illustrates several interesting cases. The starting values for y_t and x_t are taken to be 100 and 110, respectively. The base values of the parameters of the system are

$$\delta_1 = \delta_2 = 0, \ \alpha_1 = -0.05, \ \alpha_2 = 0.05, \ \beta_0 = 0, \ \gamma_0 = 0, \ \beta_1 = 1$$

and the different cases illustrated in figure 7.2 are defined by changing the values of chosen parameters. Unless otherwise specified, the other parameters revert to their original values.

Model 1: base case. There is symmetric adjustment to a long-run equilibrium that contains neither a constant nor a time trend. The parameters on the linear combination of y_t and x_t are $(1, -1)$.

Model 2: $\beta_0 = 1$. A nonzero constant is introduced into the long-run equilibrium.

Model 3: $\beta_1 = 1.05$. The weight on x_t in the equilibrium is changed so that it is no longer unity.

Model 4: $\gamma_0 = 0.01$. A deterministic time trend is introduced into the long-run equilibrium relationship.

Model 5: $\alpha_1 = -0.1$, $\alpha_2 = 0.01$. The adjustment parameters now take different values so that the adjustment to equilibrium is asymmetric.

Model 6: $\delta_1 = \delta_2 = 0.05$. Constant terms are introduced in the equations that govern the changes in y_t and x_t, respectively. These constants therefore impinge on the dynamics of y_t and x_t and not on the nature of the equilibrium.

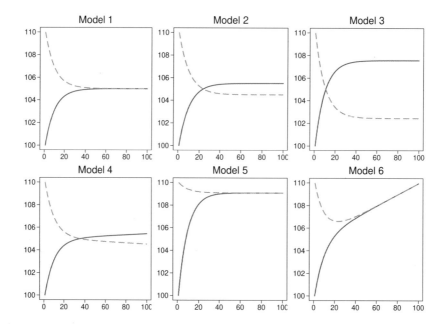

Figure 7.2. Illustrating the equilibrium adjustment paths in a bivariate error-correction model. The solid line indicates the path of y_t, and the dashed line indicates the path of x_t.

A few comments on figure 7.2 are in order. Notice that there is an adjustment to equilibrium so long as the adjustment parameters have opposite signs, reflecting the

7.3 The VECM

principle of negative feedback in a stable dynamic system. Normalizing the long-run relationship with respect to y_t requires that $\alpha_1 < 0$ in (7.2). The effect of a constant in the long-run relationship (see model 2) implies that, in equilibrium, y_t and x_t are not equal to each other as in model 1 but maintain a constant gap as $t \to \infty$. A similar effect is obtained by allowing the weight on x_t to deviate from unity (see model 3).

Model 4 is an interesting situation in which there is a time trend in the equilibrium specification. Here the distance between y_t and x_t is growing at a constant rate as $t \to \infty$. In model 5, it is clearly y_t that is doing most of the adjustment of equilibrium given that $|\alpha_1| > |\alpha_2|$. Finally, the effect of a constant in the equations for Δy_t and Δx_t does not affect the equilibrium relationship but introduces a time trend in the levels of the variables.

Although this simulation experiment is purely deterministic, it has the advantage of demonstrating that this simple bivariate model can capture a wide variety of patterns in both the equilibrium relationship between two variables and the adjustment toward this equilibrium.

7.3 The VECM

Equations (7.1)–(7.3) can be transformed into stochastic relationships by introducing error terms. The long-run or cointegrating equation is given by

$$y_t = \beta_0 + \beta_1 x_t + u_t \qquad (7.4)$$

in which u_t is a disturbance term. For simplicity, the time trend has been omitted. If y_t and x_t are both $I(1)$ variables and the linear combination given by (7.4) results in the disturbance term u_t being $I(0)$, then y_t and x_t are said to be cointegrated (Engle and Granger 1987). The β_0 and β_1 coefficients represent the cointegrating or long-run parameters. The disturbances u_t may now be referred to as equilibrium errors because they represent the stationary deviations from the long-run equilibrium in each time period.

The resulting dynamic equations for Δy_t and Δx_t are then

$$\Delta y_t = \alpha_1 (y_{t-1} - \beta_0 - \beta_1 x_{t-1}) + v_{1t}$$
$$\Delta x_t = \alpha_2 (y_{t-1} - \beta_0 - \beta_1 x_{t-1}) + v_{2t} \qquad (7.5)$$

in which v_{1t} and v_{2t} are disturbance terms. The adjustment parameters α_1 and α_2 are known as the error-correction parameters because they govern the speed of adjustment of the system to the long-run equilibrium. Inspection of the right-hand side of (7.5) reveals that the bracketed term in both equations is in fact u_{t-1}, so adjustment takes place so as to correct for the sign and magnitude of the equilibrium error from the previous period. For this reason, the system of equations (7.5) is known as the VECM.

A VECM is simply a restricted form of the VAR model discussed in chapter 4, with the restrictions following from the fact that the variables in the system are linked via a long-run cointegrating equation. Reexpressing (7.5) in levels of the variables rather than first differences yields

$$y_t = -\alpha_1\beta_0 + (1+\alpha_1)y_{t-1} - \alpha_1\beta_1 x_{t-1} + v_{1t}$$
$$x_t = -\alpha_2\beta_0 + \alpha_2 y_{t-1} + (1-\alpha_2\beta_1)x_{t-1} + v_{2t} \qquad (7.6)$$

or

$$y_t = \phi_{10} + \phi_{11} y_{t-1} + \phi_{12} x_{t-1} + v_{1t}$$
$$x_t = \phi_{20} + \phi_{21} y_{t-1} + \phi_{22} x_{t-1} + v_{2t} \qquad (7.7)$$

Note that in both the VAR in (7.7) and the VECM in (7.5), the longest lags present are the terms y_{t-1} and x_{t-1}. This relationship between VECMs and VARs implies that the information criteria used to determine the optimal lag structure of a VAR, namely, the **varsoc** command (see [TS] **varsoc**) discussed in chapter 4, can also be used to determine the optimal lag length of a VECM.[2]

There are four parameters to be estimated in the VECM of (7.5): $(\alpha_1, \alpha_2, \beta_0, \beta_1)$. In contrast, there are six parameters in the VAR of (7.7): $(\phi_{10}, \phi_{20}, \phi_{11}, \phi_{12}, \phi_{21}, \phi_{22})$. Apart from the constant terms, the parameters in (7.7) are related to those in (7.6) by the expressions

$$\phi_{11} = 1+\alpha_1, \quad \phi_{12} = -\alpha_1\beta_1, \quad \phi_{21} = \alpha_2, \quad \phi_{22} = 1-\alpha_2\beta_1$$

It follows that if y_t and x_t are $I(1)$ and cointegrated, then instead of fitting a simple VAR in differences, the appropriate specification is to impose the restrictions implied by cointegration and fit a VECM.

An interesting conundrum arises if a constant term is added to each equation of the VECM in (7.5) so that both the error-correction term and the short-run dynamics have a constant term. The resulting equations are

$$\Delta y_t = \delta_1 + \alpha_1(y_{t-1} - \beta_0 - \beta_1 x_{t-1}) + v_{1t}$$
$$\Delta x_t = \delta_2 + \alpha_2(y_{t-1} - \beta_0 - \beta_1 x_{t-1}) + v_{2t}$$

Comparing this setup with the unrestricted VAR in (7.7), it is clear that in this instance

$$\phi_{10} = \delta_1 - \alpha_1\beta_0, \qquad \phi_{20} = \delta_2 - \alpha_2\beta_0$$

so that, assuming that α_1 and α_2 may be identified, the problem remains that there is enough information to identify only two of the remaining three parameters. This problem is usually circumvented by simply apportioning values to the three intercept terms. Note, however, that this procedure must result in at least one of the constants being reported without a standard error. Usually, this constant is the intercept in the cointegrating regression.

---

2. Note, however, that as the dependent variables in the VAR are in levels and those in the VECM are in first differences, it is standard to refer to the number of lags in a VECM as $k-1$, where k is the number of lags in the corresponding unrestricted VAR.

7.4.1 Single-equation methods

A final word is necessary on the specification of VECMs concerning the introduction of time trends. From figure 7.2, the inclusion of a time trend in the cointegrating equation results in an equilibrium in which the distance between y_t and x_t grows at a constant rate. Because the inclusion of a constant introduces a time trend in the levels of the variables, including a time trend in the dynamic equations will introduce a quadratic trend in the levels of the variables. For completeness, Stata allows for these two situations, but because they are rarely encountered in practice, they will not be dealt with in detail here.

7.4 Fitting VECMs

The VECM can be expressed conveniently in matrix notation by defining the $(N \times 1)$ vectors

$$\mathbf{y}_t = \begin{bmatrix} y_{1t} \\ y_{2t} \\ \vdots \\ y_{Nt} \end{bmatrix}, \quad \boldsymbol{\delta} = \begin{bmatrix} \delta_1 \\ \delta_2 \\ \vdots \\ \delta_N \end{bmatrix}, \quad \boldsymbol{\alpha} = \begin{bmatrix} \alpha_1 \\ \alpha_2 \\ \vdots \\ \alpha_N \end{bmatrix}, \quad \boldsymbol{\beta} = \begin{bmatrix} 1 \\ -\beta_2 \\ \vdots \\ -\beta_N \end{bmatrix}, \quad \mathbf{v}_t = \begin{bmatrix} v_{1t} \\ v_{2t} \\ \vdots \\ v_{Nt} \end{bmatrix}$$

where the notation for the dependent variable has been changed deliberately to emphasize that the variables constitute a system. The VECM can now be written as

$$\Delta \mathbf{y}_t = \boldsymbol{\delta} + \boldsymbol{\alpha}(\boldsymbol{\beta}'\mathbf{y}_{t-1} - \boldsymbol{\beta}_0) + \sum_{j=1}^{k-1} \boldsymbol{\Gamma}_j \Delta \mathbf{y}_{t-j} + \mathbf{v}_t \tag{7.8}$$

where $\boldsymbol{\Gamma}_j$ is a $(N \times N)$ matrix of parameters at lag j. Additional dynamics have been introduced via lags of the dependent variables, and all lagged dependent variables appear in all equations as in a standard VAR. The VECM can be extended to allow for multiple cointegrating equations, which is appropriate when there are several equilibrium relationships, or cointegrating vectors, involving the $I(1)$ variables.

7.4.1 Single-equation methods

The original method for fitting a VECM was a simple two-stage method that was introduced in the seminal article by Engle and Granger (1987). The Engle–Granger procedure is implemented by estimating the cointegrating regression and then the dynamic equations sequentially by ordinary least squares as follows:

Long-run: Regress y_{1t} on a constant and $y_{2t} \ldots y_{Nt}$ to obtain estimates of the cointegrating parameters, and compute the residuals $\widehat{u}_t$. This regression represents an arbitrary normalization on y_{1t}, but any y_{it} could be used for normalization purposes.

Short-run: Estimate each equation of the VECM in turn by ordinary least squares. The error-correction parameter estimates, $\widehat{\alpha}_j$, are the slope parameter estimates on $\widehat{u}_{t-1}$ in these equations.

This estimator yields superconsistent estimates of the cointegrating parameters, in the sense that the estimates converge to the true population parameters at the rate T, which is double the usual $\sqrt{T}$ rate for stationary processes. On the other hand, the Engle–Granger estimator does not produce estimates that are asymptotically efficient, because it ignores the fact that all the variables are potentially endogenous. Also, the lack of dynamic terms in the long-run equation usually means that u_t will be autocorrelated. As a consequence, the ordinary least-squares estimates have nonstandard distributions rendering inference more difficult. In other words, although it is not necessary to account for short-run dynamics to obtain superconsistent estimates of the long-run parameters, the short-run dynamics must be modeled to obtain an efficient estimator with t statistics that have asymptotic distributions based on the normal distribution.

There are numerous single-equation methods that deal with the lack of efficiency of the single-equation approach. These include fully modified ordinary least squares (Phillips and Hansen 1990) and dynamic least squares (Saikkonen 1991; Stock and Watson 1993). These estimators are available as community-contributed packages for Stata, and they are not pursued any further here.

The results of implementing the Engle–Granger two-stage procedure for the crush spread data, with the relationship arbitrarily normalized with respect to the spot price of soybeans, are as follows:

```
. use http://www.stata-press.com/data/eeus/crushspread
. tsset datevec, monthly
        time variable:  datevec, 1980m1 to 2017m3
                delta:  1 month
. regress psoy poil pmeal
```

Source	SS	df	MS		Number of obs	=	447
					F(2, 444)	=	18894.80
Model	4783102.56	2	2391551.28		Prob > F	=	0.0000
Residual	56197.9218	444	126.571896		R-squared	=	0.9884
					Adj R-squared	=	0.9883
Total	4839300.49	446	10850.4495		Root MSE	=	11.25

psoy	Coef.	Std. Err.	t	P>|t|	[95% Conf. Interval]	
poil	.1831141	.0038374	47.72	0.000	.1755724	.1906558
pmeal	.6846921	.0095673	71.57	0.000	.6658893	.703495
_cons	-3.611907	1.559856	-2.32	0.021	-6.677525	-.5462901

```
. predict double egresid, resid
```

7.4.2 System estimation

```
. regress D.psoy L.D.psoy L.D.poil L.D.pmeal L.egresid, noconstant

      Source |       SS           df       MS      Number of obs   =       445
-------------+----------------------------------   F(4, 441)       =     15.68
       Model |  18700.2812         4   4675.07029  Prob > F        =    0.0000
    Residual |  131474.253       441   298.127558  R-squared       =    0.1245
-------------+----------------------------------   Adj R-squared   =    0.1166
       Total |  150174.534       445   337.470863  Root MSE        =    17.266

      D.psoy |      Coef.   Std. Err.      t    P>|t|     [95% Conf. Interval]
-------------+----------------------------------------------------------------
        psoy |
         LD. |   .5744645   .2028311     2.83   0.005     .1758287    .9731003
             |
        poil |
         LD. |   .0118074   .0471886     0.25   0.803    -.0809351    .1045499
             |
       pmeal |
         LD. |  -.3039955   .1454596    -2.09   0.037    -.5898757   -.0181153
             |
     egresid |
         L1. |  -.1274655   .0745081    -1.71   0.088    -.2739006    .0189697
```

The first striking feature of these results is that the cointegrating parameters are remarkably similar to the equilibrium suggested by the physical process of crushing. The estimated cointegrating parameters are 0.183 for soybean oil and 0.685 for soybean meal, respectively. Note that it is not appropriate to judge the significance of the constituents of the long-run relationship based on the standard errors returned by this ordinary least-squares regression, because the distribution of these parameters is not standard. In terms of the second-stage dynamic regression, the lagged equilibrium errors have a p-value of 0.088, indicating that the adjustment to equilibrium is only marginally significant at conventional levels of significance.

7.4.2 System estimation

Instead of estimating the cointegrating parameters and then the dynamic equation in a two-stage process, a fully efficient system approach was developed by Johansen (1988, 1991, 1995) in which all the parameters are estimated simultaneously by maximum likelihood. Let the vector of VECM disturbances in (7.8) be normally distributed with a null mean vector and covariance matrix $\mathbf{V}$. Based on these assumptions, the log-likelihood function for a sample of T observations is

$$\log L = -\frac{N}{2}\log 2\pi - \frac{1}{2}\log|\mathbf{V}| - \frac{1}{2(T-p)}\sum_{t=p+1}^{T}\mathbf{v}'_t \mathbf{V}^{-1}\mathbf{v}_t \qquad (7.9)$$

Given the nonlinearities arising from cointegration, an iterative gradient algorithm can be used to maximize the log likelihood in (7.9) with respect to the unknown parameters.[3]

---

3. See Martin, Hurn, and Harris (2013) for an example of the maximization of the log-likelihood function in (7.9).

Instead of using an iterative gradient algorithm to estimate the unknown VECM parameters, Johansen developed an algorithm based on solving an eigenvalue problem while maximizing the log-likelihood function (7.9). A full description of the Johansen algorithm is not provided here, but for the bivariate case of $N = 2$, an intuitive explanation of the procedure is as follows:

1. Compute residuals from four auxiliary regressions of $\{\Delta y_{1t}, \Delta y_{2t}, y_{1\,t-1}, y_{2,t-1}\}$ on the regressors $\{1, \Delta y_{1\,t-1}, \Delta y_{2\,t-1}\}$.

2. Perform an eigenvalue decomposition of the matrix of these residuals. The estimates of the cointegrating parameters, $\widehat{\beta}_0$ and $\widehat{\beta}_1$, are given by the eigenvector corresponding to the largest eigenvalue of this matrix.

3. Construct the cointegrating residuals $\widehat{u}_t = y_{1t} - \widehat{\beta}_0 - \widehat{\beta}_1 y_{2t}$, and estimate the two dynamic equations in the bivariate VECM. For one lag of the dependent variable, the regressions would be $\{\Delta y_{1t}, \Delta y_{2t}\}$ on the regressors $\{1, \widehat{u}_{t-1}, \Delta y_{1\,t-1}, \Delta y_{2\,t-1}\}$.

The optimal lag length to use in the VECM for the prices of soybean, soybean oil, and soybean meal can be chosen by using the `varsoc` command. The results are not reported here, but both the Hannan–Quinn information criterion and the Schwarz information criteria suggest an optimal lag length of 2 for the underlying VAR. The VECM can now be fit using `vec` (see [TS] **vec**). Before doing so, however, the specification of the VECM must be chosen from among the specifications available in that command's `trend()` option. The most general form of the VECM can be expressed as

$$\Delta \mathbf{y}_t = \boldsymbol{\alpha}\left(\boldsymbol{\beta}'\mathbf{y}_{t-1} + \boldsymbol{\mu} + \boldsymbol{\rho} t\right) + \sum_{i=1}^{p-1} \Gamma_i \Delta \mathbf{y}_{t-i} + \boldsymbol{\gamma} + \boldsymbol{\tau} t + \epsilon_t \quad (7.10)$$

and the five model specifications are defined as follows:

1. `trend(trend)`: estimated as shown; the cointegrating equations are trend stationary

2. `trend(rtrend)`: $\boldsymbol{\tau} = \mathbf{0}$; the cointegrating equations are trend stationary, and trends in levels are linear but not quadratic

3. `trend(constant)`: $\boldsymbol{\tau} = \boldsymbol{\rho} = \mathbf{0}$; the cointegrating equations are stationary around constant means with linear trends in levels (default)

4. `trend(rconstant)`: $\boldsymbol{\tau} = \boldsymbol{\rho} = \boldsymbol{\gamma} = \mathbf{0}$; the cointegrating equations are stationary around constant means, with no linear time trends in the data

5. `trend(none)`: $\boldsymbol{\tau} = \boldsymbol{\rho} = \boldsymbol{\gamma} = \boldsymbol{\mu} = \mathbf{0}$; the cointegrating equations, levels, and differences of the data all have means of zero

Essentially, the choice is between including a constant in both the long-run equilibrium equation and the short-run dynamic equations, or restricting the constant in the long-run relationship. Because there is no compelling evidence of linear trends in the levels of the price series, the choice of `trend(rconstant)` is adopted. In performing

7.4.2 System estimation

the estimation, a single cointegrating vector is produced. The output below suppresses listings of the short-run dynamic parameters so that only the parameter estimates for the cointegrating vector and the `alpha` parameters, representing speeds of adjustment to long-run equilibrium, are reported.

```
. vec psoy poil pmeal, lag(2) trend(rconstant) alpha noetable
Vector error-correction model
Sample:  1980m3 - 2017m3                        Number of obs   =         445
                                                AIC             =    24.04448
Log likelihood = -5334.897                      HQIC            =    24.09895
Det(Sigma_ml) =    5196185                      SBIC            =    24.18262
```

Cointegrating equations

Equation	Parms	chi2	P>chi2
_ce1	2	1539.093	0.0000

Identification: beta is exactly identified
                 Johansen normalization restriction imposed

	beta	Coef.	Std. Err.	z	P>|z|	[95% Conf. Interval]	
_ce1							
	psoy	1	.	.	.	.	.
	poil	-.220885	.0188568	-11.71	0.000	-.2578437	-.1839262
	pmeal	-.5716935	.0471671	-12.12	0.000	-.6641394	-.4792476
	_cons	-1.81055	7.578134	-0.24	0.811	-16.66342	13.04232

Adjustment parameters

Equation	Parms	chi2	P>chi2
D_psoy	1	7.817871	0.0052
D_poil	1	1.545491	0.2138
D_pmeal	1	3.010733	0.0827

	alpha	Coef.	Std. Err.	z	P>|z|	[95% Conf. Interval]	
D_psoy							
_ce1							
	L1.	-.1806605	.0646129	-2.80	0.005	-.3072994	-.0540216
D_poil							
_ce1							
	L1.	-.1720038	.1383582	-1.24	0.214	-.4431809	.0991733
D_pmeal							
_ce1							
	L1.	-.1162697	.0670086	-1.74	0.083	-.2476041	.0150647

In these results, which are produced in a dynamic multivariate setting with all variables treated as endogenous, the distribution of all the coefficients are standard, and conventional inference is possible. Once again, the estimated parameters of the cointegrating vector accord strongly with theoretical priors with the crush margin. The estimated values of 0.221 and 0.572 for oil and meal, respectively, are similar to the estimates returned by the single-equation estimation. The adjustment parameter on the error-correction term in the dynamic equation for changes in the price of soybeans of -0.181 is now statistically significant with a p-value of 0.005. The adjustment process seems to be asymmetric, with neither of the adjustment parameters in the other two dynamic equations being significantly different from zero at the 5% level. The interpretation is that it is the price of soybeans that does most of the adjusting to equilibrium in the soybean market.

Up to this point in the discussion, the existence of a single cointegrating vector has simply been assumed based on theoretical grounds. In an empirical application, however, a crucial element is to test for evidence of cointegration because the identification of cointegration is a crucial first step in modeling nonstationary variables.

7.5 Testing for cointegration

A natural way to test for cointegration is a two-step procedure consisting of estimating the cointegrating equation by least squares in the first step and testing the residuals from the cointegrating regression for stationarity in the second step. In this way, optimal weights are chosen for constituent series of the crush spread. When the weights are estimated rather than assumed to take on their theoretical values, the critical values for a unit-root test on the residuals must account for the loss of degrees of freedom in estimating the cointegrating equation. These critical values depend on the sample size and the number of deterministic terms and other regressors in the first stage regression. The community-contributed command `egranger` (Schaffer 2010) provides this test using response surface estimates of the critical values.

```
. egranger psoy poil pmeal
Engle-Granger test for cointegration          N (1st step)  =      447
                                              N (test)      =      446

                Test         1% Critical      5% Critical      10% Critical
           Statistic               Value            Value             Value

   Z(t)       -3.980              -4.326           -3.760            -3.466
Critical values from MacKinnon (1990, 2010)
```

The results are fairly straightforward. The three series are cointegrated at the 95% level of confidence when the weights are chosen optimally. The test statistic of -3.980 is greater in absolute value than the 5% critical value of -3.760 but falls short of the 1% critical value of -4.326. The null hypothesis that the crush spread is nonstationary is therefore rejected at the 5% level. Figure 7.3 plots the residuals from the first stage of the Engle–Granger procedure. There seems to be a negative trend in the residuals over the first half of the sample that disappears in the late 1990s. This result perhaps supports the marginal insignificance of the error-correction term in the dynamic equation estimated in the second step of the Engle–Granger procedure. The result is, however,

7.5 Testing for cointegration

out of line with the fact that the adjustment parameter on the error-correction term in the VECM is significant.

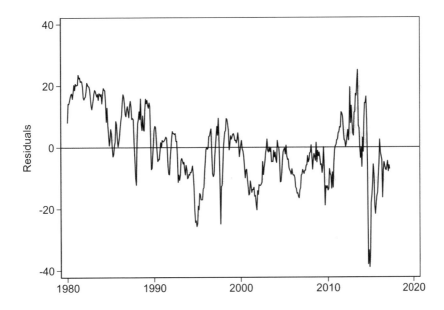

Figure 7.3. Residuals obtained from the first stage of the Engle–Granger two-stage procedure

An alternative and more general approach to testing for cointegration is based on the Johansen procedure and involves testing a sequence of hypotheses. If there are N variables being tested for possible cointegration, the maximum number of hypotheses considered is N.

Stage 1:

H_0: No cointegrating vectors
H_1: One or more cointegrating vectors

Under the null hypothesis, all the variables are $I(1)$, and there is no linear combination of the variables that exhibits cointegration. Under the alternative hypothesis, there is (at least) one linear combination of the $I(1)$ variables that yields a stationary disturbance and hence supports the existence of cointegration.

Stage 2:

H_0: One cointegrating vector
H_1: Two or more cointegrating vectors

If the null hypothesis is not rejected, the testing procedure stops and we conclude that there is one cointegrating vector. Otherwise, proceed to the next stage.

...

Stage N:

H_0: $N-1$ cointegrating vectors
H_1: All variables are stationary

At the final stage, the alternative hypothesis is that all variables are stationary rather than that there are N cointegrating equations. For there to be N linear stationary combinations of the variables, they must be stationary in the first place.

The Johansen cointegration test is a likelihood-ratio (LR) test that follows naturally from the fact that the Johansen estimator is a maximum likelihood estimator based on the log-likelihood function in (7.9). The Johansen estimator is not usually implemented in terms of a full nonlinear iterative search procedure but rather in terms of an eigenvalue decomposition problem. Likewise, the LR test is computed in terms of the estimated eigenvalues obtained from applying the Johansen estimator. For an N-dimensional system, this form of the LR statistic is

$$\text{LR} = -(T-k) \sum_{i=r+1}^{N} \log\left(1 - \widehat{\lambda}_i\right)$$

where $\widehat{\lambda}_i, i = 1, 2, \ldots, N$ are the estimated eigenvalues ordered from highest to lowest obtained from the Johansen estimator. The subscript r represents the number of cointegrating vectors. Essentially, this test can be interpreted as a test of the number of estimated eigenvalues that are nonzero. The Johansen cointegration test is also called the trace test because the trace of a matrix is the sum of its eigenvalues.

If all variables in $\mathbf{y}_t$ are $I(1)$ and there is no cointegration among the nonstationary variables, then a VAR in their first differences is the correct specification. If all the variables in y_t are $I(0)$, then a VAR in their levels is the correct specification. Conversely, if $r > 0$, then $\boldsymbol{\alpha}$ and $\boldsymbol{\beta}$ in (7.10) are $(N \times r)$ matrices of rank r. The $\boldsymbol{\beta}$ matrix contains the cointegrating vectors, while the matrix $\boldsymbol{\alpha}$ contains the speed of adjustment parameters that determine how disequilibria in the prior period will adjust the current values of each y_t.

The Johansen trace test is implemented by the vecrank (see [TS] **vecrank**) command. For the model with a restricted constant and a lag length of 2, the results are as follows:

7.6 Cointegration and structural breaks   143

```
. vecrank psoy poil pmeal, lags(2) trend(rconstant)
                      Johansen tests for cointegration
  Trend: rconstant                                         Number of obs =    445
  Sample:  1980m3 - 2017m3                                          Lags =      2
```

maximum rank	parms	LL	eigenvalue	trace statistic	5% critical value
0	9	-5346.6919	.	46.5032	34.91
1	15	-5334.8973	0.05163	22.9138	19.96
2	19	-5326.2531	0.03811	5.6254*	9.42
3	21	-5323.4404	0.01256		

The Johansen trace test suggests that there are two or fewer cointegrating vectors. This situation are often encountered in applied work, where the data support results that run counter to strong theoretical priors. This particular result will not be pursued any further here, given the strong theoretical prior that there is only one cointegrating relationship.[4]

7.6 Cointegration and structural breaks

The statistical properties of the crush spread are of interest for more than mere econometric reasons because biofuels are manufactured directly from oils and starches. In particular, biodiesel is made from vegetable oils such as palm oil, soybean oil, canola, and rapeseed, while ethanol is distilled from starchy crops including sugar, corn, wheat, and barley. Many countries, including the United States, give regulatory support to biofuel. For example, in the United States, the Renewable Fuel Standard of 2005 and the Energy Independence and Security Act of 2007 specify minimum quantities of biofuel usage annually, with a minimum 10% ethanol content in unleaded gasoline. There is empirical evidence that the petroleum and agricultural markets are becoming linked as a result of these regulations (see for example, Silvennoinen and Thorp, 2016). An interesting question for environmental policy is then whether this kind of regulation has the potential to disturb equilibrium relationships between the price of agricultural commodities that are used in the biofuel industry and their byproducts.

As already noted, casual inspection of figure 7.1 suggests that sometime after 2000, the behavior of the commodity price indices that make up the crush spread may have altered. This shift could be either an intercept shift or perhaps even a change in linear trend. The change also is reflected in the crush spread itself. This observation then raises the possibility of allowing for structural breaks in cointegration analysis.

Gregory and Hansen (1996a,b) provide a single-equation framework for assessing cointegration in the presence of breaks in the cointegrating vector. Let y_{1t} be a single dependent variable, and let $\mathbf{y}_{2t}$ be a vector of variables that, together with y_{1t}, potentially form a long-run equilibrium relationship. The general model is

---

4. Fitting and interpreting VECMs with a cointegrating rank greater than 1 and appropriate normalization schemes are explored in the exercises.

$$y_{1t} = \alpha_1 + \alpha_2 \text{DU}_{bt} + \gamma_1 t + \gamma_2 t \text{DU}_{bt} + \beta_1' \mathbf{y}_{2t} + \beta_2' \mathbf{y}_{2t} \text{DU}_{bt} + v_t$$

in which DU_{bt} is the dummy variable defined as

$$\text{DU}_{bt} = \begin{cases} 0, & t \leq T_b \\ 1, & t > T_b \end{cases}$$

and the break point is T_b or equivalently τT where τ is the break fraction corresponding to T_b. This general model allows for breaks in level, trend, and cointegrating parameters.

The procedure to test for cointegration is to compute the regression residuals from this model for all possible breaks and for all possible break fractions, τ, although in practice the model is usually simplified to look only for a specific type of break. For each case, the relevant cointegration test based on the residuals, either Dickey–Fuller or Phillips–Perron, is computed. Statistical inference is then based on the smallest value (largest negative) returned for the test statistic. In practice, the break fraction is limited to take values from 0.15 to 0.85, and only integer values of the break fraction are used. There is no distribution theory available for these statistics, so critical values are obtained by simulation.

The **ghansen** package (Pérez 2002) available from the Statistical Software Components Archive implements this method for testing for cointegration in the presence of breaks. The package deals with a single dependent variable and up to four independent variables. There are two required options, namely, `break()` and `lagmethod()`. The former option takes an argument that specifies the structural break to be of type `level`, `trend`, `regime`, and `regimetrend`, respectively. The `lagmethod()` option also takes an argument that specifies the algorithm to be used to choose the number of lags in the Dickey–Fuller test.

The **ghansen** method applied to the crush spread data for both an intercept shift and a break in trend yields the following results:

```
. use http://www.stata-press.com/data/eeus/crushspread, clear
. tsset datevec, monthly
        time variable:  datevec, 1980m1 to 2017m3
                delta:  1 month
. ghansen psoy poil pmeal, break(level) lagmethod(aic)
Gregory-Hansen Test for Cointegration with Regime Shifts
Model: Change in Level                       Number of obs   =        447
Lags  =   4  chosen by Akaike criterion      Maximum Lags    =          4
              Test        Breakpoint   Date      Asymptotic Critical Values
              Statistic                             1%          5%         10%
              
       ADF    -5.61         121       1990m1     -5.44       -4.92       -4.69
       Zt     -7.48         119       1989m11    -5.44       -4.92       -4.69
       Za     -78.73        119       1989m11    -57.01      -46.98      -42.49
```

7.6 Cointegration and structural breaks

```
. ghansen psoy poil pmeal, break(trend) lagmethod(aic)
Gregory-Hansen Test for Cointegration with Regime Shifts
Model: Change in Level and Trend              Number of obs  =     447
Lags =  2  chosen by Akaike criterion         Maximum Lags   =       4

              Test    Breakpoint   Date      Asymptotic Critical Values
           Statistic                            1%       5%      10%

    ADF     -7.79       216      1997m12      -5.80    -5.29    -5.03
    Zt      -8.57       211      1997m7       -5.80    -5.29    -5.03
    Za    -101.84       211      1997m7      -64.77   -53.92   -48.94
```

In both cases, the hypothesis of no cointegration is rejected, and the result is stronger than that obtained for the simple Engle–Granger two-stage procedure in section 7.5. The estimates of the break date for the trend-break model appear to be marginally more in line with figure 7.1. The break dates returned by the intercept-shift model appear to be a little early. Taking the break point to be July 1997 gives the cointegrating equation

```
. generate DU = 0
. replace DU = 1 if datevec >= tm(1997m7)
(237 real changes made)
. generate t  = _n
. generate Dt = t*DU
. regress psoy poil pmeal t Dt

      Source |      SS         df      MS           Number of obs  =     447
-------------+------------------------------        F(4, 442)      = 22140.93
       Model | 4815268.65       4  1203817.16       Prob > F       =   0.0000
    Residual | 24031.8337     442  54.3706645       R-squared      =   0.9950
-------------+------------------------------        Adj R-squared  =   0.9950
       Total | 4839300.49     446  10850.4495       Root MSE       =   7.3736

        psoy |    Coef.   Std. Err.     t     P>|t|    [95% Conf. Interval]
-------------+------------------------------------------------------------
        poil |  .1886017   .0025427   74.17   0.000    .1836045    .1935989
       pmeal |  .7476676    .00682   109.63   0.000     .734264    .7610711
           t | -.1551257   .0082809  -18.73   0.000   -.1714005   -.1388509
          Dt |  .0591226   .0059397    9.95   0.000    .0474491    .0707961
       _cons |  1.503949   1.202021    1.25   0.212   -.8584382    3.866337
```

Once again, the long-run parameters are in line with expectations. What is different now, however, is that the residuals from the first-stage regression, shown in figure 7.4, appear to be stationary. In particular, when comparing this plot with figure 7.3, the residuals now seem better behaved and do not have the same slight negative trend observed when there is no allowance for the trend break.

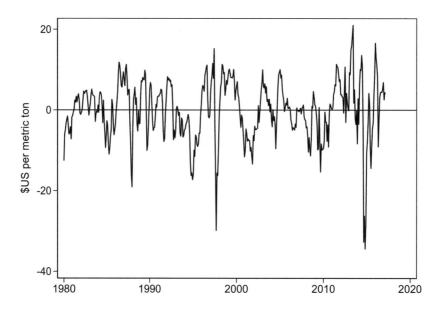

Figure 7.4. Residuals obtained from the first stage of cointegrating regression that allows for a trend break following the Gregory–Hansen approach.

The Gregory–Hansen single-equation approach to structural breaks and cointegration focuses only on the cointegrating regression. The multivariate approach seeks to generalize this treatment and work with the full VECM model. Divide the sample period into subsamples $T_0 < T_1 < \cdots < T_q$ based on breakpoints. A VAR of order k for each subsample j is then

$$\Delta \mathbf{y}_t = \Pi \mathbf{y}_{t-1} + \Pi_j t + \mu_j + \sum_{i=1}^{k-1} \Gamma_i \Delta \mathbf{y}_{t-i} + v_t$$

Note that Π and Γ_i, which relate to the stochastic component of the time series, are the same in each subsample. The parameters Π_j and μ_j, which relate to the deterministic components, vary over the sample periods.

Based on this model, Johansen, Mosconi, and Nielsen (2000) suggest several hypotheses to test. The $H_l(r)$ hypothesis is that the linear trend in each subsample is different but that there is cointegration

$$H_l(r) : \text{rank}(\Pi, \Pi_1 \ldots \Pi_q) \leq r \quad \text{or} \quad (\Pi, \Pi_1 \ldots \Pi_q) = \alpha \begin{pmatrix} \beta \\ \gamma_1 \\ \vdots \\ \gamma_q \end{pmatrix}'$$

7.6 Cointegration and structural breaks

Currently, there is no Stata implementation of these tests.

To conclude, there seems to be evidence to suggest that the crush spread, conventionally defined in terms of the parameters associated with the physical crushing process, may be regarded as the outcome of cointegration between its constituent price series. This has important implications for market participants who behave in a way that assumes the spread is stationary and adjust their exposures in derivative markets based on the belief that the spread will return to some equilibrium level. Interestingly, the behavior of the crush spread seems to have changed a little since the introduction in the 2000s of regulations relating to biofuels. In terms of the crush spread, however, the cointegration result is strengthened when this structural break is accommodated in the long-run equilibrium relationship.

Exercises

1. The environmental Kuznets curve is named after Kuznets (1955), who argued in his 1954 Presidential Address to the American Economic Association that income inequality was related to economic growth and development. Specifically, he hypothesized that income inequality rises at first in the early stages of the development process and then starts to fall with increases in economic growth because of developments in technology, environmental regulation, awareness, and education. This argument has now been made with respect to the relationship between greenhouse gas emissions and economic activity, usually measured in terms of income per capita. mw-data.dta contains annual observations on per capita gross domestic product (GDP), CO_2 emissions, and SO_2 emissions for 19 OECD countries.[5]

    a. Using data for the United States, draw a scatterplot of the logarithm of SO_2 emissions versus the logarithm of per capita GDP for 1870–2000. Use the Stata command lpoly (see [R] **lpoly**) to superimpose on the scatterplot a local polynomial fit. Is the pattern consistent with the environmental Kuznets curve hypothesis? Discuss why this pattern is to be expected in SO_2 emissions. Is the environmental Kuznets curve hypothesis potentially misleading in terms of policy prescription?
    [Hint: You may find the argument of Carson (2010) helpful.]

    b. Test log per capita GDP and SO_2 emissions in the United States for a unit root. Do you conclude that both are $I(1)$?

    c. In empirical applications, the existence of the environmental Kuznets curve is usually taken to imply that the long-run equilibrium relationship between the logarithms of SO_2 emissions (p_t) and per capita GDP (y_t) is nonlinear. Using the Engle–Granger two-step approach, estimate the cointegrating regression

    $$p_t = \alpha + \gamma t + \beta_1 y_t + \beta_2 y_t^2 + v_t$$

---

[5]. The data are used by Wagner (2015) and were obtained from the *Journal of Applied Econometrics* data archive.

in which v_t are the equilibrium errors in the event of a valid cointegrating relationship. In this specification, t is a deterministic time trend aimed at capturing improvements in technology.

d. Repeat exercises 1b and 1c for a selection of OECD countries. What do you conclude about the robustness of the relationship between emissions and GDP?

e. Comment on the problems with approach to validating the environmental Kuznets curve.
[Hint: The work of Wagner (2012, 2015) is important in this regard.]

2. twobalance.dta contains annual observations for the period 1955 to 2011 on temperature in the shallow ocean and atmosphere (gissv3$_t$), a deep ocean layer (ohc$_t$), and radiative forcing (forcing$_t$), from factors like greenhouse gases, solar irradiance, and reflective aerosols from volcanic eruptions.[6] Rename these variables U_t, L_t, and F_t, respectively. An energy balance model requires that the temperatures of the three layers satisfy two equilibrium relations. The first relationship is the link between the temperature of the upper layer (U_t) and radiative forcing (F_t) in terms of net heat flux. The second equilibrium relationship describes the heat transfer between the upper (U_t) and lower (L_t) ocean components.

a. Plot the levels and first differences of the three data series, and comment on their stationarity properties.

b. Use the generalized least-squares version of the augmented Dickey–Fuller test with optimally chosen lag length to determine if each of these variables is $I(1)$. Are your results consistent with your priors developed in exercise 2a?

c. Use the varsoc command on the underlying VAR to determine if a lag length of 2 is supported by the data.

d. Use the vecrank command with the default VECM specification, namely, trend(constant) and using two lags, to test for cointegration. How many cointegrating vectors are detected, and is this result consistent with the theory? How robust is your conclusion to the choice of lag length?

e. The energy balance model can be written as a VECM (with a single lag in the underlying VAR) with two cointegrating vectors given by

$$\begin{bmatrix} \Delta U_t \\ \Delta L_t \\ \Delta F_t \end{bmatrix} = \begin{bmatrix} \alpha_{11} & \alpha_{12} \\ \alpha_{21} & \alpha_{22} \\ \alpha_{31} & \alpha_{32} \end{bmatrix} \begin{bmatrix} \beta_{11} & 0 & 1 \\ 1 & \beta_{22} & 0 \end{bmatrix} \begin{bmatrix} U_{t-1} \\ L_{t-1} \\ F_{t-1} \end{bmatrix} + \begin{bmatrix} u_{1t} \\ u_{2t} \\ u_{3t} \end{bmatrix}$$

The first cointegrating equation models climate feedback on surface temperature, and in this equation the coefficient on radiative forcing is normalized to one, $\beta_{13} = 1$. Furthermore, the temperature in the lower ocean is not

---

6. The data for the temperature of the upper ocean layer and radiative forcing were kindly provided by Felix Pretis of Oxford University (http://www.climateeconometrics.org); see Pretis (2020). The ocean heat content data were downloaded from the National Oceanographic Data Centre.

7.6 Cointegration and structural breaks

affected by this relationship, so $\beta_{12} = 0$. In the second cointegrating equation, the coefficient on the upper layer, U_{t-1}, is normalized to 1, and forcing is excluded, so the relevant restrictions are $\beta_{21} = 1$ and $\beta_{23} = 0$. Fit the VECM with two cointegrating vectors subject to these restrictions and using the option `lags(2)`.

[Hint: The normalization restrictions are specified in the form of constraints using the `constraint` (see [R] **constraint**) command. In the vec command, use the `bconstraints()` option to specify which constraints are being applied. The `rank()` option specifies how many cointegrating vectors are to be estimated.]

f. Interpret the estimates of the long-run equilibrium response of the system.

g. Because of its physical role as a buffer between the heat content of the lower ocean and radiative forcing, upper component temperatures adjust to both cointegrating vectors. The lower component temperatures adjust only to the second cointegrating vector, so it is to be expected that $\alpha_{21} = 0$. Furthermore, radiative forcing is taken as exogenous and does not contribute to the equilibrium adjustment, so $\alpha_{31} = \alpha_{32} = 0$. Interpret the estimates obtained in exercise 2c in light of this theory.

h. Test the restrictions

$$\beta_{22} = -1, \quad \alpha_{21} = 0, \quad \alpha_{13} = 0, \quad \alpha_{23} = 0$$

by means of an LR test. What do you conclude?

[Hint: Because the restrictions involve both the `e(alpha)` and `e(beta)` matrices, the `test` (see [R] **test**) command cannot be used. Estimate the system subject to these four constraints by listing them in the `bconstraints()` and `aconstraints()` options, and use the `lrtest` (see [R] **lrtest**) command to perform the test of restrictions.]

8 Forecasting

One of the main purposes of modeling an environmental variable is to forecast its future behavior and so contribute to effective decisions by managers or public policymakers. Indeed, forecasting may be the most crucial facet of the entire exercise, given some of the challenges presently facing environmental policymakers. Forecasting extreme or severe weather systems, health-threatening air quality, and climate change are some prominent examples.

A forecast is a quantitative estimate about the most likely future value of a particular variable or a range of likely future values with a specified level of confidence. Forecasts are typically based on past and current information about the variable itself and other observable variables that are thought to be related to it. In econometric forecasting, this information is typically embodied in an empirical model whose solution shows the dependence of the variable of interest on other observable variables and unobserved random variables representing errors and disturbances. The mechanism of econometric forecasting then relies on the estimation of the equations of these models with observed data and the use of the fitted equations to create projections of future values.

Forecasting also serves a useful purpose as a means to compare and rank alternative models and to assess different methods of estimation. In carrying out such exercises, forecast errors are useful in directing attention toward the potential weaknesses in model specifications that lead to systematic errors in forecast performance. Forecast evaluation based on past successes and failures also provides a useful way of choosing between alternative models and a way to combine models using the information that is contained in past performance.

It is particularly easy to deal with forecasting in Stata. The `forecast` (see [TS] **forecast**) suite of commands enables building, solving, and forecasting both single equations (after estimation commands such as `regress`, `ivregress`, or `arima`) and systems of equations (after commands like `sureg`, `var`, or `reg3`).

8.1 Forecasting wind speed

The rising costs and negative environmental effects of traditional, nonrenewable energy sources have led to increased research regarding the viability of alternative energy sources. Of these, wind is one of the fastest growing and environmentally friendly sources of energy sources. Because there is currently no cost effective method for storing the energy that a wind turbine produces, irregular wind speeds complicate operational issues relating to wind-generated power. Power flow in watts, P, from a wind turbine is related to wind speed by the relation

$$P = \frac{1}{2}\rho\frac{\pi d^2}{4}v^3$$

where ρ is the air density in kilograms per cubic meter, d is the diameter of the turbine rotors, and v is the wind speed measured in meters per second. It is this cubic relationship between wind speed and power output that makes wind speed forecasting so important. Additional complications arise because at very low wind speeds, there is insufficient torque exerted by the wind on the turbine blades to make them rotate. The speed at which wind-driven turbines first start to rotate and generate power is called the cut-in speed and is typically between three and four meters per second. Furthermore, for winds in excess of 25 meters per second, the turbines cut out to reduce stress and prevent damage to the machinery.

Long-range forecasts and simulation of wind speed are typically the domain of large-scale numerical weather prediction models developed by meteorologists. These models are physical models that use considerations like terrain, air pressure, and temperature to model the future wind speed. From the perspective of environmental econometrics, therefore, it is shorter-term forecasting on the scale of a few minutes to several hours ahead that is of primary importance. These forecasts are important to manage the risks associated with the variability of wind speed. For example, electricity generators bid on the amount of energy they will produce prior to actual production. As a consequence, the accurate forecasting of wind speed is absolutely crucial for several reasons.

1. The stability and reliability of power system operations are highly dependent on the predicted power outputs.
2. Improved wind power forecasts result in more accurate bids on the energy market, thus avoiding penalties for not supplying the contracted power.
3. Operation and maintenance costs could be reduced with accurate wind speed forecasts.

According to Schreck, Lundquist, and Shaw (2008), forecasting errors in wind speeds of only 1% can lead to a loss of U.S. $12,000,000 over the lifetime of a 100 megawatt wind facility comprised of several wind turbines. In addition, this effect will be exacerbated if incorrect forecasts lead to excessive loading on individual wind turbines, thus reducing their average lifespan, which is currently around 20 years. Thus, the ability to forecast wind speeds accurately and precisely has become increasingly important as more wind power is introduced into electricity markets.

8.2 Introductory terminology

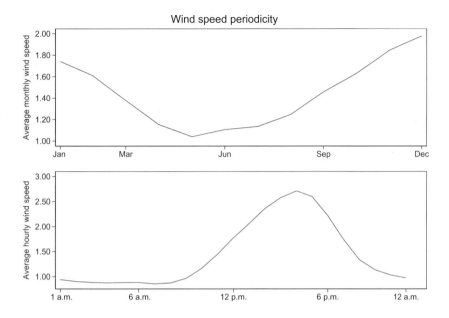

Figure 8.1. Illustrating the periodicity of wind speed. The top panel shows average monthly wind speed (annual cycle) measured at the El Bosque weather station in Santiago, Chile, while the bottom panel shows the average hourly wind speed (daily cycle) for the same station. Data are for the period January 1, 2011 to August 31, 2015.

8.2 Introductory terminology

It is useful at the outset to establish key concepts used in forecasting. Consider an observed sample of data $\{y_1, y_2, \ldots, y_T\}$ and an econometric model that is to be used to generate forecasts of y over a horizon of H periods. The forecasts of y_t are denoted by $\widehat{y}_t$. For the moment, only forecasts based on models using lagged values of y_t are considered, but generalizing to include explanatory variables is straightforward. Consider the following schematic representation:

Sample	$y_1, y_2, \ldots, y_{T-H}, y_{T-H+1}, y_{T-H+2}, \ldots, y_T$
Case 1	$y_1, y_2, \ldots, y_{T-H}, \widehat{y}_{T-H+1}, \widehat{y}_{T-H+2}, \ldots, \widehat{y}_T$
Case 2	$y_1, y_2, \ldots, y_{T-H}, y_{T-H+1}, y_{T-H+2}, \ldots, y_T, \quad \widehat{y}_{T+1}, \ldots, \widehat{y}_{T+H}$

The first row of entries represents the entire set of observations on which estimation can be based. Based on this representation, a glossary of commonly used forecasting terms is as follows.

In-sample (or ex post) forecasts: The model is fit over a restricted sample period that excludes the last H observations, $\{y_1, y_2, \ldots, y_{T-H}\}$. The model is then forecasted out of sample from y_{T-H+1} through to y_T. Because the actual values of these later observations are known, it is possible to compare the accuracy of the forecasts with the actual values. This situation is illustrated in case 1 of the schematic representation.

Out-of-sample (or ex ante) forecasts: The entire sample $\{y_1, y_2, \ldots, y_T\}$ is used to fit the model and then forecast over the horizon $T+1$ to $T+H$. This situation is illustrated in case 2.

One-step-ahead (or static) forecasts: If H is taken to be 1, then forecasts are called one-step-ahead or static forecasts. Static forecasts can be only in sample or ex post in nature apart from a single step from y_T to $\widehat{y}_{T+1}$.

Multistep-ahead (or dynamic) forecasts: If $H > 1$, forecasting requires the successive generation of $\widehat{y}_{T+1}$, $\widehat{y}_{T+2}$ up to and including $\widehat{y}_{T+H-1}$ for use in the forecasting equations. Dynamic forecasts can be either in-sample or out-or-sample forecasts, but their key feature is that actual values of y_t are not used to generate the multistep forecasts.

Point forecasts: Forecasts in which only a single figure, say, $\widehat{y}_{T+H}$, is reported for period $T+H$ is known as a point forecast. The point forecast represents the most likely value, or expected value, of y_{T+H}.

Interval forecasts: Even if a point forecast is a particularly good one and it is known that on average the forecast is correct, there is some uncertainty associated with every forecast. Interval forecasts encapsulate this uncertainty by providing a range of forecast values, $\widehat{y}_{T+H}$, within which the actual value y_{T+H} is expected to be found at some given level of confidence.

Finally, it is worth pointing out that there is a distinction between forecasting based on dynamic time-series models that are known collectively as recursive forecasts and forecasts based on more elaborate linear or nonlinear regression models. Forecasts that are based on econometric models are known as structural forecasts. Note, however, that the distinction between these two types of forecasts is often unclear because econometric models often contain both structural and dynamic time-series features.

8.3 Recursive forecasting in time-series models

This section considers a class of dynamic models for a single dependent variable known as stationary time-series models. The dynamics enter the model either through the lags of the dependent variable, through the lags of the disturbances, or through both. This type of model has been extensively studied, and the discussion here will be brief and selective.

8.3.1 Single-equation forecasts

Consider the first-order autoregressive [AR(1)] model

$$y_t = \phi_0 + \phi_1 y_{t-1} + v_t, \qquad v_t \sim \text{i.i.d. } N(0, \sigma_v^2)$$

The one-step-ahead forecasts for $y_{T+1}, y_{T+2}, \ldots, y_{T+H}$ are, respectively,

$$\widehat{y}_{T+1} = \widehat{\phi}_0 + \widehat{\phi}_1 y_T$$
$$\widehat{y}_{T+2} = \widehat{\phi}_0 + \widehat{\phi}_1 \widehat{y}_{T+1}$$
$$\vdots = \vdots \quad \vdots$$
$$\widehat{y}_{T+H} = \widehat{\phi}_0 + \widehat{\phi}_1 \widehat{y}_{T+H-1}$$

Note that unknown parameters are replaced by their estimates, and the forecast from the previous step is used to generate a recursive forecast in the next step.

The ideas generalize easily to the case of an AR(2) model given by

$$y_t = \phi_0 + \phi_1 y_{t-1} + \phi_2 y_{t-2} + v_t$$

Replacing the parameters $\{\phi_0, \phi_1, \phi_2\}$ with their sample estimators $\{\widehat{\phi}_0, \widehat{\phi}_1, \widehat{\phi}_2\}$, the recursive scheme becomes

$$\widehat{y}_{T+1} = \widehat{\phi}_0 + \widehat{\phi}_1 y_T + \widehat{\phi}_2 y_{T-1}$$
$$\widehat{y}_{T+2} = \widehat{\phi}_0 + \widehat{\phi}_1 \widehat{y}_{T+1} + \widehat{\phi}_2 y_T$$
$$\widehat{y}_{T+3} = \widehat{\phi}_0 + \widehat{\phi}_1 \widehat{y}_{T+2} + \widehat{\phi}_2 \widehat{y}_{T+1}$$
$$\vdots = \vdots \quad \vdots$$
$$\widehat{y}_{T+H} = \widehat{\phi}_0 + \widehat{\phi}_1 \widehat{y}_{T+H-1} + \widehat{\phi}_2 \widehat{y}_{T+H-2}$$

Recursive forecasts are also applicable to the general autoregressive moving-average (ARMA) class of models. Consider the ARMA(1,1) model

$$y_t = \phi_0 + \phi_1 y_{t-1} + u_t + \psi_1 u_{t-1} \tag{8.1}$$

where u_t is a disturbance term with mean 0 and variance σ_u^2. The process for creating a recursive forecast remains identical to that discussed in detail for the AR(1) case. The two principles are as follows:

1. Replace unknown parameters with consistent estimates, and replace unknown values with their predictions.
2. Regard the fitted model as the true model.

Of course, the parameters of the ARMA model are not as easy to estimate at those of an AR model, because an iterative algorithm is required. But consistent maximum likelihood estimates are relatively straightforward to obtain using the Stata `arima` (see [TS] **arima**) command. Once these are estimates are available, the `predict` (see [R] **predict**) command may be used to generate the residuals $\hat{u}_t$. Using the two forecasting principles, the recursive forecasts of the ARMA(1,1) model for up to $H > 1$ periods ahead are as follows:

$$\hat{y}_{T+1} = \hat{\phi}_0 + \hat{\phi}_1 y_T + \hat{\psi}_1 \hat{u}_{T-1}$$
$$\hat{y}_{T+2} = \hat{\phi}_0 + \hat{\phi}_1 \hat{y}_{T+1}$$
$$\hat{y}_{T+3} = \hat{\phi}_0 + \hat{\phi}_1 \hat{y}_{T+2}$$
$$\vdots = \vdots \quad \vdots$$
$$\hat{y}_{T+H} = \hat{\phi}_0 + \hat{\phi}_1 \hat{y}_{T+H-1}$$

Notice that the first-order moving-average [MA(1)] term in (8.1) appears explicitly in the forecast of $\hat{y}_{T+1}$ but disappears thereafter because a MA(q) process has a memory of q periods (in this case, one period). Just as in the case of forecasting from the AR(1) model, the forecasts quickly revert to the unconditional mean of the series, and the variance of the prediction errors grows with the forecast horizon and approaches the variance of the series.

8.3.2 Multiple-equation forecasts

This recursive method also applies to generating dynamic forecasts for a multivariate time-series model. Consider a bivariate vector AR with one lag [VAR(1)], given by

$$y_{1t} = \phi_{10} + \phi_{11} y_{1t-1} + \phi_{12} y_{2t-1} + v_{1t}$$
$$y_{2t} = \phi_{20} + \phi_{21} y_{1t-1} + \phi_{22} y_{2t-1} + v_{2t}$$

where v_{1t} and v_{2t} are disturbance terms. Given data up to time T, and replacing the parameters $\{\phi_{10}, \phi_{11}, \phi_{12}, \phi_{20}, \phi_{21}, \phi_{22}\}$ with their sample estimates, the forecast at time $T+1$ is

$$\hat{y}_{1T+1} = \hat{\phi}_{10} + \hat{\phi}_{11} y_{1T} + \hat{\phi}_{12} y_{2T}$$
$$\hat{y}_{2T+1} = \hat{\phi}_{20} + \hat{\phi}_{21} y_{1T} + \hat{\phi}_{22} y_{2T}$$

In general, the forecasts of the VAR(1) model for H periods ahead are

$$\hat{y}_{1T+H} = \hat{\phi}_{10} + \hat{\phi}_{11} \hat{y}_{1T+H-1} + \hat{\phi}_{12} \hat{y}_{2T+H-1}$$
$$\hat{y}_{2T+H} = \hat{\phi}_{20} + \hat{\phi}_{21} \hat{y}_{1T+H-1} + \hat{\phi}_{22} \hat{y}_{2T+H-1}$$

An important feature of this result is that even if forecasts are required for just one of the variables, say, y_{1t}, it is also necessary to generate forecasts of the other variables in the model. In Stata, this can be done with the `fcast compute` (see [TS] **fcast compute**) command.

The vector error-correction model (VECM), introduced in chapter 7, is a restricted VAR model. In other words, a VECM can be reexpressed in VAR form, which in turn can be used to forecast the variables of the model. Consider the following bivariate VECM model:

$$\Delta y_{1t} = \alpha_1(y_{1\,t-1} - \beta_0 - \beta_2 y_{2\,t-1}) + \gamma_{11}\Delta y_{1\,t-1} + \gamma_{12}\Delta y_{2\,t-1} + v_{1t}$$
$$\Delta y_{2t} = \alpha_2(y_{1\,t-1} - \beta_0 - \beta_2 y_{2\,t-1}) + \gamma_{21}\Delta y_{1\,t-1} + \gamma_{22}\Delta y_{2\,t-1} + v_{2t}$$

This model may be rearranged to give a restricted VAR(2) model

$$y_{1t} = -\alpha_1\beta_0 + (1 + \gamma_{11} + \alpha_1)y_{1t-1} - \gamma_{11}y_{1t-2} + (\gamma_{12} - \alpha_1\beta_2)y_{2t-1} - \gamma_{12}y_{2t-2} + v_{1t}$$
$$y_{2t} = -\alpha_2\beta_0 + (\gamma_{21} + \alpha_2)y_{1t-1} - \gamma_{21}y_{1t-2} + (1 + \gamma_{22} - \alpha_2\beta_2)y_{2t-1} - \gamma_{22}y_{2t-2} + v_{2t}$$

Alternatively, it is possible to write this system in VAR form as

$$y_{1t} = \phi_{10} + \phi_{11}y_{1t-1} + \phi_{12}y_{1t-2} + \phi_{13}y_{2t-1} + \phi_{14}y_{2t-2} + v_{1t}$$
$$y_{2t} = \phi_{20} + \phi_{21}y_{1t-1} + \phi_{22}y_{1t-2} + \phi_{23}y_{2t-1} + \phi_{24}y_{2t-2} + v_{2t}$$

in which the VAR and VECM parameters are related as follows:

$$\phi_{10} = -\alpha_1\beta_0 \qquad \phi_{20} = -\alpha_2\beta_0$$
$$\phi_{11} = 1 + \alpha_1 + \gamma_{11} \qquad \phi_{21} = \alpha_2 + \gamma_{21}$$
$$\phi_{12} = -\gamma_{11} \qquad \phi_{22} = -\gamma_{21}$$
$$\phi_{13} = -\alpha_1\beta_2 + \gamma_{12} \qquad \phi_{23} = 1 - \alpha_2\beta_2 + \gamma_{22}$$
$$\phi_{14} = -\gamma_{12} \qquad \phi_{24} = -\gamma_{22}$$

These equations give the explicit form of the reduced rank restrictions implicit in the VECM system. Once the VECM is expressed as a VAR with the VAR parameter estimates obtained from the VECM parameter estimates, forecasts are generated as for a VAR model. Stata handles these transformations automatically when the `fcast compute` command is invoked.

8.3.3 Properties of recursive forecasts

There are two basic properties related to the mean and the variance of recursive forecasts. Without loss of generality, these will be discussed in the context of an AR(1) model. Consider first the expected value of the next period's observation given information up to time T, $E_T(y_{T+1}) = \phi_0 + \phi_1 y_T$. In the same way, the conditional expectation of y_{T+2} given information to time T is

$$E_T(y_{T+2}) = (\phi_0 + \phi_1 y_{T+1} + v_{t+2}) = \phi_0 + \phi_1 E_T(y_{T+1}) = \phi_0 + \phi_0\phi_1 + \phi_1^2 y_T$$

By analogy, it follows that

$$E_T(y_{T+3}) = \phi_0 + \phi_0\phi_1 + \phi_0\phi_1^2 + \phi_1^3 y_T$$

and more generally at horizon H,

$$E_T(y_{T+H}) = \phi_0 + \phi_0\phi_1 + \phi_0\phi_1^2 + \cdots + \phi_0\phi_1^{H-1} + \phi_1^H y_T$$
$$= \phi_0(1 + \phi_1 + \phi_1^2 + \cdots \phi_1^{H-1}) + \phi_1^H y_T \qquad (8.2)$$

In other words, the forecast at $T+H$ can be written as

$$\widehat{y}_{T+H} = \widehat{\phi}_0(1 + \widehat{\phi}_1 + \widehat{\phi}_1^2 + \cdots \widehat{\phi}_1^{H-1}) + \widehat{\phi}_1^H y_T$$

The two key properties of recursive forecasts may be summarized as follows:

1. When the model is stationary, $|\phi_1| < 1$, it follows from (8.2) that

$$\lim_{H \to \infty} E_T(y_{T+H}) = \frac{\phi_0}{1 - \phi_1}$$

   In other words, the conditional mean converges to the unconditional mean of y_t. This result relies on $(1 + \phi_1 + \phi_1^2 + \cdots + \phi_1^{H-1})$ being a convergent series and also on the elimination of the second term on the right-hand side of (8.2).

2. The variances of the forecasts at different horizons are

$$\text{var}\{y_{T+1} - E_T(y_{T+1})\} = \sigma_v^2$$
$$\text{var}\{y_{T+2} - E_T y_{T+2})\} = \sigma_v^2(1 + \phi_1^2)$$
$$\vdots \quad = \quad \vdots$$
$$\text{var}\{y_{T+H} - E_T(y_{T+H})\} = \sigma_v^2\left(1 + \phi_1^2 + \phi_1^4 + \phi_1^6 + \cdots \phi_1^{2(H-1)}\right)$$

   where $\sigma_v^2 = E(v_t^2)$ is the variance of the one-step-ahead optimal forecast error. It follows that the variance of the optimal forecast is an increasing function of the forecast horizon H. This implies that the interval forecasts will become less and less precise as the forecast horizon increases.

8.4 Forecast evaluation

The discussion so far has concentrated on ex ante forecasting of a variable or variables over a forecast horizon H, beginning after the last observation in the dataset. However, it is also of interest to be able to compare the forecasts with the actual values. A common solution adopted to determine the forecast accuracy of a model is to fit the model over a restricted sample period that excludes the last H observations. The model is then forecasted out of sample over these observations, but because the actual values of these observations have already been observed it is possible to compare the accuracy of the forecasts with the actual values. Because the data are already observed, forecasts computed in this way are known as in-sample forecasts.

8.4 Forecast evaluation

There are several simple summary statistics that are used to determine the accuracy of forecasts. Define the forecast error as the difference between the actual and forecast value over the forecast horizon

$$y_{T+1} - \widehat{y}_{T+1},\ y_{T+2} - \widehat{y}_{T+2},\ \ldots,\ y_{T+H} - \widehat{y}_{T+H}$$

then it follows immediately that the smaller the forecast error, in absolute terms, the better the forecast. The most commonly used summary measures of overall accuracy of the forecasts to the actual values are

$$\text{Mean absolute error:} = \frac{1}{H}\sum_{h=1}^{H} |y_{T+h} - \widehat{y}_{T+h}|$$

$$\text{Mean absolute percentage error:} = \frac{1}{H}\sum_{h=1}^{H} \left|\frac{y_{T+h} - \widehat{y}_{T+h}}{y_{T+h}}\right|$$

$$\text{Mean squared error:} = \frac{1}{H}\sum_{h=1}^{H} (y_{T+h} - \widehat{y}_{T+h})^2$$

$$\text{Root mean squared error:} = \sqrt{\frac{1}{H}\sum_{h=1}^{H} (y_{T+h} - \widehat{y}_{T+h})^2}$$

Theil's U is a measure of relative forecast accuracy (Theil 1966). The measure is designed to compare the variance in the forecast with that of a naïve forecasting model. This measure is particularly useful for ex ante forecasts, which frequently must be compared with an artificial baseline. Suppose there is a forecast available from a naïve model, denoted $\widetilde{y}_{t+h}$. Theil's U statistic is

$$U = \sqrt{\frac{\sum_{t=T}^{T+h} \{(y_{t+h} - \widehat{y}_{t+h})/y_{t+h}\}^2}{\sum_{t=T}^{T+h} \{(\widetilde{y}_{t+h} - \widehat{y}_{t+h})/y_{t+h}\}^2}}$$

All of these measures are available from the Statistical Software Components (SSC) Archive as package `fcstats` (Baum 2017), which can also produce a time-series plot of the actual and forecast series.

Probably the most widely used formal test for comparing two different forecasts is the Diebold and Mariano test (1995). Suppose we have two competing forecasts and we compute the forecast error, $\widehat{u}_{T+h}(M_j) = \widehat{y}_{T+h} - y_{T+h}$, for the jth model, M_j. Now, define the difference

$$w_t = \widehat{u}_{T+h}(M_1) - \widehat{u}_{T+h}(M_2)$$

The Diebold–Mariano test of equal predictive accuracy is the simple t test that $E(w_t) = 0$. A Stata implementation of the test is available from the SSC Archive as package `dmariano` (Baum 2003).

An extension of the Diebold–Mariano test has been proposed by Giacomini and Rossi (2010). This extended test, `giacross`, deals with instabilities in the underlying process by using rolling regression procedures (see [TS] **rolling**). A community-contributed command by Rossi and Sekhposyan (2016), `rosssekh`, deals with the rationality of forecasts in the presence of instabilities. The features of these tests are described in Rossi and Soupre (2017). Both tests can be accessed from the SSC Archive as package `forec_instab` or from the *Stata Journal* website.

8.5 Daily forecasts of wind speed for Santiago

Consider forecasting the daily average wind speed at the Parque O'Higgins weather station in Santiago, Chile.

```
. use http://www.stata-press.com/data/eeus/santiago
. tabstat v, by(winddir) stat(N mean sd min max)
Summary for variables: v
     by categories of: winddir

 winddir |         N        mean          sd         min         max
---------+------------------------------------------------------------
     NNE |         9      1.1996     .423581     .791025    2.119838
     ENE |        31    .8839483    .3833133     .509925     2.03455
     ESE |        96    .649599    .1696369     .401125    1.584796
     SSE |       194   .7483339    .2154902    .4004417    1.544463
     SSW |      1056   1.215522    .3947056       .3871    2.120525
     WSW |       310    1.19463    .4001765    .3943958    2.016158
     WNW |         4   .9044812    .2354468     .651625    1.213442
     NNW |         4   1.549359    .8646453    .7059667    2.571279
---------+------------------------------------------------------------
   Total |      1704   1.120586    .4168381       .3871    2.571279
```

The prevailing wind direction is south-southwest, but it appears as though the northerly winds are faster, with north-northwest being the fastest direction.

Three simple models are used here to illustrate forecasting daily average wind speed. The first of these is the AR(7) model, which is fit using the Stata `arima` (see [TS] **arima**) command. To forecast the model, the Stata suite of commands relating to `forecast` is used. To create the single-equation forecasting model, we must use `estimates` (see [R] **estimates**) to store the fitted model, create the model as a named object (in this case `ar`), and then use `forecast solve` to specify that the model is to be solved over the out-of-sample period. The `prefix()` option creates a forecast series with that prefix appended to the name of the dependent variable, that is, `f_v`, so that we can compare the actual values of `v` with the model's forecasts. Likewise, a variable `sd_v` is created by the `simulate()` option, providing the time-varying standard deviation of the point forecast via simulation. This allows the computation of both point and interval forecasts over the forecast horizon. The model is fit over the period ending on July 31, 2015 and used to forecast the last month of wind speeds in the dataset, from August 1, 2015 to August 31, 2015. The results indicate that there are sizable AR coefficients out to lag 7. The actual (solid) and forecast (dashed) values are shown in figure 8.2 together with their 95% confidence intervals.

8.5 Daily forecasts of wind speed for Santiago

```
. arima v if tin(,31jul2015), ar(1/7) nolog
ARIMA regression
Sample: 01jan2011 - 31jul2015                    Number of obs    =      1673
                                                 Wald chi2(7)     =   2190.37
Log likelihood = -66.48114                       Prob > chi2      =    0.0000
```

	v	Coef.	OPG Std. Err.	z	P>|z|	[95% Conf. Interval]	
v	_cons	1.12524	.0853308	13.19	0.000	.9579944	1.292485
ARMA	ar						
	L1.	.3651914	.0216935	16.83	0.000	.3226729	.40771
	L2.	-.0064085	.0248916	-0.26	0.797	-.0551952	.0423782
	L3.	.1099003	.0251668	4.37	0.000	.0605742	.1592263
	L4.	.1440553	.0242163	5.95	0.000	.0965921	.1915184
	L5.	.0750974	.0258196	2.91	0.004	.0244919	.1257028
	L6.	.1330993	.0260303	5.11	0.000	.0820809	.1841177
	L7.	.1055986	.0232746	4.54	0.000	.0599813	.151216
/sigma		.2516527	.0034802	72.31	0.000	.2448316	.2584738

Note: The test of the variance against zero is one sided, and the two-sided
      confidence interval is truncated at zero.

```
. estimates store ar
. forecast create ar, replace
  Forecast model ar started.
. forecast estimates ar
  Added estimation results from arima.
  Forecast model ar now contains 1 endogenous variable.
. forecast solve, prefix(f_) begin(td(1aug2015)) end(td(31aug2015))
> simulate(betas, reps(500) stat(s, pref(sd_)))
  (output omitted)
. generate f_ar = f_v
. generate upper = f_v + 1.96*sd_v
(1,673 missing values generated)
. generate lower = f_v - 1.96*sd_v
(1,673 missing values generated)
```

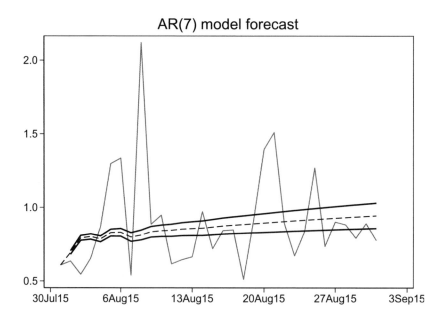

Figure 8.2. Actual (solid line) and forecasts (dashed line) of average daily wind speeds (with the 95% confidence bands shown as darker solid lines) for the month of August 2015 for the Parque O'Higgins weather station in Santiago, Chile.

The two characteristics of recursive forecasts based on time-series models, namely, the reversion to the long-run mean of the series and the increasing uncertainty of the forecast as the horizon increases, are clearly evident. The model is unable to capture the short-term variations in wind speeds nor is it able to pick the sharp gusts that are observed.

8.5 Daily forecasts of wind speed for Santiago

An ARMA(2,2) model fit to the same data yields the following parameter estimates:

```
. arima v if tin(,31jul2015), arima(2,0,2) nolog
ARIMA regression
Sample:  01jan2011 - 31jul2015              Number of obs   =       1673
                                            Wald chi2(4)    =   89359.80
Log likelihood = -44.25815                  Prob > chi2     =     0.0000
```

	v	Coef.	OPG Std. Err.	z	P>|z|	[95% Conf. Interval]	
v							
	_cons	1.131773	.1630767	6.94	0.000	.8121483	1.451397
ARMA							
	ar						
	L1.	.8382256	.1118075	7.50	0.000	.6190868	1.057364
	L2.	.1562251	.1108123	1.41	0.159	-.060963	.3734131
	ma						
	L1.	-.5077638	.1069769	-4.75	0.000	-.7174345	-.298093
	L2.	-.3309279	.0881449	-3.75	0.000	-.5036888	-.158167
	/sigma	.2483061	.0033957	73.12	0.000	.2416506	.2549617

```
Note: The test of the variance against zero is one sided, and the two-sided
      confidence interval is truncated at zero.
. estimates store arma
. forecast create arma, replace
  (Forecast model ar ended.)
  Forecast model arma started.
. forecast estimates arma
  Added estimation results from arima.
  Forecast model arma now contains 1 endogenous variable.
. drop f_v sd_v upper lower
. forecast solve, prefix(f_) begin(td(1aug2015)) end(td(31aug2015))
> simulate(betas, reps(500) stat(s, pref(sd_)))
  (output omitted)
. generate f_arma = f_v
```

Both the AR and MA components of the model are significant, although the second-order AR term could be omitted. Comparing the forecasts from the AR(7) model (f_ar) with those from the ARMA(2,2) model (f_arma) gives

```
. fcstats v f_ar f_arma if tin(1aug2015,31aug2015)
Forecast accuracy statistics for v, N = 31
              f_ar         f_arma
RMSE          .33915818    .36479128
MAE           .22585608    .23210443
MAPE          .23549075    .21835606
Theil's U     .81465736    .85868302
```

These results indicate that the AR(7) model is slightly preferable in terms of all the forecast evaluation statistics except for the mean absolute percentage error. The significance of the difference in the forecasts can be assessed using the Diebold–Mariano test.

```
. dmariano v f_ar f_arma if tin(1aug2015,31aug2015), maxlag(7)
Diebold-Mariano forecast comparison test for actual : v
Competing forecasts:  f_ar versus f_arma
Criterion: MSE over 31 observations
Maxlag = 7   Kernel : uniform

Series                        MSE
-----------------------------------
f_ar                         .115
f_arma                       .1331
Difference                  -.01804

By this criterion, f_ar is the better forecast
H0: Forecast accuracy is equal.
S(1) =     -6.05  p-value = 0.0000
```

It appears that the AR(7) model provides a slightly more accurate forecast and that the difference is statistically significant.

A final model considered here is the ARMA model that includes additional explanatory variables (ARMA-X). The AR(7) model is augmented with dummy variables for wind direction that will be taken to be the only explanatory variables. The model is fit as follows:

```
. arima v NNE ENE ESE SSE SSW WSW WNW NNW if tin(,31jul2015), ar(1/7) nolog
> noconstant

ARIMA regression

Sample:  01jan2011 - 31jul2015              Number of obs    =       1673
                                            Wald chi2(15)    =    3125.88
Log likelihood = -20.94035                  Prob > chi2      =     0.0000
```

	v	Coef.	OPG Std. Err.	z	P>|z|	[95% Conf. Interval]	
v							
	NNE	1.418766	.1338103	10.60	0.000	1.156503	1.68103
	ENE	1.193237	.087611	13.62	0.000	1.021522	1.364951
	ESE	1.006658	.0934185	10.78	0.000	.8235606	1.189755
	SSE	1.067776	.0871651	12.25	0.000	.896936	1.238617
	SSW	1.142761	.083452	13.69	0.000	.9791985	1.306324
	WSW	1.115703	.0852918	13.08	0.000	.9485339	1.282872
	WNW	1.26861	.1503819	8.44	0.000	.9738668	1.563353
	NNW	1.848985	.0851159	21.72	0.000	1.682161	2.015809

8.5 Daily forecasts of wind speed for Santiago

ARMA							
ar							
	L1.	.3408123	.023045	14.79	0.000	.295645	.3859796
	L2.	.0012823	.0246544	0.05	0.959	-.0470394	.0496039
	L3.	.104132	.0250373	4.16	0.000	.0550599	.1532042
	L4.	.1618687	.024893	6.50	0.000	.1130794	.2106581
	L5.	.0657996	.0248008	2.65	0.008	.0171911	.1144082
	L6.	.1425644	.0251298	5.67	0.000	.0933109	.1918179
	L7.	.1110456	.0236057	4.70	0.000	.0647794	.1573118
/sigma		.2448923	.0040716	60.15	0.000	.2369122	.2528724

Note: The test of the variance against zero is one sided, and the two-sided
      confidence interval is truncated at zero.

```
. estimates store armax

. forecast create armax, replace
  (Forecast model arma ended.)
  Forecast model armax started.

. forecast estimates armax
  Added estimation results from arima.
  Forecast model armax now contains 1 endogenous variable.

. drop f_v sd_v

. forecast solve, prefix(f_) begin(td(1aug2015)) end(td(31aug2015))
> simulate(betas, reps(500) stat(s, pref(sd_)))
  (output omitted)

. generate f_armax = f_v
```

All the dummy variables have been included in the model and the constant term omitted. This avoids the dummy variable trap but also clarifies the interpretation of the effect of wind direction on wind speed. The results are unequivocal and indicate the importance of the relationship between wind direction and wind speed. The problem for forecasting, however, is that the wind direction will need to be forecast to provide a forecast for wind speed. In other words, the forecast is no longer simply recursive. If we are entitled to treat wind direction as known, the improvement in the accuracy of the forecast is notable.

```
. fcstats v f_ar f_armax if tin(31jul2015,31aug2015)
Forecast accuracy statistics for v, N = 32
              f_ar         f_armax
RMSE        .33381677      .2816458
MAE         .21879808      .2024678
MAPE        .22813167      .22521312
Theil's U   .80805564      .66368599
```

This comparison is, however, somewhat unrealistic because the actual values of the wind direction have been used.

8.6 Forecasting with logarithmic dependent variables

Using a logarithmic transformation on the dependent variable of a regression equation is a common strategy. There are a variety of reasons for doing this transformation. The transformation converts a potential exponential trend into a linear trend, and it reduces the effect of heteroskedasticity. In the context of wind forecasting, a reason often given for modeling the logarithm of wind speed is that it transforms the distribution of the dependent variable, making it more symmetric and more similar to the normal distribution. The difference in the distribution between hourly wind speed and its logarithm is shown in figure 8.3.

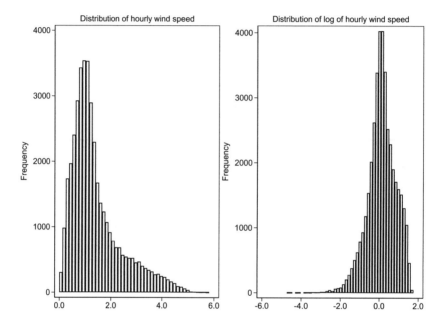

Figure 8.3. Histograms of the distribution of hourly wind speed for the El Bosque weather station in Santiago, Chile, for the period January 1, 2011 to August 31, 2015.

Irrespective of whether the reason for the transformation is valid, the facts are that it is commonly used and that it gives rise to several complications. For example, recall that the R^2 statistic measures the fraction of the variance of the dependent variable explained by the regressors. It follows, therefore, that using the R^2 to compare models is only valid if they have the same dependent variable. So it is not possible to compare the fit of two models when one of them has the dependent variable expressed in levels and the other one expresses the dependent variable as a logarithm.

8.6 Forecasting with logarithmic dependent variables

Probably a more important problem relates to prediction and forecasting. When the dependent variable is transformed by taking logs, the estimated regression can generate predicted values of $\log y_t$, but the fundamental problem being addressed usually requires predicted values of the levels of the dependent variable, y_t. Consider the model

$$\log y_t = \beta_0 + \beta_1 x_{1t} + \cdots + \beta_k x_{kt} + u_t$$

and note that

$$\begin{aligned} y_t &= \exp\left(\beta_0 + \beta_1 x_{1t} + \cdots + \beta_k x_{kt} + u_t\right) \\ &= \exp\left(\beta_0 + \beta_1 x_{1t} + \cdots + \beta_k x_{kt}\right) \exp(u_t) \end{aligned}$$

Taking expectations of this expression reveals that $\exp(\beta_0 + \beta_1 x_{1t} + \cdots + \beta_k x_{kt})$ is a biased estimate of the expected value of y_t because by Jensen's inequality

$$E\{\exp(u_t)\} \neq \exp\{E(u_t)\} = 1$$

In the context of prediction and forecasting using logged dependent variables, this problem is known as retransformation bias.

Consider, for example, fitting the ARMA-X model with seven AR terms and using wind direction dummy variables as explanatory variables to the logarithm of wind speed. A simple linear regression with no constant term to allow all the dummy variables to be included yields the following results:

```
. use http://www.stata-press.com/data/eeus/allstations, clear
. reshape wide pm co t hr vv dv station, i(datevec) j(id)
(output omitted)
. tsset datevec, delta(1 hour)
        time variable:  datevec, 01jan2011 01:00 to 31aug2015 23:00
                delta:  1 hour
. local id = 4
. rename vv`id' v
. rename dv`id' dv
. generate lv = log(v)
(10 missing values generated)
. generate dd = dofc(datevec)
. format dd %td
. summarize dd, mean
. local ll = `r(min)'
. local ul = `r(max)'
. generate mdv = .
(40,895 missing values generated)
. generate mdv_str = .
(40,895 missing values generated)
```

```
. forvalues w=`ll'/`ul' {
  2.         quietly circsummarize dv if dd==`w', ci
  3.         quietly replace mdv = `r(vecmean)' if dd==`w'
  4.         quietly replace mdv_str = `r(vecstr)' if dd==`w'
  5. }
. collapse v mdv, by(dd)
. tsset dd, daily
        time variable:  dd, 01jan2011 to 31aug2015
                delta:  1 day
. capture drop winddir
. generate winddir = int((mdv-1)/45) + 1
. label define wdlab 1 NNE 2 ENE 3 ESE 4 SSE 5 SSW 6 WSW 7 WNW 8 NNW
. label value winddir wdlab
. tabulate winddir, gen(wd)
 (output omitted)
. rename wd1 NNE
. rename wd2 ENE
. rename wd3 ESE
. rename wd4 SSE
. rename wd5 SSW
. rename wd6 WSW
. rename wd7 WNW
. rename wd8 NNW
. generate lv = log(v)
. regress lv L(1/7).lv NNE ENE ESE SSE SSW WSW WNW NNW, noconstant
```

Source	SS	df	MS		Number of obs	=	1,697
					F(15, 1682)	=	685.60
Model	296.828871	15	19.7885914		Prob > F	=	0.0000
Residual	48.5477655	1,682	.028863119		R-squared	=	0.8594
					Adj R-squared	=	0.8582
Total	345.376637	1,697	.203521884		Root MSE	=	.16989

lv	Coef.	Std. Err.	t	P>|t|	[95% Conf. Interval]	
lv						
L1.	.3097738	.0239423	12.94	0.000	.2628139	.3567337
L2.	.1114269	.0247744	4.50	0.000	.0628349	.1600188
L3.	.1225875	.0248344	4.94	0.000	.0738778	.1712971
L4.	.105232	.0249201	4.22	0.000	.0563544	.1541096
L5.	.0627009	.0248771	2.52	0.012	.0139075	.1114942
L6.	.1073216	.0248051	4.33	0.000	.0586694	.1559737
L7.	.0703484	.0236592	2.97	0.003	.0239438	.116753
NNE	-.2365278	.1701628	-1.39	0.165	-.5702809	.0972254
ENE	.154311	.0455847	3.39	0.001	.0649023	.2437197
ESE	-.048105	.0165477	-2.91	0.004	-.0805612	-.0156487
SSE	-.0153201	.0085257	-1.80	0.073	-.0320423	.0014021
SSW	.0497294	.0082055	6.06	0.000	.0336353	.0658236
WSW	.0691227	.0114159	6.05	0.000	.0467318	.0915136
WNW	.0204427	.0981754	0.21	0.835	-.172116	.2130014
NNW	.9428312	.1702945	5.54	0.000	.6088198	1.276843

8.6.1 Staying in the linear regression framework

As in the model specified in levels, all the AR terms are strongly significant, and there is strong evidence that wind direction is an important variable in the context of modeling wind speed. We compute the mean of actual wind speed and compare it with the mean of wind speed computed from the predictions of the logarithmic model. In so doing, it is important to limit the sample to exclude the missing observations generated as a result of the inclusion of seven lags of the dependent variable.

```
. scalar sig = e(rmse)
. predict lfit, xb
(7 missing values generated)
. predict uhat, resid
(7 missing values generated)
. generate vfit = exp(lfit)
(7 missing values generated)
. summarize v vfit if !mi(vfit)
```

Variable	Obs	Mean	Std. Dev.	Min	Max
v	1,697	1.424231	.4336292	.3175625	2.669042
vfit	1,697	1.406606	.3773083	.4112506	2.447183

The results reveal a negative transformation bias in the mean of the expected value of wind speed generated from the logarithmic regression (1.40 km/h) with respect to the mean of the data (1.42 km/h).

8.6.1 Staying in the linear regression framework

There are two ways to ameliorate retransformation bias in a regression framework.

1. If the disturbance term u_t is assumed to be normally distributed, then

$$u_t \sim N(0, \sigma^2) \Rightarrow E\{\exp(u_t)\} = \exp\left(\frac{\sigma^2}{2}\right)$$

This result implies that

$$\widehat{y_t} = \exp\left(\frac{\widehat{\sigma}^2}{2}\right) \exp\left(\widehat{\log y_t}\right)$$

will be an unbiased estimator of y_t, conditional on the assumption of normality. In the current example, this approach yields the following results:

```
. capture drop vfit
. generate vfit = exp(sig^2/2)*exp(lfit)
(7 missing values generated)
. summarize v vfit if !mi(vfit)
```

Variable	Obs	Mean	Std. Dev.	Min	Max
v	1,697	1.424231	.4336292	.3175625	2.669042
vfit	1,697	1.427053	.3827929	.4172286	2.482756

Using the correction based on the assumption of the normality of the disturbances contributes to a significant improvement in the transformation bias in the wind speed example. This correction can also be computed by the `levpredict` command (Baum 2009), available from the SSC Archive.

2. If u_t is not assumed to be normally distributed, then an appropriate scaling factor may be computed by regressing y_t on the fitted values, $\exp\left(\widehat{\log y_t}\right)$, using a regression equation that does not contain a constant term,

$$y_t = \alpha \exp\left(\widehat{\log y_t}\right) + e_t$$

The ordinary least-squares estimate, $\widehat{\alpha}$, is then used as the scaling factor

$$\widehat{y_t} = \widehat{\alpha} \exp\left(\widehat{\log y_t}\right)$$

to account for the retransformation bias. Implementing this correction gives the following results:

```
. capture drop vfit
. generate vfit = exp(lfit)
(7 missing values generated)
. regress v vfit, noconstant

      Source |       SS           df       MS      Number of obs   =     1,697
-------------+----------------------------------   F(1, 1696)      =  70950.81
       Model |  3673.35349         1  3673.35349   Prob > F        =    0.0000
    Residual |  87.8074145     1,696   .05177324   R-squared       =    0.9767
-------------+----------------------------------   Adj R-squared   =    0.9766
       Total |  3761.16091     1,697  2.21635881   Root MSE        =    .22754

------------------------------------------------------------------------------
           v |      Coef.   Std. Err.      t    P>|t|     [95% Conf. Interval]
-------------+----------------------------------------------------------------
        vfit |   1.010273   .0037928   266.37   0.000     1.002834    1.017712
------------------------------------------------------------------------------

. replace vfit = _b[vfit]*vfit
(1,697 real changes made)

. summarize v vfit if !mi(vfit)

    Variable |        Obs        Mean    Std. Dev.       Min        Max
-------------+---------------------------------------------------------
           v |      1,697    1.424231    .4336292   .3175625   2.669042
        vfit |      1,697    1.421056    .3811844   .4154754   2.472323
```

Once again, the correction has resulted in the elimination of the transformation bias.

It is clear that both of these simple fixes work quite well in the current example. The satisfactory results obtained by using the correction based on the normal distribution derive from the fact that the distribution of the residuals from the AR(7) regression model of the log of wind speed are indeed approximately normal. Figure 8.4 plots the histogram of the residuals with the best fitting normal distribution overlaid. Of course, as the assumption of normality is often inappropriate when working with environmental data, the latter approach is probably to be recommended. It also improves the prediction in the case where the residuals are normally distributed.

8.6.2 Generalized linear models 171

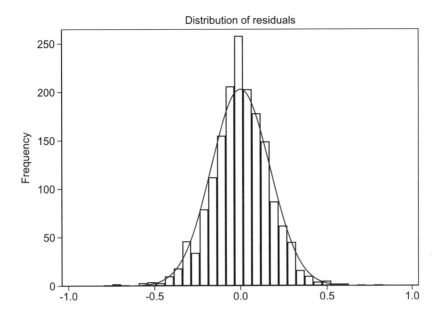

Figure 8.4. Histograms of the distribution of hourly wind speed for the El Bosque weather station in Santiago, Chile, for the period January 1, 2011 to August 31, 2015.

8.6.2 Generalized linear models

An alternative approach to eliminating transformation bias is to use a class of model known as the generalized linear model (GLM). This approach is well known in the biostatistics literature but is less well known in the mainstream econometrics literature.[1] Without loss of generality, consider the case where a dependent variable y_t depends on one explanatory variable x_t. GLMs have the form

$$g\{E(y_t|x_t)\} = g(\mu_t) = \beta_0 + \beta_1 x_t, \qquad y_t \sim F$$

The model has two important features.

1. The so-called link function, $g(\mu_t)$, which allows a linear model for the conditional mean to be related to y_t via a nonlinear function. Although the link function is nonlinear, the conditional mean is modeled as a linear combination of explanatory variables.

2. The dependent variable, y_t, is assumed to follow a distribution that belongs to the family of linear exponential distributions, F. This family encompasses a large range of probability distributions that includes the normal, binomial, Poisson, and gamma distributions.

---

1. For a thorough discussion, see Hardin and Hilbe (2018).

The general treatment of the GLM is beyond the scope of this book, but consider the case of a logarithmic link function

$$\log\{E(y_t|x_t)\} = \log(\mu_t) = \beta_0 + \beta_1 x_t$$

Because this specification is appropriate for a continuous variable y_t that takes only positive values, the appropriate choice of distribution from the exponential family is the gamma distribution. In this instance, the log-likelihood function is given by

$$\log L_t = -\frac{y_t}{\mu_t} + \log\left(\frac{1}{\mu_t}\right)$$

The GLM in the case of the logarithmic link has the advantage that predicted values are generated for the actual levels of y_t prior to the application of the link function. This approach therefore avoids any transformation bias.

The Stata command `glm` (see [R] **glm**) fit GLMs. Applied to daily wind speed data, the estimation results are as follows:

```
. glm v L(1/7).v NNE ENE ESE SSE SSW WSW WNW NNW, family(gamma) link(log) nolog
> vsquish noconstant

Generalized linear models                          Number of obs   =      1,697
Optimization     : ML                              Residual df     =      1,682
                                                   Scale parameter =   .0324629
Deviance         =   54.51636469                   (1/df) Deviance =   .0324116
Pearson          =   54.60261665                   (1/df) Pearson  =   .0324629

Variance function: V(u) = u^2                      [Gamma]
Link function    : g(u) = ln(u)                    [Log]

                                                   AIC             =   2.653374
Log likelihood   = -2236.387855                    BIC             =  -12453.87
```

	Coef.	OIM Std. Err.	z	P>|z|	[95% Conf. Interval]	
v						
L1.	.2195148	.0188632	11.64	0.000	.1825436	.2564861
L2.	.0797292	.0192239	4.15	0.000	.042051	.1174074
L3.	.0888898	.0193389	4.60	0.000	.0509862	.1267933
L4.	.0912237	.0192355	4.74	0.000	.0535229	.1289245
L5.	.0475654	.019259	2.47	0.014	.0098185	.0853123
L6.	.0919811	.0192301	4.78	0.000	.0542908	.1296714
L7.	.0598766	.0185693	3.22	0.001	.0234816	.0962717
NNE	-.9317747	.1809776	-5.15	0.000	-1.286484	-.5770651
ENE	-.4719171	.0507604	-9.30	0.000	-.5714057	-.3724286
ESE	-.7120434	.0238015	-29.92	0.000	-.7586936	-.6653933
SSE	-.6832066	.0189386	-36.07	0.000	-.7203256	-.6460875
SSW	-.632008	.022018	-28.70	0.000	-.6751624	-.5888536
WSW	-.6335215	.0262952	-24.09	0.000	-.6850591	-.581984
WNW	-.630135	.1053485	-5.98	0.000	-.8366142	-.4236558
NNW	.2482516	.1808886	1.37	0.170	-.1062836	.6027868

8.6.2 Generalized linear models

The summary statistics of the predicted values indicate that the GLM approach is more effective than either of the regression-based approximations because both the mean and the standard deviation of the predicted values are closer to those of the original data.

```
. capture drop vfit
. predict vfit
(option mu assumed; predicted mean v)
(7 missing values generated)
. summarize v vfit  if !mi(vfit)
    Variable |       Obs        Mean    Std. Dev.       Min        Max
-------------+--------------------------------------------------------
           v |     1,697    1.424231    .4336292   .3175625   2.669042
        vfit |     1,697    1.426862    .3904369   .6544176   2.447183
```

Exercises

1. `allstations.dta` contains hourly observations of wind speed measured at 11 weather stations in Santiago, Chile, saved in long format. One feature of these data is that hourly wind speed for Santiago shows strong periodicity.

   a. Load the data, reshape them to wide format, and `tsset` them as an hourly time series. Select El Bosque station, and plot hourly wind speed.

   b. Assuming that lags 1 and 24 are sufficient to capture the seasonal persistence in wind speed, v_t, fit the model

   $$v_t = \phi_0 + \phi_1 v_{t-1} + \phi_{24} v_{t-24} + \sum_{j=1}^{1} 1\delta_i Dm_{jt} + \sum_{j=1}^{2} 3h_i Dh_{jt} + u_t$$

   in which Dm_{it} and Dh_{it} are monthly and hourly dummy variables, respectively. Interpret the results.

   c. Compute the residuals from the model in exercise 1b and then compute and plot the autocorrelation function of the residuals out to a horizon of 48 hours. Discuss the features of the autocorrelation function.

   d. Now, fit the seasonal autoregressive integrated moving-average model, which combines a simple ARMA(1,1) model with the same multiplicative seasonal specification at the seasonal lag of 24 to capture the strong diurnal element. The model is formally written as an autoregressive integrated moving-average $(1, 0, 1) \times (1, 0, 1)_{24}$ model or, ignoring deterministic terms,

   $$v_t = \phi_1 v_{t-1} + \phi_{24} v_{t-24} + \varepsilon_t + \psi_1 \varepsilon_{t-1} + \psi_{24} \varepsilon_{t-24} + \phi_{24} \psi_{24} \varepsilon_{t-25}$$

   Interpret the results.

   e. Augment the model with monthly dummy variables to deal with the annual cycle. Compare the results with those obtained in exercise 1d.
   [Hint: The `arima` command does not allow factor variables, so another Stata feature will need to be used.]

f. Compute the residuals from the model fit in exercise 1e and, plot the autocorrelation function of these residuals out to a horizon of 48 hours. Compare the results with those obtained in exercise 1c, and hence make a judgment as to which model you prefer.

2. The issue of whether to combine different forecasts of the same variable has generated much interest. Is it better to rely on the best individual forecast, or is there any gain to averaging the competing forecasts? allstations.dta contains hourly observations of wind speed measured at 11 weather stations in Santiago, Chile, saved in long format.

   a. Load the data, reshape them to wide format, and choose the data from the Parque O'Higgins station.
   b. Using the command circsummarize, written by Nicholas Cox (2004), create a series of daily average wind direction. Once this series is constructed, use the collapse command to construct a daily time series of average daily wind speed and wind direction for Parque O'Higgins.
   c. Fit three models for daily wind speed:
      i. an AR(7) model;
      ii. an ARMA(1,1) model; and
      iii. an ARMA-X(1,1) model in which dummy variables for wind direction enter the equation as explanatory variables,

      using data up to July 31, 2015.
   d. Forecast daily wind speed for each day of the last month of the data, from August 1, 2015 to August 31, 2015, and evaluate the forecasts using the commonly used forecast evaluation statistics.
   e. Evaluate the significance of the differences in the forecasts using the Diebold Mariano methodology.
   f. Granger and Ramanathan (1984) suggest combining forecasts on the basis of weights ω_i obtained from the regression

   $$y_t = \omega_1 \widehat{y}_t^{(1)} + \omega_2 \widehat{y}_t^{(2)} + \omega_3 \widehat{y}_t^{(3)} + v_t$$

   in which $\widehat{y}^{(i)}$ is the forecast from the ith model. Does the combination forecast method yield superior forecasting performance?
   g. Now, assign equal weights to these forecasts, and construct the simple average

   $$\widehat{y}_t = \frac{1}{N} \sum_{i=}^{N} \widehat{y}_{1it}$$

   How does this simple combination perform relative to the other approaches?

9 Structural time-series models

The time-series models discussed in prior chapters have the common feature of being linear models in that the expected value of the dependent variable is a linear function of the lagged dependent variables or lagged disturbances. Although these models are powerful in tracking and forecasting many of the variables of interest in environmental econometrics, they cannot deal with situations in which there are multiple level shifts and stochastically varying periodic behavior.

This chapter deals with time-series models that can be written in a form known as state-space form, which specifies the dynamics of potentially unobserved state variables and how these latent states are related to observed variables. A wide variety of time-series models can be expressed in this format, including both the univariate and the multivariate time-series models presented in chapters 4 and 5. Structural time-series models that are modeled in terms of unobserved components such as trends, cycles, and seasonal effects are also expressible in state-space form.

The advantage of writing models in state-space form is that a recursive algorithm, the Kalman filter, can be applied to these models to estimate the unknown parameters of the system by maximum likelihood and also to obtain estimates of the latent state variables.

9.1 Sea level and global temperature

According to the Ocean Service of the United States National Oceanic and Atmospheric Administration,[1] global sea level has been rising over the past century, and the rate has increased in recent decades. In 2014, global sea level was 6.604 centimeters or 2.6 inches above the 1993 average. This was the highest annual average in the satellite record (1993–present). The National Oceanic and Atmospheric Administration estimates that the sea level continues to rise at a rate of about 0.3175 centimeters or 0.125 inches per year. In the United States, almost 40 percent of the population lives in relatively high-population-density coastal areas, where sea level plays a role in flooding, shoreline erosion, and hazards from storms. Globally, 8 of the world's 10 largest cities are near a coast, according to the United Nations Atlas of the Oceans. In urban settings, virtually all human infrastructure—roads, bridges, subways, water supplies, oil and gas wells, power plants, sewage treatment plants, landfills, etc.—is at risk from sea level rise.

---

1. https://oceanservice.noaa.gov/facts/sealevel.html

A particularly interesting hypothesis for environmental econometricians to investigate is the influence of global temperature on global sea level. Sea level rise involves contributions from thermal expansion of ocean water, the melting of glaciers and ice sheets, and changes in land water storage. Process-based models to capture this process are necessarily extremely complex. Given this complexity, it is not surprising that alternative approaches to such empirical models based on historical time-series data are being developed as a complementary alternative.

The basis of time-series approaches to the problem of sea level rise is based on the assumption that global mean sea level and global mean temperature are linked. As clearly illustrated in figure 9.1, the time series for both temperature and sea level are strongly trending. The econometric problem is that the hypothesized relationship between these two variables relates exactly to this trend behavior. The methods discussed in this chapter, based on expressing models in state-space form using the Kalman filter, will illustrate how time-series models can effectively contribute to this debate. In what follows, time-series models of global temperature will be examined, leading eventually to a bivariate structural time-series model of the relationship between sea level and global temperature.

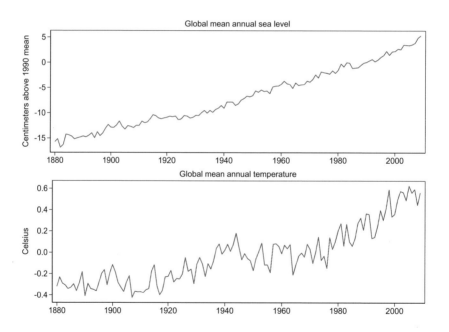

Figure 9.1. Time-series plot of mean annual global sea level and temperature.

9.2 The Kalman filter

The Kalman filter is a set of recursive equations that form predictions of the conditional means and variances of unobservable or latent state variables and then update these predictions systematically based on measurements of the observed variables in the system. It is essentially an elaborate averaging process. The prediction stage generates an estimate of the current state then the system is observed subject to noise, and a weighted average of the prediction and the measurements is computed to produce a revised estimate of the state. One of the convenient features of the recursions of the Kalman filter is that prediction errors are available in each time period t. These errors provide the basis for the maximum likelihood estimation of the unknown parameters of the system based on the assumption that the prediction errors are normally distributed. If the errors are in fact not normally distributed, Hamilton (1994) shows that the quasi–maximum likelihood estimator described in chapter 3 is still consistent and asymptotically normal, although the covariance matrix of the estimator must be estimated by the sandwich estimator for consistency.

Consider a multivariate time series, $\mathbf{y}_t$, in which there are T observations on N component elements. A state-space form for $\mathbf{y}_t$ is as follows (Harvey 1989):

$$\boldsymbol{\alpha}_t = \boldsymbol{\Phi}\boldsymbol{\alpha}_{t-1} + \mathbf{v}_t, \qquad \mathbf{v}_t \sim N(0, \mathbf{Q}) \qquad \text{(Transition equation)}$$
$$\mathbf{y}_t = \mathbf{Z}\boldsymbol{\alpha}_t + \mathbf{u}_t \qquad \mathbf{u}_t \sim N(0, \mathbf{H}) \qquad \text{(Measurement equation)}$$

in which $\boldsymbol{\alpha}_t$ is a vector of (generally) unobserved state variables, $\mathbf{v}_t$ is a vector of disturbances conformable with $\boldsymbol{\alpha}_t$, that has mean zero and covariance matrix $\mathbf{Q}$, $\mathbf{u}_t$ is an N-vector of measurement errors (disturbances) conformable with $\mathbf{y}_t$ that has zero mean and covariance matrix $\mathbf{H}$, and $\boldsymbol{\Phi}$ and $\mathbf{Z}$ are matrices of parameters. Without loss of generality and for the sake of simplicity, the other explanatory variables in the state and measurement equations have been omitted. The model here is written with time-invariant coefficients, although this restriction is not necessary.[2]

The Kalman filter cycle of prediction, measurement, log-likelihood construction, and update is implemented in terms of the following set of equations:

$$\boldsymbol{\alpha}_{t|t-1} = \boldsymbol{\Phi}\boldsymbol{\alpha}_{t-1} \qquad \text{(State prediction)}$$
$$\mathbf{P}_{t|t-1} = \boldsymbol{\Phi}\mathbf{P}_{t-1}\boldsymbol{\Phi}' + \mathbf{Q} \qquad \text{(State variance)}$$
$$\mathbf{u}_t = y_t - \mathbf{Z}\boldsymbol{\alpha}_{t|t-1} \qquad \text{(Error)}$$
$$\mathbf{F}_t = \mathbf{Z}\mathbf{P}_{t|t-1}\mathbf{Z}' + \mathbf{H} \qquad \text{(Error variance)}$$
$$\log L_t = 0.5\, N \log 2\pi - 0.5 \log |\mathbf{F}_t| - 0.5\, \mathbf{u}_t' \mathbf{F}_t^{-1} \mathbf{u}_t \qquad \text{(Log likelihood)}$$
$$\mathbf{K}_t = \widehat{\mathbf{P}}_{t|t-1} \mathbf{Z}' \widehat{F}_t^{-1} \qquad \text{(Kalman gain)}$$
$$\boldsymbol{\alpha}_t = \boldsymbol{\alpha}_{t|t-1} + \mathbf{K}_t \mathbf{u}_t \qquad \text{(State update)}$$
$$\mathbf{P}_t = (I - \mathbf{K}_t \mathbf{Z}) \mathbf{P}_{t|t-1} \qquad \text{(Variance update)}$$

where $t|t-1$ represents a quantity at time t conditional on information up to time $t-1$. The Kalman gain matrix, $\mathbf{K}_t$, is a weighting matrix that weights the relative

---
2. As of version 16, Stata is only capable of dealing with constant coefficients.

importance of the prediction and measurement in calculating the update. The larger the gain, the higher the weight given to the current measurement in the updating step. In the univariate case, the Kalman gain may be interpreted as the slope coefficient in a linear regression of the change in the state, $\boldsymbol{\alpha}_t - \boldsymbol{\alpha}_{t|t-1}$, on the prediction errors, u_t; see, for example, Martin, Hurn, and Harris (2013, 551).

If the state vector is stationary, then the filter recursions may be initiated with the unconditional distribution of the state vector,

$$\boldsymbol{\alpha}_0 = E(\boldsymbol{\alpha}_t)$$
$$\text{vec}(\mathbf{P}_0) = E(\boldsymbol{\alpha}_t \boldsymbol{\alpha}_t')$$

Typically, $\boldsymbol{\alpha}_0 = 0$ and the elements of the starting covariance matrix are given by

$$\text{vec}(\mathbf{P}_0) = [I - (T \otimes T)]^{-1} \text{vec}(\mathbf{Q})$$

where $\text{vec}(\cdot)$ indicates the column stacking of its matrix argument and $\otimes$ is the Kronecker product operator (Harvey 1989, 120–121). When the state transition equation is nonstationary, or if the initial state is not to be regarded as an arbitrary draw from the process implied by the state equation, $\boldsymbol{\alpha}_0$ will be given by a best guess. Consequently, $\mathbf{P}_0$ will be given by a positive-definite matrix summarizing the confidence in this guess. Stata provides three different options for choosing starting values, and the default option, which is based largely on the work of De Jong (1988, 1991), is usually the best option to choose.

In many applications, an important objective is to extract estimates of the unobserved state variables, $\widehat{\boldsymbol{\alpha}}_t$, once the parameters of the system have been estimated by maximum likelihood. The recursions of the Kalman filter automatically provide estimates not only of the expected values of the state but also of its variance. There are in fact three possible estimates of the state vector dependent on the information used in the construction of the estimate. These are as follows:

$$\widehat{\boldsymbol{\alpha}}_{t|t-1} = E(\boldsymbol{\alpha}_t | \mathcal{I}_{t-1}) \quad \text{(Information up to } t-1\text{)}$$
$$\widehat{\boldsymbol{\alpha}}_t = E(\boldsymbol{\alpha}_t | \mathcal{I}_t) \quad \text{(Information up to } t\text{)}$$
$$\widehat{\boldsymbol{\alpha}}_{t|T} = E(\boldsymbol{\alpha}_t | \mathcal{I}_T) \quad \text{(Information up to } T\text{)}$$

The first two equations are the familiar prediction and update equations from the recursions of the filter. The third estimate is a smoothed estimate that uses all the information in the sample to generate what is known as a fixed interval smoother estimate of the state. In practice, the fixed interval smoother requires running the Kalman filter backwards starting at T. Formally, the recursion required to construct the smoothed state is given by

$$\widehat{\boldsymbol{\alpha}}_{t|T} = \widehat{\boldsymbol{\alpha}}_{t|t} + \widehat{\mathbf{P}}_{t|t} \boldsymbol{\Phi}' \mathbf{P}_{t+1|t}^{-1} \left(\widehat{\boldsymbol{\alpha}}_{t+1|T} - \widehat{\boldsymbol{\alpha}}_{t+1|t} \right)$$

The Stata sspace (see [TS] sspace) command estimates the parameters of linear state-space models with time-invariant coefficients. Note, however, that fitting state-space models is not an entirely straightforward numerical process. Convergence problems are common, and experimentation with different optimization algorithms using the `technique()` option or specifying good starting values using the `from()` option may be necessary. Once the `sspace` model has converged, several useful `sspace postestimation` (see [TS] **sspace postestimation**) tools are available. The `predict states` command will produce a time series of the estimated latent state variables, and the three options `smethod(onestep)`, `smethod(filter)`, and `smethod(smooth)` correspond to the estimates of the states given previously.

9.3 Vector autoregressive moving-average models in state-space form

In chapters 4 and 5, autoregressive moving-average (ARMA) models were introduced and fit. The multivariate vector autoregressive moving-average (VARMA) model was introduced, but the estimation of the model was delayed because the approach to fitting this model is to use the state-space form and the Kalman filter. In this section, a series of examples of ARMA model estimation will be introduced, building up to the estimation of a VARMA(1,1) model for changes in global temperature. Changes in temperature and eventually sea levels rather than levels of the variables will be used because these models deal with stationary data, and figure 9.1 indicates that both of these variables are trending. For the moment, we simply assume that they require differencing to induce stationarity.

Consider the first-order autoregressive [AR(1)] model

$$\Delta y_t = \mu + \phi \Delta y_{t-1} + v_t$$

which may be manipulated into state space as follows:

$$\alpha_t = \phi \alpha_{t-1} + v_t$$
$$\Delta y_t = \mu + \alpha_t$$

Although the role of α_t here appears to be unnecessary additional notation, the potential of this device will become apparent when considering more complex models.

An AR(1) model for changes in global temperature can be fit as follows using `sspace`.

```
constraint 1 [D.Temp]u = 1
sspace (u L.u, state noconstant) (D.Temp u,  noerror), constraints(1) nolog
```

Notice that in Stata the notation u plays the role of α_t. The specification of `noconstant` implies that we are actually filtering a zero mean latent state as implied by theory. The first equation in `sspace` is the state equation, and the second is the measurement equation, with the following results:

```
. use http://www.stata-press.com/data/eeus/fci
. constraint 1 [D.Temp]u = 1
. sspace (u L.u, state noconstant) (D.Temp u, noerror), constraints(1) nolog
> vsquish
State-space model
Sample: 1881 - 2009                             Number of obs    =        129
                                                Wald chi2(1)     =      12.55
Log likelihood =  109.10747                     Prob > chi2      =     0.0004
 ( 1)  [D.Temp]u = 1
```

	Temp	Coef.	OIM Std. Err.	z	P>|z|	[95% Conf. Interval]	
u							
	u						
	L1.	-.2981079	.0841493	-3.54	0.000	-.4630374	-.1331783
D.Temp							
	u	1	(constrained)				
	_cons	.0064928	.0070546	0.92	0.357	-.0073341	.0203196
/state							
	var(u)	.0107787	.0013421	8.03	0.000	.0081482	.0134091

Note: Tests of variances against zero are one sided, and the two-sided
      confidence intervals are truncated at zero.

Of course, the results are easily verifiable using the arima (see [TS] arima) command. Note, however, that arima estimates standard deviations and sspace estimates variances. Also, be mindful that these commands may enforce positivity restrictions on the variance parameters in different ways, so small numerical differences may arise.

Similarly, the ARMA(1,1) model

$$\Delta y_t = \phi_1 \Delta y_{t-1} + v_t + \theta_1 v_{t-1}$$

may be written in the required form by defining

$$\boldsymbol{\alpha}_t = \begin{bmatrix} \Delta y_t \\ \theta_1 v_t \end{bmatrix}$$

It follows immediately that the required form is

$$\boldsymbol{\alpha}_t = \begin{bmatrix} \phi_1 & 1 \\ 0 & 0 \end{bmatrix} \boldsymbol{\alpha}_{t-1} + \begin{bmatrix} 1 \\ \theta_1 \end{bmatrix} v_t$$

$$\Delta y_t = \begin{bmatrix} 1 & 0 \end{bmatrix} \boldsymbol{\alpha}_t$$

9.3 Vector autoregressive moving-average models in state-space form

The Stata code to implement this model for global changes in temperature is as follows:

```
constraint 1 [u1]L.u2 = 1
constraint 2 [u1]e.u1 = 1
constraint 3 [D.Temp]u1 = 1
sspace                                      ///
    (u1 L.u1 L.u2 e.u1, state noconstant)   ///
    (u2 e.u1, state noconstant )            ///
    (D.Temp u1, noconstant),                ///
    constraints(1/3) covstate(diagonal)
```

These commands specify three constraints, two state equations, one observation equation, and two options. The first state equation defines u1, and the second defines u2. The observation equation relates D.Temp, the observed variable, to the state. The option covstate(diagonal) requires that a variance be estimated for the state.

```
. constraint 1 [u1]L.u2 = 1
. constraint 2 [u1]e.u1 = 1
. constraint 3 [D.Temp]u1 = 1
. sspace (u1 L.u1 L.u2 e.u1, state noconstant)
>        (u2 e.u1, state noconstant)
>        (D.Temp u1, noconstant),
>        constraints(1/3) covstate(diagonal) nolog vsquish
State-space model
Sample: 1881 - 2009                          Number of obs    =         129
                                             Wald chi2(2)     =      128.29
Log likelihood =   117.49262                 Prob > chi2      =      0.0000
 ( 1)   [u1]L.u2 = 1
 ( 2)   [u1]e.u1 = 1
 ( 3)   [D.Temp]u1 = 1
```

	Temp	Coef.	OIM Std. Err.	z	P>|z|	[95% Conf. Interval]	
u1							
	u1						
	L1.	.2209564	.1228463	1.80	0.072	-.0198179	.4617307
	u2						
	L1.	1	(constrained)				
	e.u1	1	(constrained)				
u2							
	e.u1	-.7132716	.0783249	-9.11	0.000	-.8667855	-.5597577
D.Temp							
	u1	1	(constrained)				
/state							
	var(u1)	.0094409	.0011756	8.03	0.000	.0071368	.0117449

Note: Tests of variances against zero are one sided, and the two-sided confidence intervals are truncated at zero.

Once again, these results are easily verified using the arima command.

Finally, consider the bivariate VARMA(1,1) model

$$\begin{pmatrix} \Delta y_{1t} \\ \Delta y_{2t} \end{pmatrix} = \begin{pmatrix} \phi_{11} & 0 \\ \phi_{21} & \phi_{22} \end{pmatrix} \begin{pmatrix} \Delta y_{1t-1} \\ \Delta y_{2t-1} \end{pmatrix} + \begin{pmatrix} \theta_{11} & 0 \\ 0 & 0 \end{pmatrix} \begin{pmatrix} v_{1t-1} \\ v_{2t-1} \end{pmatrix} + \begin{pmatrix} v_{1t} \\ v_{2t} \end{pmatrix}$$

where Δy_{1t} refers to temperature changes and Δy_{2t} refers to changes in sea level. The model makes several simplifying assumptions.

1. The AR part of temperature changes includes lagged temperature changes but not lagged sea level changes.

2. The moving-average part of temperature changes includes the lagged value of v_{1t} but not the lagged value of v_{2t}.

3. Changes in sea level are essentially treated as a vector AR(1) with no moving-average terms entering the equation.

Assumptions 1 and 2 reflect the fundamental theoretical perception that the direction of influence is from temperature to sea levels. Assumption 3 is more pragmatic because its role is merely to simplify the estimation of the model given that VARMA models are difficult to fit.

To interpret the model for **sspace**, define the state vector as

$$\begin{pmatrix} u_{1t} \\ u_{2t} \\ u_{3t} \end{pmatrix} = \begin{pmatrix} y_{1t} \\ \theta_{11} v_{1t} \\ y_{2t} \end{pmatrix}$$

$$\begin{pmatrix} u_{1t} \\ u_{2t} \\ u_{3t} \end{pmatrix} = \begin{pmatrix} \phi_{11} & 1 & 0 \\ 0 & 0 & 0 \\ \phi_{21} & 0 & \phi_{22} \end{pmatrix} \begin{pmatrix} u_{1t-1} \\ u_{2t-1} \\ u_{3t-1} \end{pmatrix} + \begin{pmatrix} 1 & 0 \\ \theta_{11} & 0 \\ 0 & 1 \end{pmatrix} \begin{pmatrix} v_{1t} \\ v_{2t} \end{pmatrix}$$

Fitting the VARMA(1,1) model yields

```
. constraint 1  [u1]L.u2 = 1
. constraint 2  [u1]e.u1 = 1
. constraint 3  [u3]e.u3 = 1
. constraint 4  [D.Temp]u1 = 1
. constraint 5  [D.Sea]u3 = 1
```

9.3 Vector autoregressive moving-average models in state-space form

```
. sspace (u1 L.u1 L.u2 e.u1, state noconstant)
>       (u2 e.u1, state noconstant)
>       (u3 L.u1 L.u3 e.u3, state noconstant)
>       (D.Temp u1, noconstant)
>       (D.Sea u3, noconstant),
>       constraints(1/5) technique(nr) covstate(diagonal) nolog vsquish
State-space model
Sample: 1881 - 2009                         Number of obs   =        129
                                            Wald chi2(4)    =     133.38
Log likelihood =  4.6934207                 Prob > chi2     =     0.0000
 ( 1)  [u1]L.u2 = 1
 ( 2)  [u1]e.u1 = 1
 ( 3)  [u3]e.u3 = 1
 ( 4)  [D.Temp]u1 = 1
 ( 5)  [D.Sea]u3 = 1
```

		Coef.	OIM Std. Err.	z	P>|z|	[95% Conf. Interval]	
u1							
	u1 L1.	.221141	.1228253	1.80	0.072	-.0195921	.4618742
	u2 L1.	1	(constrained)				
	e.u1	1	(constrained)				
u2							
	e.u1	-.7133617	.0782931	-9.11	0.000	-.8668134	-.5599099
u3							
	u1 L1.	.1971298	.4725837	0.42	0.677	-.7291172	1.123377
	u3 L1.	-.1935598	.0869001	-2.23	0.026	-.3638809	-.0232388
	e.u3	1	(constrained)				
D.Temp							
	u1	1	(constrained)				
D.Sea							
	u3	1	(constrained)				
/state							
	var(u1)	.0094409	.0011756	8.03	0.000	.0071368	.0117449
	var(u3)	.3364325	.0418912	8.03	0.000	.2543272	.4185379

Note: Tests of variances against zero are one sided, and the two-sided
      confidence intervals are truncated at zero.

Despite the fact that Stata does not have the capability of fitting a VARMA(1,1) model with a single command, the sspace facility is flexible enough to fit complex models of this kind, provided they can be cast in state-space form. From the perspective of climate modeling, however, the end result of the exercise is somewhat disappointing because the effect of temperature changes on sea level changes is found to be statistically insignificant. The estimated coefficient value of ϕ_{21} is 0.197 with a p-value of 0.677. This result is, however, to be expected because these traditional time-series methods for stationary data that rely on detrending or differencing are not particularly useful in modeling trend behavior. What is required is a framework in which both stationary and nonstationary components can be added to the model seamlessly.

9.4 Unobserved component time-series models

Structural time-series models enable time-series data to be modeled as being composed of several unobserved components: the trend component, cyclical components, a seasonal component, and an irregular component. The method aims to reconstruct the main features of the time series in terms of these constituent components. The general model is given as

$$y_t = \mu_t + \psi_t + \gamma_t + u_t$$

where μ_t, γ_t, ψ_t, and u_t are the trend, seasonal, cyclical, and irregular components, respectively. The model can, of course, be extended to account for additional observed covariates and AR components. These unobserved component models can be expressed in the state-space framework and fit via maximum likelihood using the sspace framework. There are other useful commands in this context, such as ucm (see [TS] **ucm**), for fitting structural time-series models, but knowing how to use sspace is particularly valuable because it allows nonstandard models to be dealt with fairly easily.

9.4.1 Trends

The trend represents the long-term movements in series. Harvey (1989) specifies the model with a trend component as

$$y_t = \mu_t + u_t$$

where μ_t is the trend and u_t is the irregular component, or disturbance term. Specifying the structure of μ_t is accomplished by introducing stochastic behavior into the coefficients of the deterministic time trend

$$\mu_t = \mu + \beta t$$

Setting the starting value of μ_t to be μ, the deterministic trend can be recovered from the recursion

$$\mu_t = \mu_{t-1} + \beta$$

We can introduce stochastic behavior into a deterministic trend in terms of the specification

$$\mu_t = \mu_{t-1} + \beta_{t-1} + v_{1t}, \quad v_{1t} \sim \text{i.i.d.}(0, \sigma_1^2)$$
$$\beta_t = \beta_{t-1} + v_{2t}, \quad v_{2t} \sim \text{i.i.d.}(0, \sigma_2^2)$$

This general model is known as the local linear trend, and many variations can be accommodated by imposing restrictions. Some of the more common trend specifications are listed in table 9.1.

9.4.1 Trends

Table 9.1. Common trend specifications in unobserved component time-series models

Model	Specification	Restrictions
Constant	$y_t = \mu + u_t$	$\sigma_1^2 = 0,\ \beta_t = 0$
Deterministic trend	$y_t = \mu_t + u_t$ $\mu_t = \mu_{t-1} + \beta$	$\sigma_2^2 = 0$
Local level	$y_t = \mu_t + u_t$ $\mu_t = \mu_{t-1} + v_{1t}$	$\sigma_2^2 = 0,\ \beta_t = 0$
Local level and deterministic trend	$y_t = \mu_t + u_t$ $\mu_t = \mu_{t-1} + \beta + v_{1t}$	$\sigma_2^2 = 0$
Random walk	$y_t = \mu_t$ $\mu_t = \mu_{t-1} + v_{1t}$	$\beta_t = 0,\ \sigma_u^2 = 0\ \sigma_2^2 = 0$
Random walk and drift	$y_t = \mu_t$ $\mu_t = \mu_{t-1} + \beta + v_{1t}$	$\sigma_u^2 = 0\ \sigma_2^2 = 0$
Local linear trend	$y_t = \mu_t + u_t$ $\mu_t = \mu_{t-1} + \beta_{t-1} + v_{1t}$ $\beta_t = \beta_{t-1} + v_{2t}$	none

Many of these models are designed to handle nonstationary time series. The local level, random walk, local level with deterministic trend, and random walk with drift models all allow for the modeling of I(1) series, while the local linear trend model is applicable for I(2) series. As an illustration, consider fitting the local level model

$$y_t = \mu_t + u_t$$
$$\mu_t = \mu_{t-1} + v_t$$

to the global temperature series. Fitting the model yields the following results:

```
. use http://www.stata-press.com/data/eeus/fci
. ucm Temp, model(llevel) nolog
Unobserved-components model
Components: local level
Sample: 1880 - 2009                              Number of obs    =         130
Log likelihood =   114.98779
```

	Temp	Coef.	OIM Std. Err.	z	P>|z|	[95% Conf. Interval]	
	var(level)	.0015959	.0006533	2.44	0.007	.0003154	.0028765
	var(Temp)	.0057452	.0010419	5.51	0.000	.0037031	.0077874

Note: Model is not stationary.
Note: Tests of variances against zero are one sided, and the two-sided
      confidence intervals are truncated at zero.

The first thing to note about these estimates is that there are only two parameters to estimate: the variance in the measurement equation (σ_u^2) and the variance of the local level (σ_v^2). Of particular interest is whether σ_v^2 is equal to zero because this result would mean that the trend component collapses to a constant. In this instance, the estimate returned is 0.0057, and because the p-value of the test of the null hypothesis $H_0 : \sigma_v^2 = 0$ is 0.000, the use of a local trend (as opposed to a constant) is strongly indicated. The estimated trend $\widehat{\mu}_t$ may be extracted using the `predict trend` postestimation command and issuing the `smethod(smooth)` option. The smoothed trend for global temperature is illustrated in figure 9.2.

9.4.1 Trends

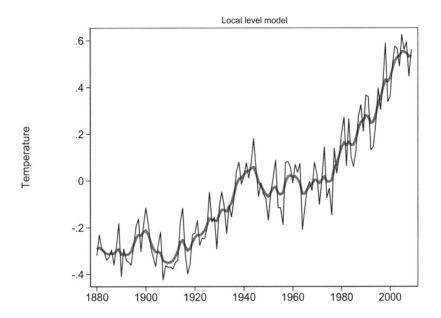

Figure 9.2. Plot of the smooth trend obtained from a local level model for mean annual global temperature.

The `predict` command may also be used to extract the irregular component, $\widehat{u}_t$. The adequacy of the proposed structural time-series model may be judged by performing a few simple tests on the standardized residuals. Figure 9.3 plots a histogram of the standardized residuals with a standard normal curve superimposed. It is clear that the model does not yield residuals that are normally distributed. Additional tests could include a formal test of normality and tests of autocorrelation to gauge the extent of the dynamic misspecification in the model.

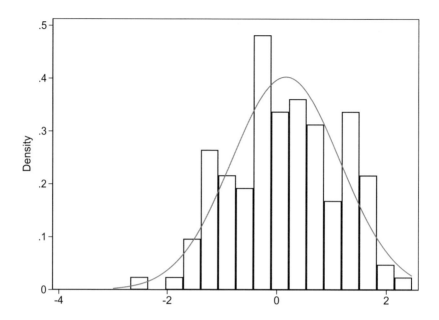

Figure 9.3. Histogram of the standardized irregular component of the local level model for mean annual global temperature overlaid with a standard normal distribution.

9.4.2 Seasonals

A seasonal component models cyclical behavior that occurs at known seasonal periodicities. Modeled in the time domain, the period of the cycle is specified as the number of time periods required for the cycle to complete: for example, 4 for quarterly seasonality, 12 for monthly seasonality. Seasonal components may be either deterministic or stochastic. Unobserved component models allow the specification of explanatory variables in *indepvars* of the command, which allows the use of factor variables, so deterministic indicator variables may be used in the ucm framework to capture deterministic seasonality.

If seasonality is considered stochastic, the variance of the seasonal component must be estimated. The specification of the stochastic seasonal component, γ_t of a structural time-series model in which the seasonal periodicity is s, is given by

$$\gamma_t = -\sum_{j=1}^{s-1} \gamma_{t-j} + v_{3t}, \qquad v_{3t} \sim \text{i.i.d. } N(0, \sigma_3^2)$$

Because the global temperature and sea level data are annual observations, no seasonal component is fit to these data.

9.4.3 Cycles

Given a time series of observations $y_t = \{y_1 \ldots y_T\}$, a basic representation of a single cycle in the data is as a cosine (or sine) wave

$$\psi_t = A\cos(\lambda t - \theta) \tag{9.1}$$

in which λ is the frequency of the cycle (in radians), A is the amplitude of the cycle, and θ is the phase which measures the displacement of the wave in terms of a shift along the time axis. The period of this cycle is given by $2\pi/\lambda$. This equation can be reexpressed as

$$\phi_t = \cos(\lambda - \theta) = \cos\lambda\cos\theta + \sin\lambda\sin\theta.$$

Using this result, (9.1) may be rewritten as

$$\begin{aligned}\psi_t &= A\cos(\lambda t - \theta)\\ &= A\cos\theta\cos(\lambda t) + A\sin\theta\sin(\lambda t)\\ &= \alpha\cos(\lambda t) + \beta\sin(\lambda t)\end{aligned}$$

in which $\alpha = A\cos\theta$ and $\beta = A\sin\theta$. Harvey (1989) now allows α and β to be time-varying by using the recursion

$$\begin{bmatrix}\psi_t\\ \psi_t^*\end{bmatrix} = \begin{bmatrix}\cos\lambda & \sin\lambda\\ -\sin\lambda & \cos\lambda\end{bmatrix}\begin{bmatrix}\psi_{t-1}\\ \psi_{t-1}^*\end{bmatrix}$$

in which $\psi_0 = \alpha$ and $\psi_0^* = \beta$.

To make the cycle stochastic, noise terms are introduced:

$$\begin{bmatrix}\psi_t\\ \psi_t^*\end{bmatrix} = \rho\begin{bmatrix}\cos\lambda & \sin\lambda\\ -\sin\lambda & \cos\lambda\end{bmatrix}\begin{bmatrix}\psi_t\\ \psi_t^*\end{bmatrix} + \begin{bmatrix}v_{4t}\\ v_{4t}^*\end{bmatrix}$$

For the model to be identified, it is usually assumed that the disturbance terms are uncorrelated but have the same variance, σ_4^2. The presence of the disturbance terms may be interpreted as an excitation of the state and to preserve stationarity a damping factor, ρ, is introduced. If $\exp(\rho) < 1$, the cycle is stationary. This stochastic-cycle model has three parameters, namely, the frequency (λ), the damping factor (ρ), describing the dispersion of the random components around that frequency, and the variance of the stochastic-cycle process (σ_4^2), which acts as a scale factor.

In this instance, the local level plus cycle model for global temperature is given by

$$y_t = \mu_t + \psi_t + u_t$$

```
. ucm Temp, model(llevel) cycle(1) iterate(20) nolog vsquish
Unobserved-components model
Components: local level, order 1 cycle
Sample: 1880 - 2009                          Number of obs    =        130
                                             Wald chi2(2)     =      45.03
Log likelihood =  118.07328                  Prob > chi2      =     0.0000
```

		OIM				
Temp	Coef.	Std. Err.	z	P>|z|	[95% Conf.	Interval]
frequency	1.510386	.2326943	6.49	0.000	1.054314	1.966458
damping	.6448814	.1806816	3.57	0.000	.2907521	.9990107
var(level)	.0013982	.0005586	2.50	0.006	.0003035	.002493
var(cycle1)	.0015356	.0011532	1.33	0.092	0	.0037959
var(Temp)	.0033713	.0009633	3.50	0.000	.0014833	.0052593

Note: Model is not stationary.
Note: Tests of variances against zero are one sided, and the two-sided confidence intervals are truncated at zero.

Convergence for this model is achieved, and the estimated frequency and damping parameters of the cycle are precisely estimated. The estimated frequency of 1.51 is small. This suggests that the estimated cycle has a relatively low frequency. The estimate of the damping factor of 0.644 is well below the upper value of 1, which is required for the stationarity of the cycle, but it does imply that the low frequency components of the model are fairly diffusely distributed about the estimated frequency. A troubling aspect of the results is that the estimated variance of the cycle, 0.0015, is statistically insignificant at the 5% level, although it is significant at the 10% level. This result casts some doubt on the model specification. The estimated variances for the local level (0.0014) and for the dependent variable (0.00347) are both significantly different from zero.

The estimated frequency of the cycle can be used to compute the period of the cycle to aid interpretation. This result is accomplished using the postestimation command `estat period`.

```
. estat period
```

cycle1	Coef.	Std. Err.	[95% Conf.	Interval]
period	4.159986	.640899	2.903847	5.416125
frequency	1.510386	.2326943	1.054314	1.966458
damping	.6448814	.1806816	.2907521	.9990107

Note: Cycle time unit is yearly.

The results obtained from `estat period` show that the estimated frequency of 1.51 translates into a period of about four years. Of course, the model is not limited to including only one stochastic cycle, but in this instance there seems to be little to support the addition of a second cycle.

Be aware that estimation of a local level plus stochastic cycle model frequently results in convergence problems because the series must have enough variation to identify the variance of the random-walk component, the variance of the idiosyncratic term, and the parameters of the stochastic-cycle component. Investigation of the Stata iteration log after a limited number of iterations often reveals which of the components is causing the problem, thus facilitating an appropriate reformulation. In particular, observe the estimate of the variance of the local level. If this estimate is being driven to zero, a deterministic trend may be appropriate.

9.5 A bivariate model of sea level and global temperature

To allow the trend in global temperature to influence sea levels, Grassi, Hillebrand, and Ventosa-Santaulària (2013) build an unobserved component time-series model for the two series as a variant of the local trend model. Denoting mean annual global temperature as y_t^T and mean annual sea level as y_t^H, the model is specified as follows. First of all, global temperature is hypothesized to follow a pure random walk,

$$y_t^T = \mu_t^T + u_t^T$$
$$\mu_t^T = \mu_{t-1}^T + \beta_{t-1}^T$$
$$\beta_t^T = \beta_{t-1}^T + \eta_t^T$$

The model for sea level is also a random walk, but the specification of its local level μ_t^H is augmented to include the local level in temperature, μ_{t-1}^T, as an explanatory variable. The equations governing sea level are therefore

$$y_t^H = \mu_t^H + u_t^H$$
$$\mu_t^H = \mu_{t-1}^H + \beta_{t-1}^H + \gamma \mu_{t-1}^T$$
$$\beta_t^H = \beta_{t-1}^H + \eta_t^H$$

Taken together, these two sets of equations can be written in state-space form as follows. The measurement equation is

$$\begin{bmatrix} y_t^H \\ y_t^T \end{bmatrix} = \begin{bmatrix} \mu_t^H \\ \mu_t^T \end{bmatrix} + \begin{bmatrix} u_t^H \\ u_t^T \end{bmatrix}$$

while the state equation is

$$\begin{bmatrix} \mu_t^H \\ \mu_t^T \\ \beta_t^H \\ \beta_t^T \end{bmatrix} = \begin{bmatrix} 1 & c & 1 & 0 \\ 0 & 1 & 0 & 1 \\ 0 & 0 & 1 & 0 \\ 0 & 0 & 0 & 1 \end{bmatrix} \begin{bmatrix} \mu_{t-1}^H \\ \mu_{t-1}^T \\ \beta_{t-1}^H \\ \beta_{t-1}^T \end{bmatrix} + \begin{bmatrix} 0 \\ 0 \\ \eta_t^H \\ \eta_t^T \end{bmatrix}$$

Using the sspace capability in Stata, the model may be specified as

```
. // Grassi model
. use http://www.stata-press.com/data/eeus/fci
. // predict out to 2100
. tsappend, add(91)
. constraint 1 [muh]L.muh = 1
. constraint 2 [muh]L.bh = 1
. constraint 3 [mut]L.mut = 1
. constraint 4 [mut]L.bt = 1
. constraint 5 [bh]L.bh = 1
. constraint 6 [bt]L.bt = 1
. constraint 7 [Sea]muh = 1
. constraint 8 [Temp]mut = 1
. sspace (muh L.muh L.mut L.bh, state noconstant noerror)
>       (mut L.mut L.bt, state noconstant noerror)
>       (bh L.bh, state noconstant) (bt L.bt, state noconstant)
>       (Sea muh, noconstant) (Temp mut, noconstant), constraints(1/8)
>       difficult nolog vsquish
State-space model
Sample: 1880 - 2009                          Number of obs    =       130
                                             Wald chi2(1)     =      2.88
Log likelihood = -3.6019597                  Prob > chi2      =    0.0899
 ( 1)  [muh]L.muh = 1
 ( 2)  [muh]L.bh = 1
 ( 3)  [mut]L.mut = 1
 ( 4)  [mut]L.bt = 1
 ( 5)  [bh]L.bh = 1
 ( 6)  [bt]L.bt = 1
 ( 7)  [Sea]muh = 1
 ( 8)  [Temp]mut = 1
```

		OIM Coef.	Std. Err.	z	P>|z|	[95% Conf. Interval]	
muh							
	muh L1.	1	(constrained)				
	mut L1.	.357974	.2110549	1.70	0.090	-.0556859	.7716339
	bh L1.	1	(constrained)				
mut							
	mut L1.	1	(constrained)				
	bt L1.	1	(constrained)				
bh							
	bh L1.	1	(constrained)				

9.5 A bivariate model of sea level and global temperature

bt						
bt L1.	1	(constrained)				
Sea						
muh	1	(constrained)				
Temp						
mut	1	(constrained)				
/state						
var(bh)	.0004753	.0005861	0.81	0.209	0	.0016242
var(bt)	7.47e-06	5.14e-06	1.45	0.073	0	.0000175
/observable						
var(Sea)	.2340919	.0359171	6.52	0.000	.1636957	.3044881
var(Temp)	.0077913	.0010201	7.64	0.000	.0057921	.0097906

Note: Model is not stationary.
Note: Tests of variances against zero are one sided, and the two-sided confidence intervals are truncated at zero.

The estimate of the parameter of interest, c, is equal to 0.358. Given the dimensions of the time series, this corresponds to a proportionality coefficient of 3.58mm/year per $C°$. This estimate is smaller than the one reported by Grassi, Hillebrand, and Ventosa-Santaulària (2013) but similar to the one found by Rahmstorf (2007) of 3.4 mm/year per $C°$. The coefficient is significant at the 10% level but not at the 5% level. This estimate leads to a substantial difference in the long-term forecast, with the long-term mean forecast to be 60cm above the 1990 mean.

Figure 9.4 shows the forecasts of sea level to 2100 for the bivariate model in which temperature interacts with sea level and the case where the interaction is switched off and a pure random walk with drift model is fit for sea level. The difference is startling. The univariate model cannot capture any influence of temperature beyond what is already contained in the sea level time series, and the long-term mean is forecast to be about 20cm above the 1990 mean.

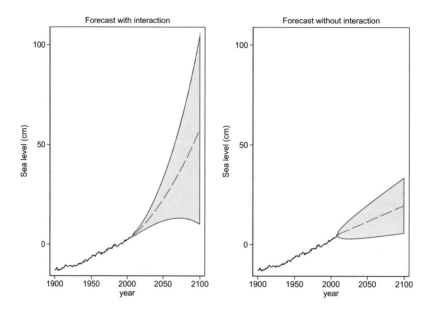

Figure 9.4. Long-term forecasts and 95% confidence intervals for mean sea level using a univariate random walk with drift model (right-hand panel) and a bivariate model in which global mean temperature interacts with mean sea level (left-hand panel).

This model of the interaction between global mean sea level and global temperature should be interpreted with care. It is presented here as an illustration of the power of the state-space modeling capability in Stata. The central point here is that state-space models provide a useful modeling tool in the presence of trending behavior. The confidence intervals around the forecasts obtained from these models are wide, perhaps indicating that there are many more influences that need to be accounted for in a fully specified model.

Exercises

1. El Niño and La Niña are the warm and cool phases of a recurring climate pattern across the tropical Pacific. The Southern Oscillation, on the other hand, refers to the connection between atmospheric pressure on the eastern and western sides of the Pacific Ocean, namely, that when pressure rises in the east, it usually falls in the west, and vice versa. Scientists eventually recognized that sea temperature and atmospheric pressure were part of the same natural phenomenon, so the joint name El Niño Southern Oscillation is used. El Niño is now understood to mean torrential flooding in the eastern tropical Pacific and severe droughts in the western tropical Pacific, while in a La Niña event, the opposite occurs. Records for sea surface temperature are not as easily obtainable as those for barometric pressure.

9.5 A bivariate model of sea level and global temperature

The latter have been recorded for over 100 years in many locations around the Pacific ocean. `soi.dta` contains monthly data on the Southern Oscillation Index, which is calculated from the monthly fluctuations in the air pressure difference between Tahiti and Darwin.

   a. The data in `soi.dta` are given in a "wide" format (see also chapter 12). Use the Stata `reshape` (see [D] **reshape**) command to form one continuous monthly series, and then plot the Southern Oscillation Index for the period 1890 to 2016. Comment on the time-series properties of the series. Note that low values of the index indicate an El Niño event. Verify that you can recognize the severe El Niño event of 1982–1983 and the slightly less severe event of 1997–1998. Hall, Skalin, and Teräsvirta (2001) report that this El Niño is blamed for the unusually vigorous rough at the 1998 United States Open Golf Tournament in San Francisco. The time-series plot also reveals the extent of the La Niña event of early 2011, during which there was severe flooding on the east coast of Australia.

   b. Fit a local-level model for the monthly Southern Oscillation Index. Use the `predict trend` postestimation command to obtain the smoothed estimate of the trend and plot of the smooth trend overlaid on the Southern Oscillation Index. Comment on the results.

   c. Fit a local-level model with a monthly seasonal component. Use the option `iterate(10)` to set the number of options to 10. Carefully inspect the iteration log, and comment on why the model is unlikely to converge.

   d. Fit a structural time-series model for the Southern Oscillation Index with a deterministic constant and one cycle. Does the estimated frequency of the cycle correspond to the well-known dominant quadrennial cycle ($\approx$ 4 year period)? The periods of the cycle component can be extracted using the `estat period` postestimation command.

   e. Li et al. (2020) examine the statistical properties of the El Niño Southern Oscillation but use a different series. Using the current data, fit a structural model with two cycles, and see if the frequencies correspond to the quadrennial cycle and the approximate biannual cycle ($\approx$ 8 month period).

2. `nile.dta` contains one of the best known and modeled time series in environmental econometrics, namely, the minimal water level of the Nile River for the years 622–1284 measured at the Roda gauge near Cairo (Beran 1994).

   a. Plot the series and comment on its properties. Can you discern any persisting cycle in the Nile data?

b. Fit a structural time-series model for the Nile data consisting of a local level, an AR(1) component, and an irregular component, given by

$$y_t = \mu_t + \phi_t + u_t \quad \text{(Measurement)}$$
$$\mu_t = \mu_{t-1} + v_{1t} \quad \text{(Local level)}$$
$$\phi_t = \rho\phi_{t-1} + v_{2t} \quad \text{[AR(1)]}$$

[Hint: You may have to use sspace rather than ucm.]

c. Using the predict trend postestimation command, extract the smoothed trend and AR(1) components and plot them. Comment on the results.

d. Using the predict irregular postestimation command, extract the standardized residuals, $\widehat{u}_t$. Draw a histogram of the standardized residuals and overlay a standard normal distribution. Comment on the plot. Provide a formal test of the normality of the standardized residuals, and interpret the results.

e. Plot the autocorrelation function of the residuals out to 30 lags. Comment on the result.

[Hint: The Nile data are often used as a tool to demonstrate the long memory properties of some environmental time series, see chapter 15.]

10 Nonlinear time-series models

The time-series models discussed in prior chapters have the common feature of being linear models in the sense that the expected value of the dependent variable is linear in the parameters. Although these models are powerful in tracking and forecasting many of the variables of interest in environmental econometrics, they cannot deal with situations in which there are asymmetric adjustments and stochastically varying cyclical behavior. The primary focus of this chapter therefore is extending the linear modeling framework to include nonlinearities.

There are two important caveats to note at the outset. First, the type of nonlinearity discussed in this chapter refers to nonlinearity in the conditional mean of the process being modeled. Issues relating to the structure of the variance are the subject matter of chapter 11. Second, in the natural world, nonlinearity is often associated with chaotic models derived from chaos theory. It is true that chaotic models are nonlinear, but not all nonlinear models are chaotic. The nonlinearities discussed in this chapter are not directly related to chaos theory, because chaotic models are frequently deterministic in nature, whereas econometric analysis emphasizes the stochastic nature of the process.

The need for nonlinear modeling in environmental econometrics arises from an increased awareness of the complexities governing natural processes, sometimes because of human activity. However, the move to a nonlinear framework does involve added complexity, so choosing to use a nonlinear model should be preceded by statistical testing. The tests that will be discussed in this chapter all test the null hypothesis of a linear model against the alternative hypothesis, which is either a specific or a nonspecific form of nonlinearity. Because there are many tests in econometrics that are named after the alternative hypothesis, such as tests for autocorrelation and heteroskedasticity, in this chapter the tests will usually be referred to as nonlinearity tests.

Once evidence is found of nonlinear behavior, a nonlinear model for the conditional mean must be specified. Predominantly, the type of nonlinearity dealt with in this chapter relates to data generating processes in which a linear process switches between several regimes according to some rule. Regime switching has often been used to motivate a nonlinear data generating process for many macroeconomic series. Examples of the nonlinear models investigated are bilinear models; threshold models, including so-called smooth transition models; and Markov switching models.

10.1 Sunspot data

Sunspots are magnetic regions on the surface of the sun with magnetic field strengths that are thousands of times stronger than the earth's magnetic field. They appear as dark spots on the surface of the sun and typically last for several days. The sunspot numbers have been of interest to climatologists who hypothesized a link between sunspot activity and climate change. The so-called Maunder Minimum[1] between 1645 and 1715 was a period in which sunspots were scarce and the winters harsh, strongly suggesting a link between solar activity and climate change. The current view is that there has been no significant long-term upward trend in solar activity since 1700. This implies that rising global temperatures since the industrial revolution cannot be attributed to increased solar activity.

The sunspot series has also been used as an explanatory variable in statistical models of climate variables. White and Liu (2008) provide evidence that the solar cycle may be the trigger for El Niño and La Niña episodes from 1900–2005, suggesting that higher solar activity implies weaker and less frequent El Niño events. This view is controversial, and no generally accepted statistical link has been established. Another line of inquiry links solar activity to rainfall in East Africa and the level of Lake Victoria.

Notwithstanding the current view concerning the tenuous link between solar activity and climate change, from a time-series perspective the sunspot data are interesting in their own right because they have proved difficult to model.

The data y_t, which are the annual averages of daily sunspot numbers from 1700 to 2017, were compiled by the Solar Influences Data Analysis Centre in Belgium. The series is plotted in figure 10.1. Not only is this series the longest directly observed index of solar activity, but it also has interesting time-series properties.[2] The autocorrelation functions (ACF) and partial autocorrelation functions (PACF) of the sunspot data are plotted in figure 10.2.

---

1. Named after the solar astronomers Annie Russell Maunder (1868–1947) and her husband, Edward Walter Maunder (1851–1928), who studied how sunspot latitudes changed with time.
2. Galileo first viewed sunspots with his telescope in 1610. From 1749, daily observations were made at the Zurich Observatory.

10.1 Sunspot data

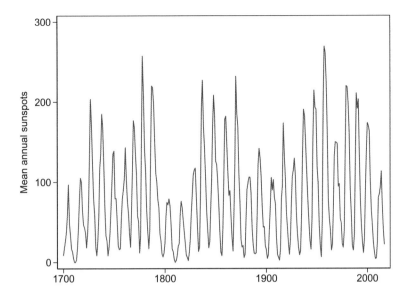

Figure 10.1. Plot of annual average sunspots series from 1700 to 2017

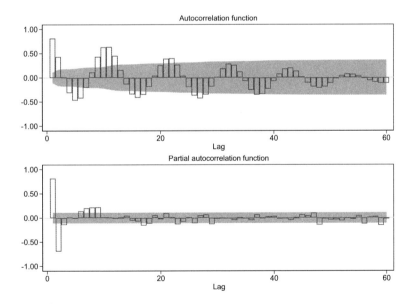

Figure 10.2. ACF (top panel) and PACF (bottom panel) of the annual average sunspot series from 1700 to 2017

The cycles in the data seen in figure 10.1 are also evident in the ACF in figure 10.2, which exhibits strong periodicity. This feature of the data has led researchers to conclude that sunspots exhibit a cycle of activity varying in duration between 9 and 14 years with an average period of approximately 11 years. The PACF also exhibits cyclical behavior out to a lag order of about 20 years. Together with this long-memory behavior, from a nonlinear modeling perspective an interesting characteristic of the series is that, in each cycle, the rise to the maximum has a steeper gradient than the fall to the next minimum. These features have given rise to many attempts to use nonlinear modeling to capture the key features of the series. These data are therefore an ideal series with which to discuss nonlinear time-series modeling methods.

10.2 Testing

As noted in the introduction, the definition of linearity refers here to the conditional mean. It is useful to adopt the conventional vector notation

$$E(y_t|\mathbf{x}_t;\boldsymbol{\beta}) = \mathbf{x}'_t\boldsymbol{\beta}_l$$

in which $\mathbf{x}_t$ is a column vector that contains the value appropriate to period t of all the explanatory variables. Note that $\mathbf{x}_t$ can contain values of y_t suitably lagged as well as current and lagged values of any explanatory variables and lagged values of any regression disturbances. All of these variables enter into the conditional mean combined linearly with a conformable column vector of parameters $\boldsymbol{\beta}_l$. For the sake of simplicity, the models discussed here are pure time-series models in which $\mathbf{x}_t$ contains lagged dependent variables (y_{t-i}) and lagged disturbances (u_{t-i}). Current and lagged values of explanatory variables will not be used, but their addition is straightforward. The variable y_t is nonlinear in mean if

$$E(y_t|\mathbf{x}_t;\boldsymbol{\beta}) = \mathbf{x}'_t\boldsymbol{\beta}_l + g(\mathbf{x}'_t;\boldsymbol{\beta}_{nl})$$

For the moment, the functional form $g(\cdot)$ is left unspecified.[3] Indeed, in most applications the nonlinear functional form is unknown at the outset, and tentative specifications are only developed after extensive search. The associated regression model is then given by

$$y_t = \mathbf{x}'_t\boldsymbol{\beta}_l + g(\mathbf{x}_t;\boldsymbol{\beta}_{nl}) + u_t \qquad u_t \sim \text{i.i.d. } N(0,\sigma^2)$$

The assumptions on the disturbance term are strict but are adopted for the sake of argument. In particular, the assumption of constant variance will be relaxed in chapter 11, which deals with time-varying variances.

As with all testing procedures, testing for nonlinearity requires a null and an alternative hypothesis. This causes an immediate difficulty in that, in some cases of interest, an alternative model has not yet been formulated, and it is a pretest for nonlinearity

---

3. This functional form can range from a simple logarithmic or polynomial function to more complex functions of the type discussed later in the chapter.

10.2 Testing

that is required. Therefore, before we discuss tests based on specific nonlinear models under the alternative, we discuss two general tests that have power against general unspecified alternative hypotheses.

One of the earliest general tests of misspecification of functional form, which can also be used to test for nonlinearity, is the RESET test (Ramsey 1969). Given the specification of a linear model under the null hypothesis, say, an AR(p) model, the procedure is as follows:

1. Fit the model under the null hypothesis

$$y_t = \beta_0 + \beta_1 y_{t-1} + \cdots + \beta_p y_{t-p} + u_t$$

   and save the fitted values $\widehat{y}_t$.

2. Estimate an auxiliary regression of y_t on the original linear regressors and powers of the fitted values

$$y_t = \beta_0 + \beta_1 y_{t-1} + \cdots + \beta_p y_{t-p} + \gamma_2 \widehat{y}_t^2 + \gamma_3 \widehat{y}_t^3 + \gamma_4 \widehat{y}_t^4 + v_t$$

   where v_t is a disturbance term. Note that there is no reason to limit the powers of the fitted values to 4, but this value is used by default in Stata.

3. Under the null hypothesis, γ_2, γ_3, and γ_4 will be statistically insignificant, and these restrictions are tested using an F test.

The idea of the RESET test is that if the powers of the fitted values help explain y_t, then the process generating y_t may involve nonlinearities. Stata refers to the RESET tests as `ovtest`, short for omitted variables. It is probably more correct to view the RESET test as a test of functional form of the relationship between the dependent and explanatory variables in the test regression.

Another important general test of nonlinearity is the so-called V23 test. Teräsvirta, Lin, and Granger (1993) suggest replacing the nonlinear function $g(\mathbf{x}_t; \boldsymbol{\beta}_{nl})$ with a third-order Taylor series expansion. The resultant expansion involves all the second and third-order cross-product terms of y_{t-i}. The equation to estimate then becomes

$$y_t = \beta_0 + \sum_{i=1}^p \beta_i y_{t-i} + \sum_{i=1}^p \sum_{j=1}^p \delta_{ij} y_{t-i} y_{t-j} + \sum_{i=1}^p \sum_{j=1}^p \sum_{k=1}^p \delta_{ijk} y_{t-i} y_{t-j} y_{t-k} + v_t \quad (10.1)$$

where v_t is once again a disturbance term. The significance of the second- and third-order cross products can be tested with an F test after estimating (10.1) directly.

To implement these tests using the sunspot data, the first step is to fit the model under the null hypothesis. This requirement raises the issue of the appropriate number of lags to use in the estimation. Using the Stata `varsoc` command with a maximum lag length of 12 produces the following results:

```
. use http://www.stata-press.com/data/eeus/annualsunspots
. tsset datevec, yearly
        time variable:  datevec, 1700 to 2017
                delta:  1 year
. varsoc sunspots, maxlag(12)
  Selection-order criteria
  Sample: 1712 - 2017                          Number of obs     =        306
```

lag	LL	LR	df	p	FPE	AIC	HQIC	SBIC
0	-1697.58				3881.28	11.1018	11.1067	11.114
1	-1531.5	332.15	1	0.000	1319.47	10.0229	10.0326	10.0472
2	-1427.89	207.22	1	0.000	674.755	9.35223	9.36683	9.38873
3	-1424.43	6.9126	1	0.009	664.01	9.33617	9.35564	9.38485
4	-1424.43	.00901	1	0.924	668.345	9.34268	9.36701	9.40352
5	-1424.38	.10053	1	0.751	672.508	9.34889	9.37809	9.4219
6	-1421.84	5.0877	1	0.024	665.758	9.3388	9.37286	9.42398
7	-1415.85	11.963	1	0.001	644.433	9.30624	9.34517	9.40358
8	-1408.44	14.836	1	0.000	617.962	9.26429	9.30809	9.3738
9	-1400.73	15.409*	1	0.000	591.471*	9.22047*	9.26913*	9.34215*
10	-1400.7	.05512	1	0.814	595.246	9.22682	9.28035	9.36068
11	-1400.69	.03335	1	0.855	599.09	9.23325	9.29165	9.37927
12	-1400.68	.02226	1	0.881	602.981	9.23971	9.30298	9.3979

```
  Endogenous:  sunspots
   Exogenous:  _cons
```

All the information criteria suggest that 9 lags is a reasonable choice. This lag length is in line with the evidence in figure 10.2, which shows that there is significant structure in the autocorrelations and partial autocorrelations out to lag 9. Applying the RESET test to the sunspot data after fitting an AR(9) model gives the following output:

```
. regress sunspots L(1/9).sunspots
```

Source	SS	df	MS		Number of obs	=	309
					F(9, 299)	=	200.59
Model	1026696.65	9	114077.405		Prob > F	=	0.0000
Residual	170043.912	299	568.708734		R-squared	=	0.8579
					Adj R-squared	=	0.8536
Total	1196740.56	308	3885.52129		Root MSE	=	23.848

sunspots	Coef.	Std. Err.	t	P>|t|	[95% Conf. Interval]	
sunspots						
L1.	1.168556	.0563697	20.73	0.000	1.057624	1.279487
L2.	-.4193821	.0879323	-4.77	0.000	-.5924267	-.2463375
L3.	-.1341714	.0908782	-1.48	0.141	-.3130134	.0446705
L4.	.104604	.0908838	1.15	0.251	-.0742489	.2834568
L5.	-.0724476	.0909361	-0.80	0.426	-.2514034	.1065082
L6.	.0053406	.0908059	0.06	0.953	-.1733589	.1840402
L7.	.0219882	.0908585	0.24	0.809	-.156815	.2007915
L8.	-.0527488	.0879303	-0.60	0.549	-.2257895	.1202919
L9.	.2221535	.0562627	3.95	0.000	.1114324	.3328746
_cons	12.65165	4.174331	3.03	0.003	4.436858	20.86644

```
. estat ovtest
Ramsey RESET test using powers of the fitted values of sunspots
      Ho:  model has no omitted variables
              F(3, 296) =        9.12
              Prob > F =       0.0000
```

The significance of the autocorrelation at lag 9 provides support for including 9 lags under the null hypothesis. The RESET statistic is $F_{(3,296)} = 9.12$ with a p-value of 0.000, which indicates the strong rejection of the null hypothesis of linearity in favor of the unspecified nonlinear model under the alternative hypothesis. Limited experimentation will show that the significance of the RESET test statistic is fairly robust to the choice of lag length.

The V23 test for the sunspot data can be implemented using the v23 command distributed with this text and specifying the desired lag structure. Implementing the V23 test after fitting an AR(9) model under the null hypotheses yields the following results:

```
. v23 sunspots, lag(9)
H0: linear specification of sunspots = AR(9)
Second Order Cross Products:  F(45,89) =    0.95 (p = 0.5626)
Third Order Cross Products:   F(165,89) =   1.29 (p = 0.0912)
V23 Test:                     F(210,89) =   2.02 (p = 0.0001)
```

The V23 test produces the test statistic $F_{(210,89)} = 2.02$ with a p-value of 0.0001, which also indicates a strong rejection of the null hypothesis of linearity. There is therefore solid evidence from both the RESET and V23 tests to show that the sunspot time series has nonlinear features that a simple autoregressive (AR) model cannot capture. Note, however, that the full V23 model that includes all the second- and third-order cross-product terms is particularly expensive in terms of degrees of freedom. The test regression has 210 coefficients as opposed to just 3 for the RESET test. Notwithstanding this disadvantage, if sample size is not an issue, the V23 test has been shown to have reasonable power against unspecified alternatives (Lee, White, and Granger 1993). A further advantage of the V23 approach is that some information can be provided in terms of the separate tests on the second- and third-order cross products. In this instance, the results of the V23 test suggest that it is the third-order terms that capture most of the nonlinear behavior.

10.3 Bilinear time-series models

Bilinear time-series models constitute a class of nonlinear models where the linear autoregressive moving-average (ARMA) model is augmented by the product of the AR and moving-average (MA) terms (Granger and Anderson 1978). Consider the ARMA(1,1) model

$$y_t = \phi y_{t-1} + v_t + \psi v_{t-1} \tag{10.2}$$

where v_t is an independently and identically distributed (i.i.d.) disturbance term with mean zero and constant variance. A simple bilinear model introduces nonlinearity through an additional term, which is simply the product of the AR(1) and MA(1) components of the model

$$y_t = \phi y_{t-1} + v_t + \psi v_{t-1} + \gamma y_{t-1} v_{t-1}$$

This additional component is known as the bilinear term, which has the effect of generating a range of interesting nonlinear features that cannot be generated from the linear ARMA(1,1) model in (10.2). For example, in the current model specification, the process y_t will mostly behave linearly but will suddenly exhibit seemingly spontaneous outbursts of activity. This bursting behavior is driven by the nonlinear product term in (10.2) and will become more pronounced as γ grows.

Simulated time series of size $T = 500$ from the bilinear model

$$y_t = 0.4 y_{t-1} + v_t + 0.2 v_{t-1} + \gamma y_{t-1} v_{t-1}, \qquad v_t \sim \text{i.i.d. } N(0,1)$$

are given in figure 10.3 for alternative values of the bilinear parameter γ. For $\gamma = 0.4$, y_t exhibits the typical pattern of a linear ARMA(1,1) model, but for larger values of this parameter, $\gamma = 1$, the model exhibits bursting behavior.

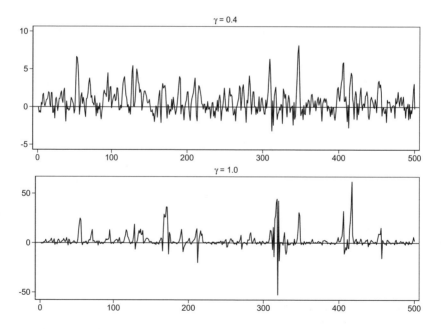

Figure 10.3. Simulated bilinear time series of length $T = 500$ for different values of the bilinear parameter, γ. The AR parameter is set at 0.4 and the MA parameter at 0.2.

A bilinear time-series model can be fit using maximum likelihood methods, but the bursting behavior of the model can sometimes cause difficulties with convergence. A

10.3 Bilinear time-series models

less efficient but perhaps more stable approach to fitting the bilinear model is to use a two-stage procedure suggested by Durbin (1959) for fitting MA models. Consider the bilinear model given by

$$y_t = \phi y_{t-1} + v_t + \gamma y_{t-1} v_{t-1}, \qquad v_t \sim \text{i.i.d. } N(0, \sigma^2)$$

Step 1: Fit a high-order linear AR model by ordinary least squares, and compute the estimated residuals, $\hat{v}_t$.

Step 2: Fit the bilinear model using $\hat{v}_{t-1}$ to construct the bilinear term.

Using this two-step estimation procedure, a simple test of bilinearity can be performed by computing the F statistic of the restriction that the coefficient of $y_{t-1}\hat{v}_{t-1}$ is zero. The test is easily extended to include a longer lag structure and additional bilinear terms.

Consider the following bilinear model for the sunspot data estimated by Battaglia and Orfei (2002).

$$y_t = \sum_{i=1}^{3} \phi_i y_{t-i} + v_t + \sum_{i=1}^{3}\sum_{j=1}^{4} \gamma y_{t-i} v_{t-i} \tag{10.3}$$

In implementing Durbin's two-step procedure, a lag length of 12 is adopted in the first stage. The use of `varsoc` for these data in the previous section indicated a choice of 9 lags. The choice of 12 lags is long enough to ensure that there are several redundant lags in the specification as required by the Durbin procedure while maintaining adequate remaining degrees of freedom. The results from fitting the bilinear model are as follows:

```
. use http://www.stata-press.com/data/eeus/annualsunspots
. tsset datevec, yearly
        time variable:  datevec, 1700 to 2017
                delta:  1 year
. quietly regress sunspots L(1/12).sunspots
. quietly predict double v, residuals
```

```
. regress sunspots L(1/3).sunspots c.L(1/3)sunspots#c.L(1/4)v, vsquish
```

Source	SS	df	MS	Number of obs	=	302
				F(15, 286)	=	150.83
Model	1031447.31	15	68763.1539	Prob > F	=	0.0000
Residual	130384.848	286	455.891077	R-squared	=	0.8878
				Adj R-squared	=	0.8819
Total	1161832.16	301	3859.9075	Root MSE	=	21.352

sunspots	Coef.	Std. Err.	t	P>|t|	[95% Conf. Interval]	
sunspots						
L1.	2.1854	.1141701	19.14	0.000	1.96068	2.41012
L2.	-1.824648	.1802456	-10.12	0.000	-2.179424	-1.469872
L3.	.4658218	.0995358	4.68	0.000	.269906	.6617375
cL.sunspots#						
cL.v	-.002381	.001273	-1.87	0.062	-.0048865	.0001246
cL.sunspots#						
cL2.v	.0041495	.0024304	1.71	0.089	-.0006342	.0089332
cL.sunspots#						
cL3.v	.0050011	.0023037	2.17	0.031	.0004667	.0095354
cL.sunspots#						
cL4.v	.006709	.0025324	2.65	0.009	.0017245	.0116934
cL2.sunspots#						
cL.v	-.0075401	.0028042	-2.69	0.008	-.0130595	-.0020207
cL2.sunspots#						
cL2.v	-.0070296	.0037585	-1.87	0.062	-.0144274	.0003683
cL2.sunspots#						
cL3.v	-.0098637	.0032095	-3.07	0.002	-.0161809	-.0035464
cL2.sunspots#						
cL4.v	-.009376	.004058	-2.31	0.022	-.0173634	-.0013887
cL3.sunspots#						
cL.v	-.0017487	.002298	-0.76	0.447	-.0062718	.0027745
cL3.sunspots#						
cL2.v	.0042801	.0019987	2.14	0.033	.0003461	.0082141
cL3.sunspots#						
cL3.v	.0058513	.0019277	3.04	0.003	.0020569	.0096457
cL3.sunspots#						
cL4.v	.0053556	.0023263	2.30	0.022	.0007767	.0099345
_cons	16.93875	3.201727	5.29	0.000	10.63682	23.24069

The AR(3) terms in the linear part of the specification are all highly significant, and several of the bilinear terms are also significant. The fitted model can now be used to test for nonlinearity by computing the F test of zero restrictions on all the bilinear terms.

10.3 Bilinear time-series models

```
. testparm c.L(1/3)sunspots#c.L(1/4)v

 ( 1)  cL.sunspots#cL.v = 0
 ( 2)  cL.sunspots#cL2.v = 0
 ( 3)  cL.sunspots#cL3.v = 0
 ( 4)  cL.sunspots#cL4.v = 0
 ( 5)  cL2.sunspots#cL.v = 0
 ( 6)  cL2.sunspots#cL2.v = 0
 ( 7)  cL2.sunspots#cL3.v = 0
 ( 8)  cL2.sunspots#cL4.v = 0
 ( 9)  cL3.sunspots#cL.v = 0
 (10)  cL3.sunspots#cL2.v = 0
 (11)  cL3.sunspots#cL3.v = 0
 (12)  cL3.sunspots#cL4.v = 0

       F( 12,   286) =   12.06
            Prob > F =   0.0000
```

The $F_{(12,286)}$ test statistic is 12.06 with a p-value of 0.000. This test result indicates that a linear AR specification would suffer from neglected nonlinearity, and the bilinear terms help to alleviate this problem. The actual and fitted values obtained from the model are plotted in figure 10.4. Although the fitted values track the actual sunspot series quite well, it can be seen that the model slightly underfits at the turning points.

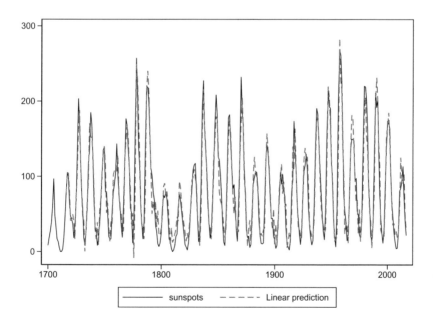

Figure 10.4. Actual and fitted values of the bilinear model in (10.3). The parameter estimates are obtained using the Durbin two-stage ordinary least-squares approach.

Although bilinear processes offer an avenue for modeling nonlinear time series, the practical question remains that the lag structure of the model that provides a reasonable fit to the data must be determined. As Poskitt and Tremayne (1986) point out, using information criteria for this purpose generally has differing results dependent on the choice of information criterion. Even more problematic is the sheer number of alternative models that must be fit to reach a decision. These difficulties, together with the lack of a theoretical structure to aid interpretation of the results, means that bilinear models have not been widely used in applied econometrics. The inclusion of the discussion of bilinear models is motivated instead by the fact that the bilinear model may be regarded as a simplification of the ideas developed by Volterra (1930) and Wiener (1958) that are often regarded as the origin of nonlinear time-series analysis.

10.4 Threshold autoregressive models

Models with more than one regime have a long history in econometrics. Quandt (1958) considered a switching regression model in which the regression equation, including the error term, switches according to a random variable. Autoregressive models in which it is assumed that the dependent variable, y_t, is governed by more than a single regime at any point in time are known as threshold autoregressive (TAR) models. Univariate TAR models are usually attributed to the work of Tong and Lim (1980).

Without loss of generality, consider a pure time-series model for y_t in which the lag order is limited to one. In the case of two regimes where each submodel has an AR(1) form, the model is expressed as

$$y_t = \alpha_0 + \alpha_1 y_{t-1} + (\beta_0 + \beta_1 y_{t-1})I_t + v_t \qquad (10.4)$$

in which the variable I_t is the indicator function defined by

$$I_t = \begin{cases} 1 & s_t \geq k \\ 0 & s_t < k \end{cases}$$

where k is some constant parameter known as the threshold, and s_t is a transition variable. The fundamental property of the TAR model is that the switch from one regime to another is instantaneous at the point k. In the case of a pure time-series model, as is the case considered here, s_t is taken to be y_{t-d}, where $d > 0$ is an integer known as the delay parameter, so that

$$I_t = \begin{cases} 1 & y_{t-d} \geq k \\ 0 & y_{t-d} < k \end{cases} \qquad (10.5)$$

In these circumstances, the model is known as a self-exciting threshold autoregressive (SETAR) model. Although this is a particularly simple specification, it is straightforward to extend the model to allow for additional lags and also additional explanatory variables. The essential idea underlying the class of TAR models is the piecewise linearization of nonlinear models over the state space by the introduction of the thresholds,

10.4 Threshold autoregressive models

implying that these models are locally linear. Although fairly simple in construction, the SETAR model can capture a wide variety of nonlinear behavior often encountered in practice such as periodicity of the kind observed in the sunspot data.

Performing a test for threshold nonlinearity raises a few problems. Intuitively, the way to proceed is to test (10.4) to see if it reduces to the linear AR model

$$y_t = \alpha_0 + \alpha_1 y_{t-1} + v_t$$

This approach is not as simple as it seems. A linear model would require that $\beta_1 = 0$, but imposing these restrictions under the null hypothesis means that the threshold parameter k is not identified. As a consequence, it is difficult to derive the distribution of the test statistic under the null hypothesis. If the threshold k and the delay parameter d are known, then a test for linearity against the alternative of a threshold nonlinearity in a model of the form in (10.4) would take the following form:

1. Fit the model under the null hypothesis by regressing y_t on y_{t-1}, and compute the ordinary least-squares residuals $\widehat{v}_t$.
2. For a given threshold k and delay d, construct $I_t(y_{t-d} > k)$.
3. Regress $\widehat{v}_t$ on $\{y_{t-1}, y_{t-1} I_t\}$.
4. Either compute the F statistic with the restriction that the coefficient on $y_{t-1} I_t$ is zero, or compute TR^2 from the auxiliary regression, where T is the sample size, and compare with a χ_1^2 distribution. The null hypothesis is rejected for large values of the test statistic.

Of course, this statistic will be valid only for this particular model. The general problem of testing in the presence of unidentified parameters was considered by Davies (1987) and in the context of threshold models by Hansen (1996).

The value of the threshold, k, is generally not known and must be estimated as a parameter. The model is nonlinear in parameters and cannot be fit directly by ordinary least squares. It has been shown that searching over potential values of the threshold so as to minimize the sum of squared errors from the fitted model yields a consistent estimate of the threshold, although this procedure does require that the SETAR model be specified before the search for the threshold takes place. The Stata command **threshold** (see [TS] **threshold**) fits threshold models including the SETAR specification. The model may have multiple thresholds, and it is possible to either specify a known number of thresholds or let **threshold** find the optimum number of thresholds. In SETAR modeling, it is also possible to use **threshold** to find the optimum delay parameter, d.

Many different flavors of SETAR models have been proposed in the literature analyzing the sunspot data. The model implemented here is

$$y_t = \begin{cases} \alpha_0 + \alpha_1 y_{t-1} + \alpha_2 y_{t-2} + v_{1t} & \text{if } y_{t-d} \leq k \\ \beta_0 + \beta_1 y_{t-1} + \beta_2 y_{t-2} + v_{2t} & \text{if } y_{t-d} > k \end{cases} \quad (10.6)$$

Using $d = \{2, 3, 4\}$ and searching between approximately the 10th and 90th percentiles of the sunspot series, the output obtained is as follows:

```
. quietly threshold sunspots, threshvar(L2.sunspots) regionvars(L(1/2).sunspots)
. estimates store Model1
. quietly threshold sunspots, threshvar(L3.sunspots) regionvars(L(1/2).sunspots)
. estimates store Model2
. quietly threshold sunspots, threshvar(L4.sunspots) regionvars(L(1/2).sunspots)
. estimates store Model3
. estimates table Model1 Model2 Model3, stats(ssr aic bic hqic)
> star(0.1 0.05 0.01)
```

Variable	Model1	Model2	Model3
Region1			
sunspots			
L1.	1.6733258***	1.5770358***	1.3540467***
L2.	-1.1051646***	-1.0577137***	-.68563658***
_cons	28.08215***	32.287646***	32.365009***
Region2			
sunspots			
L1.	1.0712287***	1.0259305***	.79664163***
L2.	-.25985826***	-.20112898***	-.10492356
_cons	1.4788222	-3.107741	4.6169736
Statistics			
ssr	161625.47	163485.04	176861.28
aic	1982.9852	1981.3498	2000.7906
bic	2005.5196	2003.8652	2023.2869
hqic	1991.9875	1990.3456	2009.7797

legend: * p<.1; ** p<.05; *** p<.01

The information criteria are all in agreement that model 2 with a delay of $d = 3$ is the best model. Fitting the model with $d = 3$ to identify the threshold yields

10.4 Threshold autoregressive models

```
. threshold sunspots, threshvar(L3.sunspots) regionvars(L(1/2).sunspots)
Searching for threshold: 1
(Running 252 regressions)
................................................   50
................................................  100
................................................  150
................................................  200
................................................  250
..
Threshold regression
                                           Number of obs    =         315
Full sample:    1703 - 2017                AIC              =   1981.3498
Number of thresholds =  1                  BIC              =   2003.8652
Threshold variable: L3.sunspots            HQIC             =   1990.3456
```

Order	Threshold	SSR
1	60.7000	1.635e+05

sunspots	Coef.	Std. Err.	z	P>|z|	[95% Conf. Interval]	
Region1						
sunspots						
L1.	1.577036	.0611753	25.78	0.000	1.457134	1.696937
L2.	-1.057714	.0999884	-10.58	0.000	-1.253687	-.8617401
_cons	32.28765	2.758035	11.71	0.000	26.882	37.6933
Region2						
sunspots						
L1.	1.02593	.0625521	16.40	0.000	.9033306	1.14853
L2.	-.201129	.0663453	-3.03	0.002	-.3311634	-.0710946
_cons	-3.107741	4.244403	-0.73	0.464	-11.42662	5.211136

The optimal threshold value returned by the search is $k = 60.7$, which is slightly larger than the values reported by Tong and Lim (1980) and Battaglia and Orfei (2002), although they use annual data over a shorter sample period in their work and include a much more complex dynamic structure. In regime 1, when the annual sunspot number is below the threshold, there are equal and opposite signed AR effects, a result that suggests the process returns to the threshold in one period. The dynamic structure for regime 2 is more complex. The large positive first-order autocorrelation coefficient is offset by a much smaller negative second-order autocorrelation coefficient, suggesting that the process persists a little longer in regime 2. This tentative interpretation seems to be supported by the plot of the actual and fitted values in figure 10.5.

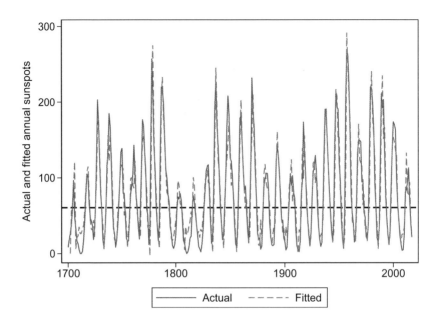

Figure 10.5. The actual and fitted values of obtained from fitting the model in (10.6) using the sunspot data. The delay parameter is $d = 3$. The estimate of the threshold of 60.7 is shown as the horizontal dashed line.

10.5 Smooth transition models

Threshold time-series models assume that the dependent variable, y_t, is governed by a single regime at each point in time. Without loss of generality, consider a pure time-series model for y_t in which the lag order is limited to 1. In the case of two regimes where each submodel has an AR(1) form, the model is expressed as

$$y_t = \alpha_0 + \alpha_1 y_{t-1} + (\beta_0 + \beta_1 y_{t-1})w_t + v_t$$

in which the variable w_t represents a time-varying weighting function with the property $0 \leq w_t \leq 1$. Although this is a particularly simple specification, it is straightforward to extend the model to allow for additional lags and also additional explanatory variables, x_t.

Rather than the abrupt change between regimes implied by the indicator function defined in (10.5), it may be that the change is more gradual. Bacon and Watts (1971) argued that instead of having an abrupt shift from one regime to the other, one could make the transition smooth. In the econometrics literature, Goldfeld and Quandt (1972, 263–264) independently presented the smooth transition regression model as a solution to the estimation problem in the switching regression model. The AR counterpart of

10.5 Smooth transition models

the smooth transition model was introduced by Chan and Tong (1986) but is mainly associated with the work of Teräsvirta (1994b).

There are two popular versions of the smooth transition model.

1. Logistic smooth transition (LSTAR)

$$w_t = \frac{1}{1 + \exp\{-\gamma(s_t - k)\}}, \qquad \gamma > 0$$

2. Exponential smooth transition (ESTAR)

$$w_t = 1 - \exp\{-\gamma(s_t - k)^2\}, \qquad \gamma > 0$$

In both models, k is the threshold, $\gamma > 0$ governs the speed of the transition from one regime to another, and s_t is known as the transition variable. In all the applications considered here, the transition variable, s_t, is taken to be the lagged dependent variable y_{t-d}. As in previous sections, d is known as the delay.

Figure 10.6 plots the transition functions for these two models with $\gamma = 5$, $k = 0$, and the transition variable, y_{t-d}, taken to lie in $[-2, 2]$. It is immediately apparent that the type of transitions implied by the two functions are quite different. The LSTAR model associates the two regimes with small and large values of the transition variable. As γ gets larger, the change becomes almost instantaneous and approaches the simple threshold model. When $\gamma = 0$, the LSTAR model reduces to the linear model. On the other hand, the two regimes in the ESTAR model are associated with small and large absolute values of y_{t-d}. A drawback of the ESTAR specification is that for both $\gamma \to 0$ and $\gamma \to \infty$, the function becomes a constant (0 or 1, respectively), and the model is then a linear one.

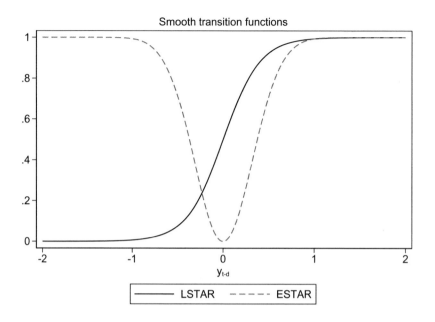

Figure 10.6. Alternative threshold functions used in TAR models: $\gamma = 5$, $k = 0$, and s_t taken to lie in the interval $[-2, 2]$

Performing a test of linearity when the alternative is a smooth transition model raises the same problems of unidentified parameters under the null hypothesis as for the simple threshold model. Imposing the restrictions required for a linear model would mean that the parameters γ and k are not identified. Luukkonen, Saikkonen, and Teräsvirta (1988) show that in the LSTAR case with $s_t = y_{t-d}$, the local behavior of the weighting function in the vicinity of $\gamma = 0$ is approximated by the cubic polynomial

$$w_t \approx \delta_0 + \delta_1 y_{t-d} + \delta_2 y_{t-d}^2 + \delta_3 y_{t-d}^3 \tag{10.7}$$

Substituting this cubic for w_t in (10.4) and simplifying yields

$$y_t = \phi_0 + \phi_1 y_{t-1} + \varphi_1 y_{t-1} y_{t-d} + \varphi_2 y_{t-1} y_{t-d}^2 + \varphi_3 y_{t-i} y_{t-d}^3 + u_t$$

where the disturbance term u_t is a function of v_t in (10.4) and the approximation error associated with using the approximation in (10.7). When d is known, the null hypothesis requires that $\varphi_1 = \varphi_2 = \varphi_3 = 0$. Testing this hypothesis guarantees power against both LSTAR and ESTAR simultaneously.

Teräsvirta (1994b) suggests the following modeling cycle for smooth transition models.

Step 1: Choose the appropriate linear model for y_t.

10.5 Smooth transition models

Step 2: Assuming without loss of generality that the simple AR(1) model adopted earlier is appropriate, the model in (10.7) provides the relevant test regression.

Step 3: Assuming that y_{t-d} is the correct transition variable, test linearity for different values of the delay parameter, d, by testing $H_0 : \varphi_1 = \varphi_2 = \varphi_3 = 0$. Assuming that linearity is rejected, take d to be that value of the parameter for which the p-value of the test is minimized. The reason for this is that if there is a correct d among the alternatives considered, the power of the test is maximized against it.

Step 4: Choose between LSTAR and ESTAR transition functions using the following sequence of tests:

- Test $H_{03}:\ \varphi_3 = 0$.
- Test $H_{02}:\ \varphi_2 = 0 | \varphi_3 = 0$.
- Test $H_{01}:\ \varphi_1 = 0 | \varphi_2 = 0 | \varphi_3 = 0$.

If the p-value for rejecting H_{02} is less than that of the two other tests, choose an ESTAR model. Otherwise, choose an LSTAR model. The foundation for this suggestion is the work of Saikkonen and Luukkonen (1988), who derive a test against the ESTAR alternative. They show the approximation in (10.7) need only use terms up to y_{t-d}^2 to capture the dynamics of the alternative.

The sunspot numbers have frequently been transformed prior to the estimation of various time-series models. One of the more popular transformations is that proposed by Ghaddar and Tong (1981). If y_t represents the original sunspot series, the proposed transformation is given by

$$x_t = 2\left(\sqrt{y_t + 1} - 1\right) \tag{10.8}$$

Ghaddar and Tong (1981) choose this transformation after a search, trying to make the distribution of the data being modeled more accurately approximate a Gaussian distribution. As an illustration, this transformation will be adopted for the smooth transition approach to modeling the sunspot data.

As in the previous sections, a simple application of `varsoc` based on a maximum lag order of 12 will show that even when transformed, the appropriate linear model for the sunspot data is an AR(9). The next step is to determine the appropriate delay parameter. The analysis is slightly more complicated given the AR(9) under the null, but the additional terms are easily dealt with using Stata's factor notation.

```
. use http://www.stata-press.com/data/eeus/annualsunspots, clear
. tsset datevec
        time variable:  datevec, 1700 to 2017
                delta:  1 unit
. generate y = 2*(sqrt(sunspots+1)-1)
. forvalues d = 1/4 {
  2.        capture drop st
  3.        capture drop st2
  4.        capture drop st3
  5.        quietly generate st  = L`d´.y
  6.        quietly generate st2 = st^2
  7.        quietly generate st3 = st^3
  8.        quietly regress y L(1/9)y c.L(1/9)y#c.st c.L(1/9)y#c.st2 c.L(1/9)y#c.st3
  9.        quietly testparm c.L(1/9)y#c.st c.L(1/9)y#c.st2 c.L(1/9)y#c.st3
 10.        display "Delay  " `d´  "   p-value    " r(p)
 11. }
Delay 1   p-value    .02053377
Delay 2   p-value    2.206e-06
Delay 3   p-value    .00023901
Delay 4   p-value    .00133907
```

The optimal value of the d is found to be 2. Step 3 requires the sequential testing for the appropriateness of the LSTAR or ESTAR are as follows:

```
. capture drop st
. capture drop st2
. capture drop st3
. quietly generate st  = L2.y
. quietly generate st2 = st^2
. quietly generate st3 = st^3
. quietly regress y L(1/9)y c.L(1/9)y#c.st c.L(1/9)y#c.st2 c.L(1/9)y#c.st3
. quietly testparm c.L(1/9)y#c.st3
. display "F03 = "  "  r(F)  "   p-value    " r(p)
F03 =   .81863884    p-value   .59940713
. quietly regress y L(1/9)y c.L(1/9)y#c.st c.L(1/9)y#c.st2
. quietly testparm c.L(1/9)y#c.st2
. display "F02 = "  "  r(F)  "   p-value    " r(p)
F02 =   2.240294     p-value   .01977215
. quietly regress y L(1/9)y c.L(1/9)y#c.st
. quietly testparm c.L(1/9)y#c.st
. display "F01 = "  "  r(F)  "   p-value    " r(p)
F01 =   5.8753797    p-value   1.556e-07
```

The strongest rejection is for H_{01}, indicating that an LSTAR transition function is appropriate.

10.5 Smooth transition models

The problem now becomes one of fitting an appropriate LSTAR model for the sunspot data based on a delay parameter $d = 2$. Assuming a normal disturbance term, $v_t \sim$ i.i.d. $N(0, \sigma^2)$, and assuming further that $s_t = y_{t-d}$ with d is known, the conditional log-likelihood function at time t is

$$\log L_t = -\frac{1}{2}\ln 2\pi\sigma^2 - \frac{1}{2}\frac{(y_t - \alpha_0 - \alpha_1 y_{t-1} - (\beta_0 - \beta_1 y_{t-1})w_t)^2}{\sigma^2}$$

which is to be maximized. Even when d is known, this estimation is not trivial, as was first pointed out by Haggan and Ozaki (1981) in the case of an ESTAR model and by Bates and Watts (1988) and Seber and Wild (1989) in the case of LSTAR dynamics. Consequently, there are several practical steps to take in fitting smooth transition models. First, it is useful to scale γ by the standard deviation of the transition variable σ_y^2. In this standardized form, a useful starting value is $\gamma = 1$. Thereafter, it is useful to rely on nonlinear least-squares estimation to obtain estimates of the γ and c for given values of the other parameters and use a zigzag procedure to find the final parameter estimates.

Because there is no theory available to help with the specification of the dynamics, the model used by Teräsvirta (1994a) will be adopted here for illustrative purposes. The model is given by

$$w_t = \left[1 + \exp\left\{-\frac{\gamma}{\sigma_x^2}(x_{t-2} - k)\right\}\right]^{-1}, \quad \gamma > 0$$

$$x_t = \alpha_1 x_{t-1} + \alpha_2 x_{t-2} + \alpha_7 x_{t-7}$$
$$+ (\beta_0 + \beta_1 x_{t-1} + \beta_2 x_{t-2} + \beta_3 x_{t-3} + \beta_7 x_{t-7} + \beta_8 x_{t-8} + \beta_9 x_{t-9} + \beta_{11} x_{t-11})w_t$$

where x_t refers to the sunspot numbers transformed according to (10.8). Taking the delay parameter as $d = 2$, the starting values of $\gamma = 4$ and $k = 8$ for the transition function are taken from Teräsvirta (1994a). Starting values for dynamic terms, α_i and β_i, are easily obtained from a regression in which the transition function is evaluated at its starting values. The full model fits obtained from the Stata `nl` (see [R] **nl**) command are as follows:

```
. quietly summarize L2.y
. global ss = 1/r(sd)
. // transition function
. generate w = 1/( 1 + exp(-4.7*$ss*(L2.y - 7.8)))
(2 missing values generated)
. // OLS starting values
. quietly regress y L.y L2.y L7.y w c.L1.y#c.w c.L2.y#c.w c.L3.y#c.w
> c.L7.y#c.w c.L8.y#c.w c.L9.y#c.w c.L11.y#c.w, noco
. global a1   = _b[L1.y]
. global a2   = _b[L2.y]
. global a7   = _b[L7.y]
. global b0   = _b[w]
. global b1   = _b[cL1.y#c.w]
```

```
. global b2  = _b[cL2.y#c.w]
. global b3  = _b[cL3.y#c.w]
. global b7  = _b[cL7.y#c.w]
. global b8  = _b[cL8.y#c.w]
. global b9  = _b[cL9.y#c.w]
. global b11 = _b[cL11.y#c.w]
. generate l1y  = L.y
(1 missing value generated)
. generate l2y  = L2.y
(2 missing values generated)
. generate l3y  = L3.y
(3 missing values generated)
. generate l7y  = L7.y
(7 missing values generated)
. generate l8y  = L8.y
(8 missing values generated)
. generate l9y  = L9.y
(9 missing values generated)
. generate l11y = L11.y
(11 missing values generated)
. dropmiss, any obs force
(11 observations deleted)
. nl (y = {a1}*l1y + {a2}*l2y + {a7}*l7y
> + ({b0}+{b1}*l1y+{b2}*l2y+{b3}*l3y+{b7}*l7y+{b8}*l8y+{b9}*l9y+{b11}*l11y)*
> (1/( 1 + exp(-{gam} *$ss *(l2y - {c})))))),
> variables(y) nolog vsquish
> init(gam 4 c 8 a1 $a1 a2 $a2 a7 $a7 b0 $b0 b1 $b1 b2 $b2
> b3 $b3 b7 $b7 b8 $b8 b9 $b9 b11 $b11)
```
  (output omitted)

10.5 Smooth transition models

```
. estimates replay

active results
```

Source	SS	df	MS			
				Number of obs =		307
Model	79644.085	13	6126.46811	R-squared =		0.9792
Residual	1694.1799	294	5.76251679	Adj R-squared =		0.9783
				Root MSE =		2.400524
Total	81338.265	307	264.94549	Res. dev. =		1395.617

y	Coef.	Std. Err.	t	P>|t|	[95% Conf. Interval]	
/a1	1.566445	.0732736	21.38	0.000	1.422238	1.710653
/a2	-.9056647	.1531199	-5.91	0.000	-1.207015	-.6043147
/a7	.2854469	.03896	7.33	0.000	.2087709	.3621228
/b0	3.340136	1.079076	3.10	0.002	1.216444	5.463828
/b1	-.7014796	.1209454	-5.80	0.000	-.9395081	-.463451
/b2	.9247369	.1925651	4.80	0.000	.5457561	1.303718
/b3	-.321229	.0696922	-4.61	0.000	-.4583877	-.1840702
/b7	-.2633054	.0760153	-3.46	0.001	-.4129084	-.1137023
/b8	-.1530382	.1010696	-1.51	0.131	-.3519499	.0458734
/b9	.1773336	.0826892	2.14	0.033	.0145958	.3400714
/b11	.1232929	.0497386	2.48	0.014	.0254041	.2211817
/gam	4.956032	2.312002	2.14	0.033	.4058596	9.506204
/c	9.925958	.8467419	11.72	0.000	8.259514	11.5924

The estimates of the dynamic terms are all statistically significant, providing support for this particular choice of dynamic specification. Teräsvirta (1994a) finds that the data support the restrictions $\alpha_2 = -\beta_2$ and $\alpha_7 = -\beta_7$. Testing these restrictions gives the following output:

```
. test /a2+/b2=0

 ( 1)  [a2]_cons + [b2]_cons = 0

       F(  1,   294) =    0.02
            Prob > F =    0.8751
. test /a7+/b7=0

 ( 1)  [a7]_cons + [b7]_cons = 0

       F(  1,   294) =    0.13
            Prob > F =    0.7214
```

In both cases, the data do not reject these restrictions, suggesting that they could be imposed in the estimation. At one extreme, when the transition function approaches 0, the AR(3) appears to be stationary but persistent because the sum of the α_i is approximately equal to 0.93. At the other when the transition function approaches 1, the dynamics are more complex. The transition parameter γ is relatively small, indicating that the function switches between 0 and 1 relatively smoothly. This result is suggestive that the LSTAR model is perhaps to be favored over the simple SETAR model. Figure 10.7 plots the actual and fitted values obtained from the LSTAR model.

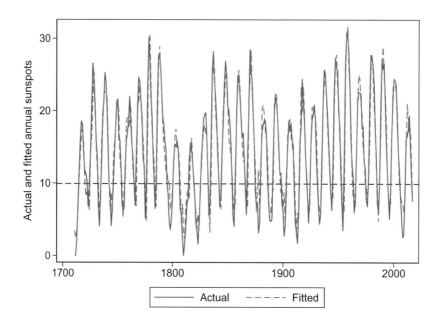

Figure 10.7. Actual and fitted values from the fitted LSTAR model for sunspot numbers proposed by Teräsvirta (1994a). The estimated threshold is shown as the horizontal dashed line.

By comparison with figure 10.5, which plots the actual and fitted values of the SETAR model, not much improvement appears to have been gained by the extra complexity of the LSTAR model. What is noticeable, however, is a visible improvement in fit at the beginning of the sample and also in the troughs immediately after 1800. The LSTAR model appears to capture the turning points of the troughs and the quick return to the threshold slightly better.

10.6 Markov switching models

The Markov switching model due to Hamilton (1989) is widely used to model nonlinearities in time series. This model has much in common with the threshold time-series models discussed previously, except in those models, switching is deterministic, whereas in the Markov switching model, it is stochastic. Let w_t, represent a stochastic weighting variable that switches between regimes according to

$$w_t = \begin{cases} 1: & \text{Regime 1} \\ 0: & \text{Regime 2} \end{cases}$$

10.6 Markov switching models

Based on this binary indicator, consider the model

$$y_t = \alpha + \beta w_t + z_t \sqrt{\gamma + \delta w_t} \qquad z_t \sim N(0,1) \qquad (10.9)$$

It follows that the behavior of y_t in the two states is given by

$E(y_t\|w_t = 1) = \alpha + \beta$	$\text{var}(y_t\|w_t = 1) = \gamma + \delta$	Regime 1
$E(y_t\|w_t = 0) = \alpha$	$\text{var}(y_t\|w_t = 0) = \gamma$	Regime 2

Unlike the threshold time-series models, the weighting function is not identified by expressing it in terms of lagged observed variables but is unobservable. Instead, w_t is specified in terms of the parameters p and q, which represent the conditional probabilities of staying in each regime

$$\Pr(w_t = 1 \mid w_{t-1} = 1, y_{t-1}, y_{t-2}, \ldots) = p$$
$$\Pr(w_t = 0 \mid w_{t-1} = 0, y_{t-1}, y_{t-2}, \ldots) = q$$

The unknown parameters of the model are $\theta = \{\alpha, \beta, \gamma, \delta, p, q\}$, and estimation by maximum likelihood requires the specification of the joint conditional distribution, which is denoted $f(y_t, w_t \mid y_{t-1}, y_{t-2}, \ldots; \theta)$, a problem that is complicated by the fact that w_t is not observed. The full details of the estimation procedures used in Markov switching models will not be spelled out here, but a good starting point is to consult the Stata documentation, [TS] **mswitch**, which also contains a full set of references.

A sketch of the three steps to constructing the log-likelihood function for this simple model is as follows:

1. Specify the rule for determining w_t at time t.

    The major assumption that is made is that w_t is a first-order Markov process, so its dynamics depend only on its value in the previous period. Let $w_{1,t+1|t}$ represent w_t taking the value 1 at time $t+1$ conditional on its value at time t and similarly for $w_{0,t+1|t}$. The first-order Markov process assumption means that the evolution of the weights is completely described by

$$\begin{pmatrix} w_{1,t+1|t} \\ w_{0,t+1|t} \end{pmatrix} = \begin{pmatrix} p & 1-q \\ 1-p & q \end{pmatrix} \begin{pmatrix} w_{1,t|t} \\ w_{0,t|t} \end{pmatrix}$$

At $t=1$, it is necessary to be able to evaluate $w_{1,1|0}$ and $w_{0,1|0}$, which are set equal to their respective stationary probabilities

$$w_{1,1|0} = \frac{1-q}{1-p+1-q}, \qquad w_{0,1|0} = \frac{1-p}{1-p+1-q} \qquad (10.10)$$

2. Specify the conditional distribution of y_t in each regime.
   From (10.9), in which the disturbances are assumed to be normal, it follows that

   $$f(y_t|w_t=1, y_{t-1}, \ldots; \theta) = \frac{1}{\sqrt{2\pi(\gamma+\delta)}} \exp\left(-\frac{(y_t - \alpha - \beta)^2}{2(\gamma+\delta)}\right)$$

   $$f(y_t|w_t=0, y_{t-1}, \ldots; \theta) = \frac{1}{\sqrt{2\pi\gamma}} \exp\left(-\frac{(y_t - \alpha)^2}{2\gamma}\right)$$

   where $f(y_t|w_t=1, y_{t-1}, \ldots)$ is the probability density function of y_t conditional on $w_t = 1$ and all past values of y_t.

3. Construct the log-likelihood function.
   The log-likelihood function now takes the form of the mixture of the two conditional distributions from step 2 with the weights given by the values of w_t obtained in step 1, $w_{1,t+1|t}$ and $w_{0,t+1|t}$, respectively. The log-likelihood function for observation t is given by

   $$\ln L_t = w_{1,t|t-1} f(y_t|w_t=1, y_{t-1}, \ldots; \theta) + w_{0,t|t-1} f(y_t|w_t=0, y_{t-1}, \ldots; \theta)$$

An important extension to this simple model is to allow for k lags

$$y_t = \alpha + \beta w_t + \sum_{i=1}^{k} \phi_i (y_{t-i} - \alpha - \beta w_{t-i}) + \sqrt{\gamma + \delta w_t}\, z_t$$

Further extensions consist of specifying y_t as a vector of variables and allowing for time-varying conditional probabilities p and q by expressing these probabilities as functions of explanatory variables.

In Stata, Markov switching models are fit using the `mswitch` command, which fits dynamic regression models that exhibit different dynamics across unobserved states using state-dependent parameters. Two types of model are available: Markov switching dynamic regression (`mswitch dr`) models, which allow a quick adjustment after the process changes state, and Markov switching autoregression (`mswitch ar`) models, which allow a more gradual adjustment. Given the importance of the AR parameters in modeling the sunspot numbers observed in previous sections, a two-state Markov switching model is proposed that has two AR lags in each regime as well as regime-specific constants and variances.

10.6 Markov switching models

```
. use http://www.stata-press.com/data/eeus/annualsunspots, clear
. tsset datevec
        time variable: datevec, 1700 to 2017
                delta: 1 unit
. mswitch ar sunspots, ar(1/2) arswitch varswitch
Performing EM optimization:

Performing gradient-based optimization:

Iteration 0:   log likelihood = -1447.7668
Iteration 1:   log likelihood = -1445.1909
Iteration 2:   log likelihood = -1444.8145   (not concave)
Iteration 3:   log likelihood = -1444.4902   (not concave)
Iteration 4:   log likelihood = -1444.2062
Iteration 5:   log likelihood = -1443.8271
Iteration 6:   log likelihood =  -1443.745
Iteration 7:   log likelihood = -1443.7445
Iteration 8:   log likelihood = -1443.7445

Markov-switching autoregression
Sample: 1702 - 2017                          Number of obs    =        316
Number of states =    2                      AIC              =     9.2009
Unconditional probabilities: transition      HQIC             =     9.2484
                                             SBIC             =     9.3198

Log likelihood = -1443.7445
```

sunspots	Coef.	Std. Err.	z	P>|z|	[95% Conf. Interval]	
State1						
ar						
L1.	1.590482	.1767283	9.00	0.000	1.244101	1.936863
L2.	-1.201039	.2393344	-5.02	0.000	-1.670126	-.7319523
_cons	45.44981	6.533081	6.96	0.000	32.64521	58.25442
State2						
ar						
L1.	1.451524	.0374072	38.80	0.000	1.378207	1.524841
L2.	-.6803299	.0356886	-19.06	0.000	-.7502781	-.6103816
_cons	70.82314	5.452706	12.99	0.000	60.13603	81.51024
sigma1	42.8272	7.183368			30.82821	59.49643
sigma2	14.96783	1.20295			12.78641	17.52141
p11	.3310541	.1431854			.1223185	.6373343
p21	.1266113	.0300844			.078386	.1981279

The header reports the sample size, fit statistics, number of states, and default method used for computing the unconditional state probabilities, which are from the stationary distribution as in (10.10). The expectation maximization algorithm was used to find the starting values for the optimization algorithm used to estimate the model parameters.[4] Finally, the header reports that the transition method was used to compute the unconditional state probabilities as a function of the transition probabilities.

The results of the estimation are a little easier to interpret once the `mswitch postestimation` (see [TS] **mswitch postestimation**) command `predict` is used to obtain time series of the predicted probabilities of being in a given state. These estimates of the probabilities may be one-step-ahead predictions (the default), filtered estimates using past and contemporaneous information, or smoothed estimates that use all sample information. The smoothed probabilities of being in state 1 and state 2 are shown in figure 10.8.

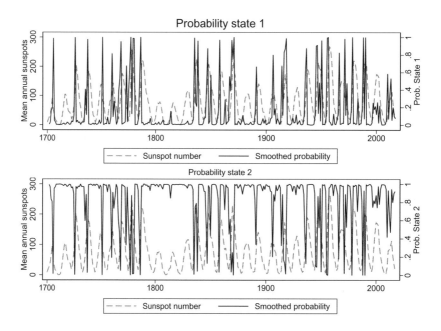

Figure 10.8. The smoothed probabilities of being in state 1 and state 2 for the two-state Markov switching model of mean annual sunspot numbers.

The estimated states are most easily interpreted in terms of variances. State 1 is a high-variance state with a mean annual sunspot number of 45.45 and a variance of

---

4. The expectation maximization algorithm due to Dempster, Laird, and Rubin (1977) is used in situations where there are unobserved variables. The algorithm performs an expectation step, which creates a function for the expectation of the log likelihood evaluated using the current estimate for the parameters, and a maximization step, which computes parameters maximizing the expected log likelihood found in the expectation step.

10.6 Markov switching models

42.83. State 2 is a low-variance state. Although it has a higher mean annual sunspot number of 70.82, the variance of 14.97 is significantly lower than that in state 1. So state 2 refers to periods of relatively low but stable sunspot numbers, while state 1 refers to periods of high variation that correspond to the quick rise and fall of numbers to peak amplitude. Figure 10.8 gives the immediate impression that state 2 is more persistent than state 1. To aid more detailed interpretation of the actual transition probabilities, the matrix of transition probabilities and the expected duration of remaining in each state can be calculated using `mswitch postestimation` commands `estat transition` and `estat duration`.

```
. estat transition
Number of obs = 316
```

Transition Probabilities	Estimate	Std. Err.	[95% Conf. Interval]	
p11	.3310541	.1431854	.1223185	.6373343
p12	.6689459	.1431854	.3626657	.8776815
p21	.1266113	.0300844	.078386	.1981279
p22	.8733887	.0300844	.8018721	.921614

```
. estat duration
Number of obs = 316
```

Expected Duration	Estimate	Std. Err.	[95% Conf. Interval]	
State1	1.494889	.3199756	1.139365	2.757361
State2	7.898189	1.876704	5.047244	12.75739

The parameter $p11$ is the estimated probability of staying in state 1 in the next period given that the process is in state 1 in the current period. The estimate of 0.33 implies that state 1 is not particularly persistent. This conjecture is borne out by the fact that the `estat duration` results estimate the expected duration of the process in state 1 at about 1.5 years. The probability of staying in state 2 is $p22 = 0.87$, which implies that state 2 is more persistent than state 1, with the expected duration in state 2 of 7.8 years. Note that the probability of being in state 1 is given by $p11 + p12$, while the probability of being in state 2 is $p21 + p22$. The smoothed probability of being in the low-variance state 2 obtained from the fit Markov switching model highlights the period around 1800, which is known as the Dalton minimum.[5] This period of low solar activity also corresponded to a period of low global average temperatures, but the relationship between solar activity and global temperature is not well understood.

To conclude, as with the other models used in this chapter to model the sunspot numbers, the Markov switching model yields results that are sensible and interpretable.

---

5. The Dalton minimum was a period of low sunspot count from about 1790 to 1830 representing low solar activity, named after the English meteorologist John Dalton.

Exactly which nonlinear model to choose is therefore a difficult task. One advantage of Markov switching models, probably accounting for their popularity, is that the notion of regimes has a physical interpretation that appeals to applied researchers. However, Markov switching models have long been recognized to suffer from a discrepancy between in-sample and out-of-sample performance. In-sample analysis of Markov switching models often leads to appealing results, but out-of-sample performance, in contrast, is frequently inferior to simple benchmark models. For an interesting analysis of this problem, see Boot and Pick (2018). It seems that the choice boils down to the ultimate goal of the exercise. If understanding in-sample performance is important, then Markov switching models seem the logical choice. If accurate forecasting is also required, then perhaps some other nonlinear model should be considered.

Exercises

1. `sst34.dta` contains monthly data on the NINO3.4 index from January 1870 to December 2017. NINO3.4 is one of several El Niño/Southern Oscillation indicators based on sea surface temperatures. It is the average sea surface temperature anomaly in the region bounded by 5°N to 5°S, from 170°W to 120°W. This region is where changes in local sea surface temperature are important for shifting the large volume of rainfall typically located in the far western Pacific.

    a. The data are supplied in wide form with each row comprising 12 monthly observations. Use the `reshape long` command to reformat the data into a single monthly time series.

    b. An El Niño (La Niña) event is identified if the five-month running-average of the NINO3.4 index is above $+0.4°C$ (is below $-0.4°C$) for at least six consecutive months. Compute and plot the 5-month running average of NINO3.4 index for the period January 1980 to December 2016, together with $\pm 0.4°C$ bands and hence identify the El Niño and La Niña events over the period.

    c. Determine the optimum lag order for a linear autoregression of NINO3.4 based on a maximum possible lag order of 12.

    d. Consider fitting a simple threshold model using a lagged value of the NINO3.4 index as the threshold variable and with the optimum lag order fixed at six lags. Using the method suggested by Teräsvirta (1994b) test for the optimum delay parameter d in the threshold model.

    e. Irrespective of your results in exercise 1c assume that $d = 2$. Fit a simple threshold model that uses y_{t-d} as the threshold variable, sets the value of the threshold at 0, and imposes the restriction that there is only one unrestricted constant in the model. Interpret the results. Now, fit a similar model using the Stata `threshold` command, and comment on the differences in specification.

    f. Once again using the method suggested by Teräsvirta (1994b), test to determine whether an LSTAR model is appropriate for the data.

10.6 Markov switching models

g. Fit an LSTAR model, and provide a plot of the transition function. Interpret the results.

2. `pm_daily.dta` contains daily data for the period January 1, 2009 to December 31, 2014 on particulate matter, $PM_{2.5}$, from the Cerrillos station in Santiago, Chile.

   a. Plot the ACF and PACF of the $PM_{2.5}$ time series using a lag horizon of 24 days. Interpret the results.
   b. Using the Stata `varsoc` (see [TS] **varsoc**) command, compute the optimal lag order of a linear autoregression based on a maximum lag order of 24 days. Are your results consistent with those obtained in exercise 2a.
   c. The World Health Organization recommends that the maximum daily average of $PM_{2.5}$ is around 25 $\mu g/m^3$. Examine the mean and standard deviation of the data both above and below this recommended threshold.
   d. Assuming that there are two states (good air quality and bad air quality), use the command `mswitch dr` to fit a Markov switching model for $PM_{2.5}$ based on the optimum lag order identified in exercise 2b. Are the results consistent with the underlying assumption of the two states.
   e. Use a Wald test to test the null hypothesis that all lags greater than 2 in both states are equal to zero.
   f. Irrespective of the result in exercise 2e, fit a dynamic Markov switching model with only two lags of $PM_{2.5}$ in each state. Compute the predicted probabilities of being in the good and bad states, and overlay these probabilities on time-series plots of the series (you will need two figures, one for each state). Interpret the results.
   g. Compute the transition probabilities and expected durations of the good and bad states, and comment on the results.

11 Modeling time-varying variance

It is difficult to provide a simple definition of environmental risk because the concept relates to such a wide variety of phenomena, such as climate change related to greenhouse gas emissions, catastrophic environmental disasters, and more subtle issues such as the unintended consequences of environmental regulations. One particularly interesting aspect of environmental risk, however, is related directly to finance theory. Investors have been increasingly concerned with the environmental records of firms in making their investment choices. There is growing pressure for investors to invest in green companies, and there is a growing market for "green bonds". In the context of finance, the natural definition of environmental risk involves the uncertainty relating to the returns from investing in green companies by comparison with the returns to more conventional investment strategies.

The central paradigm of modern finance theory is the risk/reward tradeoff: risks must be taken to generate rewards, or returns to investment. An optimal strategy maximizes rewards and minimizes risk. In finance theory, the variance of the reward or its square root, volatility, is an appropriate measure of risk associated with an investment. The Dow Jones sustainability indices (DJSI)[1] are constructed by selecting constituent firms in terms of their environmental and social records. These indices, available for worldwide firms, can be used as a benchmark of green firms' financial performance. Environmental risk in a financial sense may then be defined as the uncertainty associated with the returns to these sustainability indices (Hoti, McAleer, and Pauwels 2007). Since the seminal article of Engle (1982), however, these risks have been regarded as time-varying, and a vast literature has developed that is aimed at understanding this aspect of financial risk. This chapter uses the time-series techniques developed in the field of financial econometrics to explore the time-varying risk associated with investing in green firms. In so doing, it draws distinctions between the various types of risk inherent in the investment decision.

11.1 Evaluating environmental risk

A stock index is an aggregate summary measure of a collection of stocks combined in a particular way to create a portfolio. The value of the index then represents the value of an underlying portfolio and is expressed in terms of an average price that has been normalized in some way. The DJSI are the first indices developed to track the stock performance of the world's leading companies in terms of economic, environmental,

---

1. See http://www.sustainability-indices.com for details.

and social criteria. There are three regional DJSI indices of interest: North America, Europe, and the Asia Pacific region. These may be aggregated into the World DJSI, which summarizes the performance of all the regional indices.

These indices provide a benchmark for investors who integrate sustainability considerations into their portfolios. The DJSI include stocks based on total sustainability scores calculated from an annual corporate sustainability assessment, and the indices are rebalanced quarterly. Sources of information for such assessments come from online questionnaires, company documentation, publicly available information, policies, reports, and direct contacts with a variety of firms.

The usual aggregate stock market indices from the U.S. market—the Dow Jones Industrial Average, the S&P 500, and the Russell 2000—may be used to provide a conventional investment strategy against which to measure the performance of the DJSI. Figure 11.1 plots the North America, Europe, and Asia Pacific indices as well as the conventional S&P 500 index. Although the scales of the indices are different, it is immediately apparent that the time-series behavior of the regional DJSI indices are broadly similar to that of traditional indices that proxy returns on the market. There are sharp falls in the value of all the indices corresponding to the global financial crisis of 2008–2009, the Greek debt crisis of late 2011, and the stock market selloff starting in August 2015 prompted by the crash in the Chinese stock market.

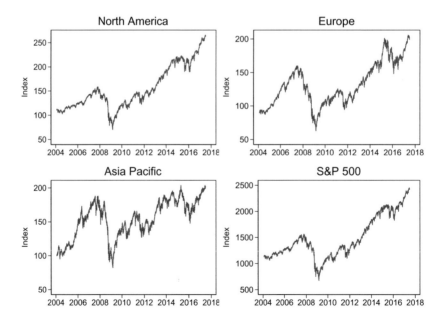

Figure 11.1. Plot of the daily North America, Europe and Asia Pacific DJSI and the S&P 500 index for the period February 1, 2004 to June 17, 2017

The real question of interest is whether investing in the basket of stocks represented by the DJSI indices is inherently riskier than investing in a more conventional portfolio of stocks. As a first step in answering this question, it is useful to consider the behavior of the variance of returns to green investment and in particular how the variance evolves over time. Another approach will be to decompose the riskiness of green investments into systematic and idiosyncratic risk. Systematic risk refers to risk inherent to the entire market or market segment and not just a particular stock or industry. This type of risk is impossible to avoid completely. Idiosyncratic risk, on the other hand, is risk that is particular to a certain stock or investment that can be reduced by diversification. In later sections, the riskiness of green investments will be examined from the perspective of these components of total risk.

11.2 The generalized autoregressive conditional heteroskedasticity model

The return on the decision to invest in green firms accounts for the capital gain or loss, is defined as the price change over the holding period of the asset, as well as the cumulative impact of the stream of cash flows (if any) that take place over the holding period. If P_t is the price of the asset at time t, the excess return[2] on the asset, r_t, is defined relative to the return on a risk-free asset,[3] r_{ft}, by

$$r_t = (\log P_t - \log P_{t-1}) - r_{ft} \qquad (11.1)$$

where the difference in logarithms of price, or the log price relative, is approximately the percentage change in the asset's value over that period.

The riskiness of the investment is therefore defined in terms of the variance of excess returns. A proxy for the variance at any point in time is taken to be the squared excess return.[4] Figure 11.2 plots both the excess returns and the squared excess returns to the DJSI World Index, computed using (11.1). The fall in the level of excess returns during the first half of the sample reflects a sharp increase in the daily risk-free rate.[5] The most important feature of the behavior of these returns is that the assumption of a constant variance over the sample period appears unwarranted. Excess returns on the DJSI World Index exhibit what is termed volatility clustering, which is a particular form of heteroskedasticity. Periods of tranquility are followed by bursts of high variation in the series. Clustering is most evident in the behavior of the squared returns series where clusters of high variability are clearly visible. These movements imply that a time-varying structure in the variance of the returns should be modeled to appropriately capture their behavior.

---

2. For a full discussion of simple, log, and excess returns, see, for example, Hurn et al. (2020).
3. The risk-free asset is usually taken to be a government bond.
4. The volatility of an investment is defined as the standard deviation of excess returns.
5. The daily risk-free factor, a standard measure in financial applications, is available from Kenneth French's website http://mba.tuck.dartmouth.edu/pages/faculty/ken.french/data_library.html.

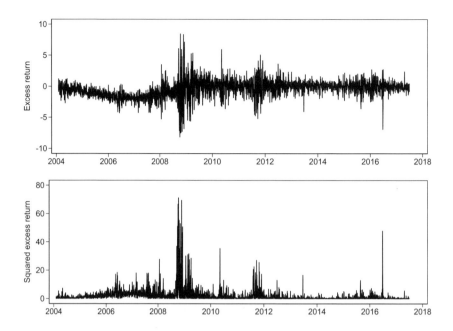

Figure 11.2. Plot of the daily percentage excess returns on the DJSI World Index (top panel) and squared excess returns (bottom panel) for the period February 1, 2004 to June 17, 2017

In a seminal article, Engle (1982) introduced the concept of autoregressive conditional heteroskedasticity (ARCH) to model time-varying variance using a simple linear model of the trajectory of the variance. A generalization of that model due to Bollerslev (1986) is known as generalized autoregressive conditional heteroskedasticity, or GARCH, and takes the form

$$r_t = \mu_0 + u_t$$
$$u_t \sim N(0, h_t)$$
$$h_t = \gamma_0 + \gamma_1 u_{t-1}^2 + \gamma_2 h_{t-1} \qquad (11.2)$$

The fundamental property of the model in (11.2), known as the GARCH(1,1) model, is that the conditional variance h_t is time varying with an autoregressive (AR) component (h_{t-1}) and a component driven by unexpected events proxied by the squared disturbance in the previous period (u_{t-1}^2). Note that the lack of inclusion of additional terms in the mean equation for r_t other than the constant term (μ_0) is for illustrative purposes only because additional variables can be easily included in both the mean equation and the variance equation. Furthermore, the assumption of only one lag on both the squared error term ($\mathbf{u}_t^2$) and the conditional variance (h_{t-1}) is only for ease of exposition. That is, additional terms in $u_{t-2}^2, u_{t-3}^2 \ldots$ could be added to the variance equation as well as additional AR terms $h_{t-2}, h_{t-3} \ldots$. In most empirical applications, the GARCH(1,1)

11.2 The generalized autoregressive conditional heteroskedasticity model

specification is usually adopted perhaps because of its simplicity but also because of the work of Hansen and Lunde (2005), who found that the forecasts of conditional variance obtained from this simple model were difficult to beat.

In this model, the parameter γ_2, where $(0 < \gamma_2 < 1)$, determines how past shocks affect the conditional variance at time t. The initial impact of the previous shock on h_t is γ_1. The effect of this shock feeds into the next period's conditional variance with strength $\gamma_1 \gamma_2$. This process continues with the effects at period τ given by $\gamma_1 \gamma_2^{\tau-1}$. Therefore, the larger γ_2 is, the larger the impact of the original u_{t-1}^2 shock.

The unconditional variance of r_t is obtained in the limit of this progression, effectively when h_t ceases to change. The unconditional volatility implied by the model is given by

$$h_{LR} = \frac{\gamma_0}{1 - \gamma_1 - \gamma_2}$$

The unconditional variance is defined as long as $(\gamma_1 + \gamma_2) < 1$. If this condition is not satisfied, the conditional variance will have a unit root. GARCH models with a unit root are referred to as integrated GARCH, following the discussion of nonstationary models in chapter 6. One way to proceed in fitting integrated GARCH models is to impose the unit-root restriction so that the conditional variance in (11.2) becomes

$$h_t = \gamma_0 + \gamma_1 u_{t-1}^2 + (1 - \gamma_1) h_{t-1}^2$$

Based on the model in (11.2), a test for ARCH involves regressing $\widehat{u}_t^2$ on $\{1, \widehat{u}_{t-1}^2\}$ and computing TR^2 from this regression, where T is the sample size. Under the null hypothesis, the statistic is distributed as χ_1^2. Intuitively, what is being tested is the explanatory power of $\widehat{u}_{t-1}^2$. Small values of the test statistic suggest that this regressor is not significant, implying that only a constant term is required. The test is available using the `regress postestimation` (see [R] **regress postestimation**) command `estat archlm`. Using the residuals from the estimation of (11.3) for the DJSI World Index, the ARCH test results for one lagged value are

```
. use http://www.stata-press.com/data/eeus/djsi
. quietly regress rwd
. estat archlm
LM test for autoregressive conditional heteroskedasticity (ARCH)
```

lags(p)	chi2	df	Prob > chi2
1	117.163	1	0.0000

H0: no ARCH effects   vs.   H1: ARCH(p) disturbance

There is strong evidence of ARCH(1) effects in the DJSI World excess returns series. The test may be carried out for higher-order ARCH(p) effects by using the `lags()` option.

The model in (11.2) is estimated by maximum likelihood methods. Based on the assumption that $\mathbf{u}_t$ is normally distributed, the log-likelihood function for observation t is

$$\log L_t = -\frac{1}{2}\log(2\pi) - \frac{1}{2}\log h_t - \frac{1}{2}\frac{(r_t - \mu_0)^2}{h_t}$$

$$= -\frac{1}{2}\log(2\pi) - \frac{1}{2}\log(\gamma_0 + \gamma_1 u_{t-1}^2 + \gamma_2 h_{t-1}) - \frac{1}{2}\frac{(r_t - \mu_0)^2}{(\gamma_0 + \gamma_1 u_{t-1}^2 + \gamma_2 h_{t-1})}$$

In Stata, the command arch (see [TS] **arch**) is used to fit ARCH and GARCH models. Estimation of a GARCH(1,1) model for the excess returns on the DJSI World Index proceeds as follows:

```
. arch rwd, arch(1) garch(1) nolog vsquish
ARCH family regression
Sample: 2 - 3377                                Number of obs   =      3,376
Distribution: Gaussian                          Wald chi2(.)    =          .
Log likelihood = -5132.562                      Prob > chi2     =          .
```

	rwd	Coef.	OPG Std. Err.	z	P>|z|	[95% Conf. Interval]	
rwd	_cons	-.2135001	.0159742	-13.37	0.000	-.2448091	-.1821912
ARCH							
arch L1.		.0762575	.0063215	12.06	0.000	.0638677	.0886473
garch L1.		.9181287	.0068197	134.63	0.000	.9047624	.931495
_cons		.0101209	.0022222	4.55	0.000	.0057656	.0144763

```
. predict hwd, variance
```

The results are typical of those often encountered when fitting GARCH(1,1) models using the returns to financial assets. The major feature of the estimates is the strong persistence of the conditional variance inherent in the estimate $\widehat{\gamma}_2 = 0.918$. This point will be returned to when discussing the prediction of the conditional variance obtained from these estimates. The sum $\widehat{\gamma}_1 + \widehat{\gamma}_2 = 0.918 + 0.076 = 0.994$ is less than 1, indicating that the variance process is nearly a unit-root process. This result is also common in many financial applications. A plot of the estimated conditional variance is shown in figure 11.3. The effects of the global financial crisis of 2008–2009 and key events in the Greek debt crisis in late 2011 are clearly evident in the behavior of the conditional variance series.

11.2 The generalized autoregressive conditional heteroskedasticity model   235

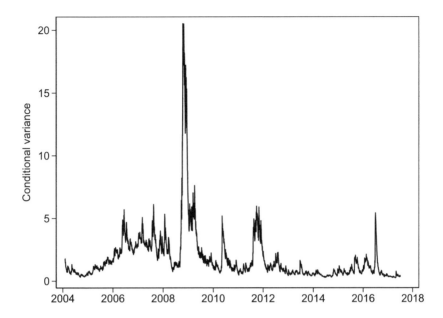

Figure 11.3. Plot of the conditional variance of returns on the DJSI World Index for the period February 1, 2004 to June 17, 2017

Based on these estimates, the predicted conditional variance of excess returns is obtained using the `arch postestimation` (see [TS] **arch postestimation**) command `predict` with the `var` option. If the `dynamic` option is used with `predict`, then dynamic forecasts of the conditional variance are obtained using the recursive structure of time-series forecasts discussed in chapter 4. The important feature to be remembered is that recursive time-series forecasts converge to the long-run conditional mean. This is also true of the forecast from a GARCH(1,1) model, which converges to the long-run or unconditional volatility implied by the model, given by

$$h = \frac{\gamma_0}{1 - \gamma_1 - \gamma_2}$$

The convergence of the dynamic forecast of the conditional variance to the unconditional variance of the process is easily demonstrated. Using the `extremes` command (Cox 2003), the largest occurrences of the conditional variance can be identified and the maximum value used as a starting point for a forecast of the conditional variance.

```
. extremes hwd date, n(3) high

  obs:         hwd        date

 1188.    20.30956   17oct2008
 1197.    20.51848   30oct2008
 1187.    20.54214   16oct2008
```

Figure 11.4 demonstrates this convergence of the forecast series for DJSI World Index returns. A GARCH(1,1) model is fit and then out-of-sample predictions are made for the conditional variance starting on October 16, 2008. The forecast converges to the long-term mean despite the fact that it starts well above the unconditional variance. The fact that convergence to the long-run (unconditional) value occurs quite slowly over a 24-month period alludes to the point made earlier about the level of persistence in GARCH models. It is clear from figure 11.4 that the estimated persistence of conditional volatility is too pronounced. Indeed, the actual estimated conditional variance series drops off a lot more quickly than the forecast.

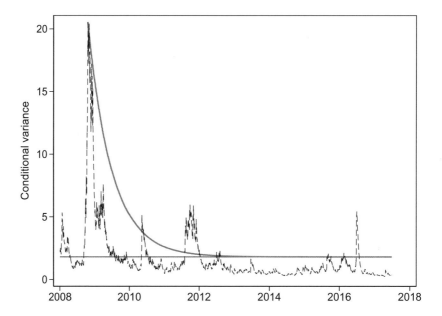

Figure 11.4. Plot of the forecast conditional variance of returns on the World DJSI from a GARCH(1,1) fitted for the period February 1, 2004 to June 17, 2017. The conditional variance is forecast starting on October 17, 2008.

11.3 Alternative distributional assumptions

The GARCH model in (11.2) specifies that the distribution of u_t is normal with zero mean and (conditional) variance h_t. The empirical distribution of many financial series is decidedly not normal, because these distributions often have "fat tails" or excess kurtosis, suggesting that large changes in prices or returns are much more prevalent than they would be if those quantities were normally distributed. Fat-tailed error distributions are handled in Stata by means of the t distribution and the generalized error distribution (GED).

For the t distribution, the log-likelihood function for each observation is given by

$$\log L_t = -\frac{1}{2}\log\{\pi(\nu-2)\} - \frac{1}{2}\log h_t + \log\Gamma\left(\frac{\nu+1}{2}\right) - \log\Gamma\left(\frac{\nu}{2}\right)$$
$$- \left(\frac{\nu+1}{2}\right)\log\left(1 + \frac{(r_t - \mu_0)^2}{h_t(\nu-2)}\right)$$

where h_t is the conditional variance of the GARCH(1,1) model and $\Gamma(k)$ is the gamma function—the generalized factorial function for any real k. The parameter ν, the degrees of freedom parameter, allows for distributions with higher peaks and fatter tails than the normal distribution (leptokurtosis). In fitting the model, Stata enforces the restriction that $\nu > 2$. When $\nu \approx 30$, the t distribution is well approximated by the normal.

For the GED distribution, the log-likelihood function for each observation is given by

$$\log L_t = \log s - \log \lambda - \frac{s+1}{d}\log 2 - \log\Gamma(s^{-1}) - \frac{1}{2}\left|\frac{r_t - \mu_0}{\lambda h_t}\right|^s$$
$$\lambda = \left(\frac{\Gamma(s^{-1})}{2^{2/s}\Gamma(3s^{-1})}\right)^2$$

The shape parameter, s, of the GED distribution, controls tail thickness. The GED may be used to model either thinner tails (platykurtosis) or fatter tails (leptokurtosis) than the normal distribution. In fitting the model, Stata enforces the restriction that $s > 0$. When $s = 2$, the GED is the same as the normal distribution, with larger values corresponding to fatter tails.

To illustrate, the GARCH(1,1) model is reestimated with all three distributions using the `dist()` option on the `arch` command, as displayed in table 11.1. The results show that the estimates of γ_1 and γ_2 are positive as required and significantly different from zero. As for the model based on the normal distribution, the sum $\widehat{\gamma}_1 + \widehat{\gamma}_2$ remains close to 1 for the t and GED distributions, indicating that the conditional variance is estimated to be very persistent. There is thus strong memory in the conditional variance of the DJSI.

Both of the distributional parameters of the models based on the t distribution and the GED are significantly different from zero. Note, however, that Stata estimates these parameters so as to enforce the restrictions that $\nu > 2$ in the t distribution and

$s > 0$ in the GED distribution. The postestimation command `nlcom` may be used to obtain estimates of the actual parameter and its standard error. In the case of the t distribution, Stata estimates the parameter as $\log(\nu - 2)$, so an estimate of ν and its standard error may be obtained using `nlcom` as follows:

```
. eststo: arch rwd, arch(1) garch(1) dist(t)
 (output omitted)
. nlcom (nu: exp( _b[lndfm2:_cons]) + 2)
       nu:  exp( _b[lndfm2:_cons]) + 2
```

rwd	Coef.	Std. Err.	z	P>|z|	[95% Conf. Interval]
nu	21.67502	4.775773	4.54	0.000	12.31467 31.03536

The estimate of the degrees-of-freedom parameter of the t distribution, $\hat{\nu}$, is 21.68 with an associated standard error of 4.78. A similar procedure for the GED distribution, this time recognizing that Stata estimates $\log s$, yields 1.87 with a standard error of 0.052. Neither estimate provides compelling evidence against the assumption of normality of the errors in the model of column 1 of table 11.1.

Table 11.1. GARCH(1,1) estimates of excess returns on the DJSI World Index under different distributional assumptions

	Normal	t distribution	GED
μ_0	−0.214	−0.206	−0.210
	(−13.37)	(−12.57)	(−13.01)
γ_0	0.0101	0.00723	0.00932
	(4.55)	(2.73)	(3.68)
γ_1	0.0763	0.0734	0.0760
	(12.06)	(9.19)	(10.55)
γ_2	0.918	0.923	0.919
	(134.63)	(112.99)	(120.06)
$\log(\nu - 2)$		2.979	
		(12.27)	
$\log s$			0.626
			(22.64)
N	3376	3376	3376
$\log L$	−5132.6	−5123.2	−5130.6

NOTE: t statistics in parentheses.

11.4 Asymmetries

In the GARCH(1,1) model in (11.2), u_{t-1} measures a shock in period $(t-1)$. For no-news days, $u_{t-1} = 0$, while a positive (negative) value of u_{t-1} represents good (bad) news. An important property of this GARCH specification is that shocks of the same magnitude, positive or negative, result in the same increase in volatility h_t when they are squared. This implies that positive news, $u_{t-1} > 0$, has the same effect on the conditional variance as negative news of a similar magnitude, $u_{t-1} < 0$, because it is only the absolute size of the news that matters when the shock is squared.

In the case of stock markets, an asymmetric response to news about a firm is often observed: negative shocks with $u_{t-1} < 0$ have a larger effect on the conditional variance, as theory predicts, than positive shocks of the same size. A negative shock raises the firm's debt-equity ratio, increasing financial leverage and consequently increasing the riskiness of the stock. This so-called leverage effect implies that bad news causes a greater increase in conditional variance than good news.

To address this asymmetry in the context of GARCH modeling, two popular specifications that relax the restriction of symmetric responses to news have been developed. The threshold ARCH specification (Zakoïan 1994; Glosten, Jagannathan, and Runkle 1993) of the conditional variance equation is

$$h_t = \gamma_0 + \gamma_1 u_{t-1}^2 + \gamma_2 h_{t-1} + \lambda u_{t-1}^2 I_{t-1}$$

where I_{t-1} is the indicator variable defined as

$$I_{t-1} = \begin{cases} 1: & u_{t-1} \geq 0 \\ 0: & u_{t-1} < 0 \end{cases}$$

To make the asymmetry in the impact of news on the conditional variance explicit, this model can also be written as

$$h_t = \begin{cases} \gamma_0 + (\gamma_1 + \lambda) u_{t-1}^2 + \gamma_2 h_{t-1}: & u_{t-1} \geq 0 \\ \gamma_0 + \gamma_1 u_{t-1}^2 + \gamma_2 h_{t-1}: & u_{t-1} < 0 \end{cases}$$

The leverage effect in equity markets leads us to expect $\lambda < 0$ so that negative news, $u_{t-1} < 0$, is associated with a higher effect on volatility than positive news of the same magnitude. Refitting the GARCH(1,1) model of the DJSI World Index excess returns using the tarch(1) option yields the following results:

```
. arch rwd, arch(1) garch(1) tarch(1) nolog vsquish
initial values not feasible
(note:  default initial values infeasible; starting ARCH/ARMA estimates from 0)
ARCH family regression
Sample: 2 - 3377                                  Number of obs   =        3,376
Distribution: Gaussian                            Wald chi2(.)    =            .
Log likelihood = -5113.088                        Prob > chi2     =            .
```

	rwd	Coef.	OPG Std. Err.	z	P>|z|	[95% Conf. Interval]	
rwd							
	_cons	-.2071389	.016305	-12.70	0.000	-.2390961	-.1751817
ARCH							
	arch						
	L1.	.0983809	.0088827	11.08	0.000	.0809711	.1157907
	tarch						
	L1.	-.0541757	.0092382	-5.86	0.000	-.0722823	-.036069
	garch						
	L1.	.9113761	.007759	117.46	0.000	.8961687	.9265834
	_cons	.0184243	.0029215	6.31	0.000	.0126983	.0241502

As expected, the estimate of the threshold ARCH parameter, $\widehat{\lambda}$, is negative and statistically significant, indicating the presence of asymmetry.

As an alternative asymmetric model for the conditional variance, the exponential GARCH specification (Nelson 1991) of the conditional variance is

$$\log h_t = \gamma_0 + \gamma_1 \left| \frac{u_{t-1}}{\sqrt{h_{t-1}}} \right| + \lambda \frac{u_{t-1}}{\sqrt{h_{t-1}}} + \gamma_2 \log h_{t-1}$$

An important advantage of the exponential GARCH model is that the conditional variance is guaranteed to be positive at each point in time because the conditional variance is expressed in terms of $\log h_t$. In this specification, the scaled shock $\mathbf{u}_t$ is included in terms of both its absolute value and its signed value. The parameter λ captures potential asymmetry in the effect of u_{t-1} on $\log h_t$. It is expected that $\lambda < 0$, so a negative news shock is associated with a larger increase in the conditional variance than a positive news shock of the same magnitude. We illustrate by refitting the GARCH(1,1) model of DJSI World Index excess returns using the earch(1) and egarch(1) options.

11.4 Asymmetries

```
. arch rwd, earch(1) egarch(1) nolog vsquish
ARCH family regression
Sample: 2 - 3377                                Number of obs  =      3,376
Distribution: Gaussian                          Wald chi2(.)   =          .
Log likelihood = -5124.28                       Prob > chi2    =          .
```

	rwd	Coef.	OPG Std. Err.	z	P>|z|	[95% Conf. Interval]	
rwd	_cons	-.2004438	.0155985	-12.85	0.000	-.2310163	-.1698713
ARCH	earch L1.	-.0206024	.0057997	-3.55	0.000	-.0319697	-.0092351
	earch_a L1.	.1533041	.0119707	12.81	0.000	.129842	.1767662
	egarch L1.	.9825784	.0025794	380.93	0.000	.9775228	.987634
	_cons	.0020755	.0016569	1.25	0.210	-.0011721	.005323

The coefficient labeled L1.earch_a is the equivalent of γ_1 in the specification above. It refers to the effect of the unsigned shock on the conditional variance. The coefficient labeled L1.earch is the coefficient expressing an asymmetric effect, which is λ in the specification above. Its significant negative value implies that a negative shock will inflate the conditional variance, while a positive shock will dampen it.

To conclude this examination of the univariate GARCH model, the central message is that the total risk from investing in green stocks as selected by the DJSI exhibits a similar pattern of time variation to that of investment in a broad market index. This time variation means that it is not sensible to speak simply of one strategy being inherently riskier than another without reference to a particular time period. Figure 11.5 plots the difference between the conditional variance of the World DJSI Index and that of the S&P 500 index also estimated using a GARCH(1,1) model. Positive values indicate that the conditional variance (and estimated risk) of the World DJSI is higher than that of the S&P 500, and vice versa. Interestingly, there are periods where the green investment strategy appears to be considerably less risky, particularly during the periods of turmoil of the Great Recession and global financial crises of 2008–2009. To better understand these relationships, a closer investigation of systematic risk is necessary.

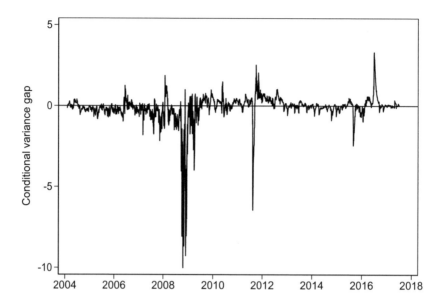

Figure 11.5. Plot of the difference between the conditional variances of the returns on the World DJSI and the conditional variance of the returns on the S&P 500 index for the period February 1, 2004 to June 17, 2017. Positive values indicate times when the sustainable index has a higher conditional variance than that of the market.

11.5 Motivating multivariate volatility models

Up to this point, the chapter has been concerned with the time-varying volatility of log returns to green investments. No further distinction has been made concerning the composition of this volatility. The capital asset pricing model (CAPM)[6] relates the excess return on an asset to the excess return on the market, r_{mt}, in terms of a simple linear regression:

$$r_t = \alpha + \beta r_{mt} + u_{it} \qquad (11.3)$$

where u_t is a disturbance term. In the CAPM, the parameter β measures the exposure of the excess returns on the asset to the excess returns on the market, both defined relative to a risk-free rate of interest r_{ft}. The CAPM provides a convenient method of decomposing the total risk of holding an asset into systematic risk (risk that is inherent to the entire market) and idiosyncratic risk (risk inherent in the characteristics of the asset itself). Formally, this decomposition is achieved by squaring both sides of (11.3) and then taking expectations:

---

6. The 1990 Nobel Memorial Prize in Economic Sciences was awarded to William F. Sharpe, Harry Markowitz, and Merton Miller for this contribution to the field of financial economics.

11.5 Motivating multivariate volatility models

$$\underbrace{E\{(r_{it})^2\}}_{\text{Total risk}} = \underbrace{E\{(\alpha + \beta r_{mt})^2\}}_{\text{Systematic risk}} + \underbrace{E(u_{it}^2)}_{\text{Idiosyncratic risk}}$$

using the result that $E(r_{mt}u_t) = 0$, that is, the return on the market is independent of the firm-specific disturbance $\mathbf{u}_t$. Evaluating the riskiness of green investments requires the evaluation of both total risk and systematic risk.

Table 11.2 reports the results of fitting the CAPM in (11.3) using the excess returns on the S&P 500 as the measure of excess returns on the market. The results suggest that the North America DJSI tracks the market closely, while those for Europe and Asia Pacific are reasonably defensive investments in terms of exposure to market risk. On the other hand, all the DJSI investment portfolios would earn negative abnormal returns, as given by the estimated α parameters, which are statistically significant. In terms of idiosyncratic risk, the Asia Pacific index shows the highest estimate of h, the conditional variance, followed by that of Europe. Merely considering total risk of the environmental portfolio obscures possibly important aspects of the risk profile of such an investment strategy. While the Asia Pacific DJSI is the riskiest of the indices, it could potentially be used as a defensive investment against large swings in market risk.

Table 11.2. Estimates of the CAPM for DJSI indices for the period February 1, 2004 to June 17, 2017

	(1) North America	(2) Europe	(3) Asia Pacific
β	0.948 (409.89)	0.723 (58.36)	0.354 (19.97)
α	−0.0198 (−5.95)	−0.121 (−6.82)	−0.288 (−11.33)
N	3376	3376	3376
h	0.184	0.983	1.404

NOTE: t statistics in parentheses.

The CAPM encapsulates the risk characteristics of an asset in terms of its β, which is given by

$$\beta = \frac{\text{cov}(r_{it}, r_{mt})}{\text{var}(r_{mt})} \qquad (11.4)$$

in which r_{it} is the excess return to the relevant DJSI and r_{mt} is the return to the market, proxied here by the S&P 500 index. However, restricting β to be a constant may be unrealistic given the overwhelming evidence of the univariate GARCH models, implying that the conditional variances of the DJSI and S&P 500 indices are time varying. If β is a time-varying parameter, the unconditional expectations in (11.4) are replaced by conditional expectations:

$$\beta_t = \frac{E_{t-1}\left\{r_{it} - E_{t-1}(r_{it}),\ r_{mt} - E_{t-1}(r_{mt})\right\}}{E_{t-1}\left[\left\{r_{mt} - E_{t-1}(r_{mt})\right\}^2\right]} \tag{11.5}$$

where the numerator is the conditional covariance between the excess returns on the DJSI index and the market based on information at time $(t-1)$ and the denominator is the time-varying conditional variance of the market excess return.

To estimate β_t, it is necessary to model the conditional variance and covariance. While the univariate GARCH model provides a suitable specification of the conditional variances of excess returns on the indices, it cannot provide an estimate of a time-varying conditional covariance.

The search for a workable approach to the estimation of time-varying conditional covariances starts by considering the correlation coefficient defined as

$$\rho = \frac{\text{cov}(r_{it}, r_{mt})}{\sqrt{\text{var}(r_{it})}\sqrt{\text{var}(r_{mt})}}$$

This expression may be rearranged in terms of the covariance to yield

$$\text{cov}(r_{it}, r_{mt}) = \rho\sqrt{\text{var}(r_{it})}\sqrt{\text{var}(r_{mt})} \tag{11.6}$$

so that the covariance is the product of the correlation ρ and the standard deviations $\sqrt{\text{var}(r_{it})}$ and $\sqrt{\text{var}(r_{mt})}$. This expression suggests that an estimate of the time-varying covariance can be constructed by simply replacing the unconditional variances $\text{var}(r_{it})$ and $\text{var}(r_{mt})$ with their respective conditional variances based on univariate GARCH specifications. To complete the estimation of (11.6), ρ is estimated using the sample correlation between the excess returns on the DJSI and the S&P 500.

Using the daily excess returns to the DJSI Asia Pacific and the S&P 500 for the period February 1, 2004 to June 17, 2017, figure 11.6 gives the conditional covariance series and the estimate of the time-varying β. The latter series exhibits some cyclical behavior but stays fairly close to the constant β estimate represented by the solid line.

11.6 Multivariate volatility models

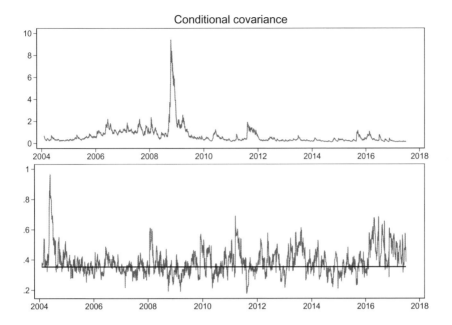

Figure 11.6. Conditional covariances (top panel) and implied time-varying β (lower panel) for the DJSI Asia Pacific with the constant β estimate superimposed. The market portfolio is the S&P 500 index. The excess returns are daily for the period February 1, 2004 to July 17, 2017.

Although this is a simple approach, the major insight it provides is crucial in developing workable multivariate GARCH (MGARCH) models. A natural extension is to consider a multivariate model that can produce both the conditional variances as well as conditional covariances of the multivariate dynamic process.

11.6 Multivariate volatility models

A simple multivariate specification is

$$\mathbf{r}_t = \mathbf{u}_t, \qquad v_t \sim N(0, \mathbf{H}_t)$$

where $\mathbf{r}_t = [r_{1t}, r_{2,t}, \ldots, r_{Nt}]'$ is now defined as a $(N \times 1)$ vector of excess returns and $\mathbf{u}_t = [u_{1t}, u_{2t}, \ldots, u_{Nt}]'$ as a $(N \times 1)$ vector of disturbance terms with covariance matrix $\mathbf{H}_t$. The matrix $\mathbf{H}_t$ is a $(N \times N)$ symmetric positive-definite matrix with the variances, h_{it}, on the main diagonal and covariances, h_{ijt}, as the off-diagonal elements. For the case of two assets

$$\mathbf{H}_t = \begin{pmatrix} h_{1t} & h_{12t} \\ h_{21t} & h_{2t} \end{pmatrix}$$

where $h_{12t} = h_{21t}$ is the conditional covariance.

Based on the assumption that the disturbance vector $\mathbf{u}_t$ follows the multivariate normal distribution, the log-likelihood function at time t is

$$\log L_t = -\frac{N}{2}\log(2\pi) - \frac{1}{2}\log|\mathbf{H}_t| - \frac{1}{2}\mathbf{u}_t'\mathbf{H}_t^{-1}\mathbf{u}_t$$

from which it becomes apparent that the specification of $\mathbf{H}_t$ is central to the problem.

11.6.1 The vech model

The vech model is a direct generalization of the univariate GARCH model to multiple dimensions. This model takes its name from the matrix operator vech($\cdot$), which represents the column stacking of the unique elements of a symmetric matrix. The diagonal vech (DVECH) specification is

$$\text{vech}(\mathbf{H}_t) = \mathbf{C} + \mathbf{A}\,\text{vech}(u_{t-1}u_{t-1}') + \mathbf{D}\,\text{vech}(\mathbf{H}_{t-1})$$

To fix ideas, for the case of $N = 2$ variables, $\mathbf{H}_t$ is given by

$$\begin{bmatrix} h_{1t} \\ h_{12t} \\ h_{2t} \end{bmatrix} = \begin{bmatrix} c_1 \\ c_2 \\ c_3 \end{bmatrix} + \begin{bmatrix} a_{11} & a_{12} & a_{13} \\ a_{21} & a_{22} & a_{23} \\ a_{31} & a_{32} & a_{33} \end{bmatrix} \begin{bmatrix} u_{1\,t-1}^2 \\ u_{1\,t-1}u_{2\,t-1} \\ u_{2\,t-1}^2 \end{bmatrix}$$
$$+ \begin{bmatrix} d_{11} & d_{12} & d_{13} \\ d_{21} & d_{22} & d_{23} \\ d_{31} & d_{32} & d_{33} \end{bmatrix} \begin{bmatrix} h_{1\,t-1} \\ h_{12\,t-1} \\ h_{2\,t-1} \end{bmatrix}$$

The vech model does not guarantee that $\mathbf{H}_t$ will be positive definite at each t. Even in the bivariate case, there are many parameters that need to be estimated.[7] The Stata command `mgarch dvech` (see [TS] **mgarch dvech**) fits a diagonal version of the vech model in which all the parameter matrices are restricted to be diagonal matrices, which reduces the dimensionality of the model considerably at the cost of ruling out interactions across the equations.

Fitting a DVECH(1,1) model for excess returns on the DJSI Asia Pacific and the S&P 500 involves specifying the mean equations, which in this case only contain a constant term. Finance theory suggests that in efficient markets, equity prices follow a random walk and are not forecastable from prior information. Therefore, there is no rationale to include earlier values of returns in the mean equations.

---

7. For the $N = 2$ case, there are 21 parameters in the model, and for $N = 4$ variables, this number grows to 210.

11.6.1 The vech model

```
. mgarch dvech (rap rsp = ), arch(1) garch(1) vsquish nolog
Diagonal vech MGARCH model
Sample: 2 - 3377                                  Number of obs    =      3,376
Distribution: Gaussian                            Wald chi2(.)     =          .
Log likelihood = -10454.15                        Prob > chi2      =          .
```

		Coef.	Std. Err.	z	P>|z|	[95% Conf.	Interval]
rap							
	_cons	-.1235773	.0210184	-5.88	0.000	-.1647726	-.082382
rsp							
	_cons	-.1393511	.0177953	-7.83	0.000	-.1742294	-.1044729
/Sigma0							
	1_1	.0188292	.0046849	4.02	0.000	.009647	.0280114
	2_1	.0014703	.0007731	1.90	0.057	-.0000449	.0029856
	2_2	.0149639	.0033208	4.51	0.000	.0084553	.0214725
L.ARCH							
	1_1	.0720134	.0081648	8.82	0.000	.0560108	.088016
	2_1	.0317646	.0041354	7.68	0.000	.0236594	.0398699
	2_2	.0822245	.0095404	8.62	0.000	.0635257	.1009234
L.GARCH							
	1_1	.9178574	.0092416	99.32	0.000	.8997442	.9359707
	2_1	.9623763	.0047014	204.70	0.000	.9531618	.9715907
	2_2	.907757	.010423	87.09	0.000	.8873283	.9281857

The parameters of the equations that construct the conditional variance matrix are all statistically significant. The parameters labeled Sigma0 are the constants labeled as vector **C** in the specification, while the L.ARCH and L.GARCH parameters are the diagonal elements of matrices **A** and **D**, respectively. Of particular importance is the fact that the equation for the covariance h_{12t} (the 2_1 estimates) is well determined, suggesting that there is an advantage to fitting the fully specified model rather than relying on the simple procedure described above. This is confirmed by the plot of the (smoothed) time-varying β for the DJSI Asia Pacific in figure 11.7. This estimate is computed by using the variance option on the postestimation predict (see [R] **predict**) command. This time series of covariances is then used to construct the time-varying β given in (11.5). By contrast with figure 11.6, where the estimate appears to be a noisy version of the constant β estimate, the estimate based on the DVECH model shows that the DJSI Asia Pacific index has become more defensive over the sample period. Far from being riskier than a conventional investment in the market index, the Asia Pacific Index appears to act as a hedge against systematic risk as proxied by the excess returns on the S&P 500 index.

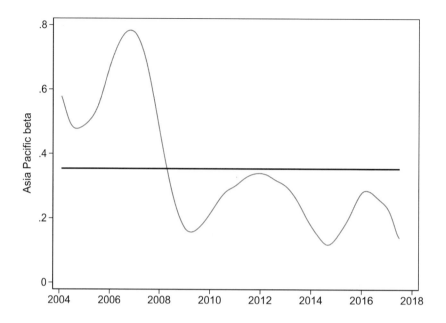

Figure 11.7. The time-varying β for the DJSI Asia Pacific computed from a DVECH MGARCH model with the constant β estimate superimposed. The market portfolio is the S&P 500 index. The excess returns are daily for the period February 1, 2004 to July 17, 2017.

11.6.2 The dynamic conditional correlation model

Even with the restriction of diagonal parameter matrices, the vech model often encounters convergence problems. A solution to the parameter dimensionality problem proposed by Engle (2002) is the dynamic conditional correlation (DCC) model, which is now the most widely adopted MGARCH specification in empirical work.[8]

The DCC model is based on specifying the conditional covariance matrix as

$$\mathbf{H}_t = \mathbf{S}_t \mathbf{R}_t \mathbf{S}_t \qquad (11.7)$$

where $\mathbf{R}_t$ is a $(N \times N)$ conditional correlation matrix and $\mathbf{S}_t$ is a diagonal matrix containing the conditional standard deviations

$$\mathbf{S}_t = \begin{bmatrix} \sqrt{h_{11t}} & & 0 \\ & \ddots & \\ 0 & & \sqrt{h_{NNt}} \end{bmatrix}$$

---

8. Stata's `mgarch` (see [TS] **mgarch**) command also supports the constant conditional correlation approach and the Tse and Tsui (2002) varying conditional correlation approach. The latter model is an alternative to the DCC approach.

11.6.2 The dynamic conditional correlation model

This expression is the multivariate analogue of (11.6), in which the covariance between two assets is constructed from the product of the correlation and the two standard deviations. For the DCC model, the conditional variances are assumed to have univariate GARCH representations. Without loss of generality, assume that each of the N equations (for each $i = 1, \ldots, N$) for the conditional variances take the GARCH(1,1) form

$$h_{iit} = \alpha_{i0} + \alpha_{i1} u_{i\,t-1}^2 + \beta_{i1} h_{ii\,t-1} \tag{11.8}$$

The conditional correlation matrix $\mathbf{R}_t$ in (11.7) is defined as[9]

$$\mathbf{R}_t = \text{diag}(\mathbf{Q}_t)^{-1/2} \, \mathbf{Q}_t \, \text{diag}(\mathbf{Q}_t)^{-1/2}$$

The matrix $\mathbf{Q}_t$ represents a pseudocorrelation matrix that has a GARCH(1,1) specification given by

$$\mathbf{Q}_t = (1 - \alpha - \beta)\mathbf{Q} + \lambda_1 \mathbf{z}_{t-1} \mathbf{z}_{t-1}' + \lambda_2 \mathbf{Q}_{t-1} \tag{11.9}$$

with unknown scalar parameters α and β. The $z_{it} = u_{it}/\sqrt{h_{it}}$ series is a standardized disturbance. The definition of the intercept in the dynamics of $\mathbf{Q}_t$ as $(1 - \alpha - \beta)\mathbf{Q}$ simplifies estimation as the intercept no longer has to be estimated separately. The matrix $\mathbf{Q}$ is the unconditional matrix of standardized disturbances given by

$$\mathbf{Q} = \frac{1}{T} \sum_{t=1}^{T} \begin{bmatrix} z_{1t}^2 & z_{1t} z_{2t} & \cdots & z_{1t} z_{Nt} \\ z_{2t} z_{1t} & z_{2t}^2 & \cdots & z_{2t} z_{Nt} \\ \vdots & \vdots & \ddots & \vdots \\ z_{Nt} z_{1t} & z_{Nt} z_{2t} & \cdots & z_{Nt}^2 \end{bmatrix}$$

The unknown parameters of the DCC MGARCH model consist of the univariate GARCH parameters of the conditional variance equations in (11.8) and the parameters α and β in (11.9). Stata's mgarch dcc (see [TS] **mgarch dcc**) command implements the model. Given the parsimony of the specification, the DCC model is easier to fit than the DVECH model and can be applied to more assets.

---

9. An alternative specification of the evolution of the correlation matrix $\mathbf{R}_t$ is provided by the varying conditional correlation model mentioned above.

The full DCC(1,1) model for the excess returns on the three DJSI indices together with the S&P 500 index can now be fit:

```
. mgarch dcc (rna reu rap rsp = ), arch(1) garch(1) nolog vsquish
Dynamic conditional correlation MGARCH model
Sample: 2 - 3377                              Number of obs    =     3,376
Distribution: Gaussian                        Wald chi2(.)     =         .
Log likelihood = -13015.36                    Prob > chi2      =         .
```

	Coef.	Std. Err.	z	P>|z|	[95% Conf. Interval]	
rna						
_cons	-.1314114	.0170373	-7.71	0.000	-.1648038	-.098019
ARCH_rna						
arch						
L1.	.0571033	.0042059	13.58	0.000	.04886	.0653466
garch						
L1.	.9322538	.0051116	182.38	0.000	.9222352	.9422723
_cons	.0141006	.0021981	6.41	0.000	.0097924	.0184088
reu						
_cons	-.1476127	.0197785	-7.46	0.000	-.1863779	-.1088475
ARCH_reu						
arch						
L1.	.0716128	.00653	10.97	0.000	.0588143	.0844112
garch						
L1.	.9170451	.0075772	121.03	0.000	.902194	.9318961
_cons	.0213833	.0040442	5.29	0.000	.0134568	.0293099
rap						
_cons	-.1202483	.0207159	-5.80	0.000	-.1608506	-.0796459
ARCH_rap						
arch						
L1.	.0907578	.0107492	8.44	0.000	.0696898	.1118257
garch						
L1.	.8986642	.0117819	76.27	0.000	.875572	.9217564
_cons	.02569	.0060148	4.27	0.000	.0139011	.0374788
rsp						
_cons	-.1308017	.017481	-7.48	0.000	-.1650638	-.0965396
ARCH_rsp						
arch						
L1.	.057604	.0043748	13.17	0.000	.0490295	.0661785
garch						
L1.	.9314217	.0053471	174.19	0.000	.9209416	.9419018
_cons	.015464	.0024103	6.42	0.000	.0107399	.0201882
corr(rna,reu)	.78743	.0281209	28.00	0.000	.732314	.8425459
corr(rna,rap)	.4057181	.0626349	6.48	0.000	.2829559	.5284804
corr(rna,rsp)	.9893248	.0014965	661.10	0.000	.9863917	.9922579
corr(reu,rap)	.4729414	.0578947	8.17	0.000	.3594699	.5864128
corr(reu,rsp)	.7761467	.0296118	26.21	0.000	.7181086	.8341847
corr(rap,rsp)	.380403	.0641606	5.93	0.000	.2546506	.5061553

11.6.2 The dynamic conditional correlation model

```
/Adjustment
    lambda1     .020521    .001435    14.30   0.000    .0177085   .0233335
    lambda2    .9738858   .0018217   534.60   0.000    .9703153   .9774563
```

Despite being a system of $N = 4$ excess returns with six distinct covariance terms, the maximum-likelihood estimation procedure readily converges. The important point to note from these estimates is that the parameters governing the adjustment of the correlation matrix $\mathbf{R}_t$, λ_1 and λ_2, are statistically significant, indicating that the correlations between the assets meaningfully change over time. Note that the joint test of the restrictions $\lambda_1 = \lambda_2 = 0$ cannot be performed using a simple Wald F test, because if $\lambda_1 = 0$, the parameter λ_2 becomes unidentified. Testing the assumptions of the DCC model is the subject of ongoing research (Harvey and Thiele 2016; Silvennoinen and Teräsvirta 2016).

Figure 11.8 plots the time-varying βs for all three regional DJSI indices. It is apparent that the risk profiles of the three indices are different. Investing in the basket of stocks that make up DJSI North America is almost equivalent to a tracking portfolio for the S&P 500, but the DJSI Europe and Asia Pacific indices are far more defensive in terms of their exposure to systematic risk. While the idiosyncratic risk of the latter indices may exceed that of the DJSI North America index, their role as hedges in times of market turbulence may suit many investors. It is therefore not particularly sensible to think of green investments necessarily being riskier than broad market indices. Rather, one should consider whether the risk characteristics of the green indices match the desired risk profile and preferences of the investor.

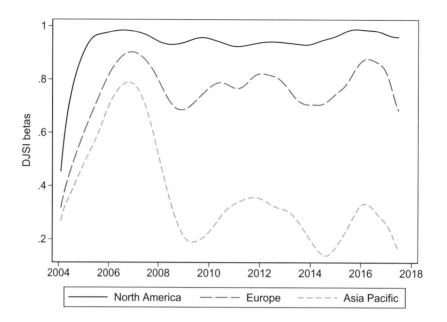

Figure 11.8. Time-varying β estimates for the DJSI regional indices computed from a DCC MGARCH model with the constant β estimate superimposed. The market portfolio is the S&P 500 index. The excess returns are daily for the period February 1, 2004 to July 17, 2017.

Exercises

1. `commodities.dta` contains daily data for the period March 31, 1983 to May 30, 2014 on the log returns on futures indices for several commodities and also the West Texas crude futures index.

    a. Plot the log returns on the corn futures index `corn` and the squared log returns. Comment on the volatility properties of the series.
    b. Fit a GARCH(1,1) model of log returns of the corn futures index with an AR(1) specification for the mean and based on the assumption of the normality of the errors. Interpret the results. Plot the conditional variance implied by the model. Comment on the results.
    c. Refit the GARCH model, but allow for asymmetries in the conditional variance. Interpret the results.

11.6.2 The dynamic conditional correlation model

   d. Corn is an important input into the production of biofuels. Consequently, there is concern that environmental legislation to promote the use of biofuels has linked volatility in corn markets to volatility in oil markets. Refit a GARCH(1,1) model for `corn`, but allow log returns to West Texas crude futures, `wti` to enter the conditional variance equation. Interpret the results.
   e. Fit a bivariate MGARCH DVECH model for log returns on corn and coffee futures. Allow for an AR(1) specification for the mean for both commodity returns. Plot the estimated conditional covariance, and provide explanations for the observed pattern.
   f. Fit a trivariate MGARCH DCC model for log returns on corn, coffee, and West Texas crude futures. Plot the conditional correlation between corn and West Texas crude and coffee and West Texas crude. Comment on the results.

2. A recent development in environmental econometrics has been the increased interest in derivative securities, which take as their underlying variable some environmental quantity such as wind speed, pollution, temperature, or rainfall. Probably the most developed of these derivatives are temperature derivatives defined on a nonlinear transformation of the daily temperature. *Cumulative* cooling degree days (CDD) over a period of N days given by

$$C_N = \sum_{k=1}^{N} \max(0, T_k - 18)$$

where T_k is the mean temperature on the kth day and 18 degrees Celsius ($\approx 65^o$ Fahrenheit) follows general convention. The buyer of a vanilla European temperature call option pays an upfront premium and receives a payout if the value C_N exceeds the strike value, D, at the maturity of the option. The tick value of a call option with strike value D and duration N days is therefore

$$\mathcal{T}_N = \max(C_N - D, 0)$$

where a payoff will be made if the actual number of CDD exceeds the strike value, set when the option contract begins. The actual monetary payoff from the contract is the product of the tick value and the tick size, defined as the cash value of a tick. Because temperature cannot be traded, weather derivatives cannot be priced in the same manner as traditional options on commodities or equity shares. A useful way of addressing the problem is to fit a model to the time series of average temperature and then simulate the distribution of the index on which the derivative is written (Campbell and Diebold 2005). `bristemp.dta` contains daily maximum and minimum temperature, for Brisbane, Australia, for the period January 1, 1887 to August 31, 2007.

   a. Plot *average* daily temperature for the period January 1, 2000 to August 31, 2007. Verify that there is a strong periodic pattern in the temperature and also that the variance of temperature may be higher in the hotter and colder months.

b. Fit a GARCH(1,1) model for temperature of the following form, based on the assumption that $\mathbf{u}_t$ follows a t distribution,

$$T_t = \mu_0 + \mu_1 T_{t-1} + \phi \cos\left(\frac{2\pi t}{365}\right) + \theta \sin\left(\frac{2\pi t}{365}\right) + \mathbf{u}_t$$

$$h_t = \alpha_0 + \alpha_1 u_{t-1}^2 + \beta_1 h_{t-1} + \varphi \cos\left(\frac{2\pi t}{365}\right) + \vartheta \sin\left(\frac{2\pi t}{365}\right)$$

where t is a repeating value that cycles through $1, \ldots, 365$ (February 29 is dropped from the dataset). Interpret the results.

c. Plot the estimated conditional variance. Does the result accord with intuition?

d. Consider pricing a three-month CDD call option for the period January 1 to March 31, 2007. Using the **forecast** (see [TS] **forecast**) suite of commands, simulate temperature outcomes over the period 1000 times. These simulations of average temperature over the life of the option provides realizations of cumulative CDDs $\{\widehat{\mathbb{C}}_{N1}, \widehat{\mathbb{C}}_{N2}, \ldots, \widehat{\mathbb{C}}_{NK}\}$. Plot a kernel estimate of the simulated distribution of CDDs.

e. Given a strike value $D = \$570$, the expected tick value (ETV) of the call option is given by

$$\text{ETV} = \frac{1}{K} \sum_{j=1}^{K} (\widehat{\mathbb{C}}_{Nj} - 570) \mathbb{I}_{(\widehat{\mathbb{C}}_{Nj} - 570)}$$

where $\mathbb{I}_{(.)}$ is the indicator function that takes the value 1 if $(\widehat{C}_{Nj} - 570) > 0$. A fair premium to purchase the call option would be one for which this ETV is zero. What is the price of the option?

f. For January, February, and March of 2007, the actual number of cumulative CDDs recorded was 611, meaning that the payoff to the call with a strike value of 570 days would have been $41. Would a call option at the simulated price generated in exercise 2e be in the money at maturity?

12 Longitudinal data models

Earlier chapters have made use of cross-sectional data where each observation represents a unit and time-series data where each observation represents a time period. In the case where there are repeated observations on the same units over time, the data are termed *longitudinal* or *panel* data. In longitudinal data, each variable has both an i subscript indicating the unit and a t subscript referring to time on a defined time-series calendar. In a strongly balanced panel, the data contain N units with T observations available on each unit so that the panel dataset has $N \times T$ observations. In an unbalanced panel, some units have fewer than T observations. This reflects the fact that some units may not have existed (or their data recorded) during the entire set of time periods.

The term panel data was originally coined in the 1960s to refer to large-scale surveys in the United States that were set up to collect data on the same families over time, aimed at obtaining a better understanding of the dynamics of the distribution of income and employment. Examples included the Panel Study of Income Dynamics, which was created in 1968, and the National Longitudinal Survey of Labor Market Experience, which was initiated in 1966. For a recent survey of the origins of panel data, see Sarafidis and Wansbeek (2020) and the references therein. This chapter discusses an important class of econometric techniques applicable to the analysis of longitudinal data, or panel data, whose multidimensional attributes support the consideration of many interesting hypotheses.

12.1 The pollution haven hypothesis

The relationship between environmental policy, the location of production, and subsequent trade flows is important, particularly if a reduction in trade barriers enables polluting firms to locate production activity in jurisdictions with less stringent environmental regulation. This phenomenon, known as the pollution haven hypothesis (PHH), predicts that pollution havens will attract polluting industries so that some measures of economic activity will be related to regulatory stringency.

A well-known dataset produced by Keller and Levinson (2002) covers 48 contiguous U.S. states from 1977 to 1994. The United States is the only country that has collected pollution abatement cost data for a significant period of time in the form of the Pollution Abatement Costs and Expenditures survey, which publishes manufacturers' pollution abatement costs at the four-digit industry level. The measure of economic activity favored is foreign direct investment, particularly for the chemical sector. Figure 12.1 illustrates the distribution of foreign direct investment (FDI) in the chemical industry by state, as well as average pollution abatement costs at the state level. It is evident that there are wide variations in both series and correlations between them. These data are used to illustrate several econometric techniques for the modeling of longitudinal data.

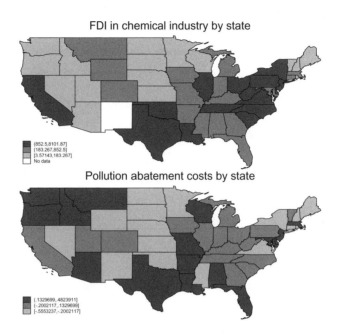

Figure 12.1. Logarithm of foreign direct investment by state for the continental U.S. (top panel) and the logarithm of average pollution abatement costs by state for the continental U.S., 1977–1994 averages (bottom panel)

This sector is chosen because it is reasonable to assume that the chemical industry will respond to regulatory incentives given the pollution-intensive nature of production. This is not necessarily because the chemical industry is the most pollution intensive but perhaps because it is one of the most heavily regulated of all manufacturing industries. In addition to the regulation of its products, it is also subject to several requirements aimed at minimizing releases of chemical substances during manufacturing and processing. Some industries are less geographically mobile than others because of transportation costs, plant fixed costs, or agglomeration economies. Consequently, these less mobile industries will be insensitive to differences in regulatory stringency between countries because they are unable to easily relocate their facilities. Cross-industry regressions that average over multiple industries could conceal the effect of environmental regulations on trade in the more footloose industries.

For all but the most heavily regulated industries, the costs of compliance with U.S. environmental regulation comprise a relatively small portion of total production costs.

In this dataset, environmental costs average around one percent of total material costs. Thus, the stringency of environmental regulations may not be a significant determinant of comparative advantage for most U.S. industries because it may be dwarfed by other determinants of industry location such as labor costs or infrastructure. However, environmental costs comprise a larger share of the total cost for a few pollution-intensive industries such as chemical manufacturing, petroleum, and primary metals. One explanation for why industries with high average pollution abatement costs may be less sensitive to increases in those costs over time is that the more pollution-intensive industries may also be less "footloose", or less able to readily relocate their production.

As is typical in the empirical literature, simple correlations between net imports and environmental regulations fail to uncover a strong relationship. However, there has been little empirical support for the proposition that environmental regulations affect trade. In a survey article, Jaffe et al. (1995) conclude that although environmental regulations impose large and significant costs on polluting industries, these costs have not appreciably affected patterns of international trade. The lack of empirical support for the proposition that environmental costs affect trade flows has been a puzzle in the trade and environment literature.

12.2 Data organization

The Stata command `xtset` or `tsset` (see [XT] **xtset** or [TS] **tsset**) (with two arguments) is used to define data as a panel. The first argument, `i`, is an integer variable identifying the unit, which can take on any positive value. The `i` values need not be sequential, because the variable is used only to group the observations related to that unit. The second argument, `t`, is an integer variable from a time-series calendar. Options to `xtset` or `tsset` define both the time units of the data (such as yearly, quarterly, monthly, or daily) and possibly the periodicity, when data may be associated with a time period but not collected every time period. For instance, the U.S. decennial census takes place every 10 years, so these are annual data with `delta(10)`, telling Stata that the observation preceding the 2000 census is the 1990 census, and so on.

The `tsset` or `xtset` command given with no arguments will report if the data have been declared as a panel, and if so, provide the names of the `i` and `t` variables, and if the panel is *strongly balanced* or *unbalanced*. Some econometric techniques and tests require that a panel be balanced, and the Stata command `spbalance` (see [SP] **spbalance**) can make any dataset into what Stata describes as "strongly balanced" by dropping periods for which fewer than N units are available. This should be done with caution, however, because applying the constraint that every unit is present in every time period leads to *survivorship bias*. The bias arises because those units who remain in the panel are survivors relative to those who leave the panel, and those two groups are likely to have different characteristics. Even though `spbalance` can be used to create a balanced panel, think carefully about why the units are missing before using it. If the missing data are entirely random, then balancing the panel will not induce any bias. If the missing data are related in some systematic fashion, the resulting balanced panel will not be representative of the population, leading to biased findings.

12.2.1 Wide and long forms of panel data

Because panel data have two subscripts, there are two ways in which these data may be represented for data manipulation and statistical analysis, known as the "wide" and "long" forms of the data. In the "wide" form, a particular measurement will be stored as multiple variables, each with T observations. For instance, foreign direct investment levels for each U.S. state, FDI_{it}, might be recorded as variables fdi_MA, fdi_NY, fdi_CA, fdi_TX, and so on. Likewise, if there are annual measures of pollution abatement costs for those states, they would be stored as variables rac_MA, rac_NY, and so on. This form of storage for panel data is quite natural when you are downloading time series for several units (countries, industries, provinces, cities, etc.) from an external source. It is also the appropriate form of storage for the seemingly unrelated regression estimator provided by sureg (see [R] **sureg**).

However, many data manipulation tasks and econometric procedures applied to panel data require that the data be stored in the "long" form. In the long form, the variables associated with a particular measurement are stacked into a single variable.[1] There is then a single variable, fdi, in which the first T observations refer to moving average, the next T to NY, and so on, and likewise for the rac variable. With the data in the long form, many data manipulation tasks become simpler. For instance, computing annual growth rates from a level series requires only one command:

```
generate dlfdi = log(fdi) - log(L.fdi)
```

In the wide format, this command would have to be given separately for each state's variables. On the other hand, some data management tasks are more readily handled in the wide format. For instance, say that each state's pollution abatement costs per capita should be expressed as a deviation from the average (or median) per capita abatement costs across states. In this case, it would be simpler to use egen (see [D] **egen**) to compute the average (or median) series and then use a foreach (see [P] **foreach**) command to compute new variables as deviations from that series before transforming the data to long format:

```
egen acavg = rowmean(racMA-racTX)
foreach v of varlist rac* {
    generate racdev`v´ = rac`v´ - acavg
}
```

Many of Stata's econometric techniques applicable to panel data, generally found in the *Stata Longitudinal-Data/Panel-Data Reference Manual*, require that panel data be stored in the long form. Several estimators and tests also require that the panel is balanced.

---

1. This should be recognized as the vec operator in linear algebra, implemented in Stata as vec().

12.2.2 Reshaping the data

Because panel data may often not be in the preferred form, a tool to transform long data into wide data or vice versa is given by Stata's reshape (see [D] **reshape**) command. The command takes an additional keyword, long or wide, indicating the target form. That is, if the data are currently stored in the wide form, they can be reshaped into the long form. Consider a simple example of three variables (lfdi, lpop, and len) being recorded for three U.S. states (IA, MI, and NC) over seven years:

```
. use http://www.stata-press.com/data/eeus/phhwide
. keep year *IA *MI *NC
. keep year lfdi?? lpop* len*
. replace year = year + 1900
variable year was byte now int
(18 real changes made)
. keep if year<1984
(11 observations deleted)
. format l* %6.3f
. list, noobs sep(0)
```

year	lfdiIA	lpopIA	lenIA	lfdiMI	lpopMI	lenMI	lfdiNC	lpopNC
1977	4.700	14.885	1.001	5.513	16.032	1.179	3.738	14.258

lenNC
0.820

year	lfdiIA	lpopIA	lenIA	lfdiMI	lpopMI	lenMI	lfdiNC	lpopNC
1978	4.990	14.886	1.128	5.617	16.037	1.278	3.761	14.263

lenNC
0.944

year	lfdiIA	lpopIA	lenIA	lfdiMI	lpopMI	lenMI	lfdiNC	lpopNC
1979	5.236	14.886	1.247	5.841	16.042	1.401	3.932	14.265

lenNC
1.125

year	lfdiIA	lpopIA	lenIA	lfdiMI	lpopMI	lenMI	lfdiNC	lpopNC
1980	5.375	14.885	1.545	5.609	16.041	1.609	4.007	14.266

lenNC
1.504

year	lfdiIA	lpopIA	lenIA	lfdiMI	lpopMI	lenMI	lfdiNC	lpopNC
1981	5.587	14.883	1.681	5.945	16.036	1.758	3.892	14.272

lenNC
1.675

year	lfdiIA	lpopIA	lenIA	lfdiMI	lpopMI	lenMI	lfdiNC	lpopNC
1982	5.598	14.876	1.742	5.989	16.025	1.831	4.078	14.274

lenNC
1.726

year	lfdiIA	lpopIA	lenIA	lfdiMI	lpopMI	lenMI	lfdiNC	lpopNC
1983	5.620	14.870	1.785	6.047	16.018	1.890	4.043	14.276

lenNC
1.803

The variables are named systematically, with the attribute (for instance, lfdi, the logarithm of foreign direct investment) followed by the two-letter state abbreviations for Iowa, Michigan, or North Carolina. When each year's variables from this wide form are stacked into a single column (for instance, for lfdi in the long form, each observation is identified by two variables: the state code and the year to which it pertains.

The reshape command is used to transform the data, indicating that year should be the main row identifier (i()) with state, a string variable as the secondary identifier (j()):

```
. reshape long lfdi lpop len, i(year) j(state) string
(note: j = IA MI NC)

Data                              wide    ->    long

Number of obs.                       7    ->      21
Number of variables                 10    ->       5
j variable (3 values)                     ->    state
xij variables:
                    lfdiIA lfdiMI lfdiNC ->    lfdi
                    lpopIA lpopMI lpopNC ->    lpop
                      lenIA lenMI lenNC  ->    len
```

In this example, lfdi lpop len are *stubs*: the part of the variable names in the wide-form variables that are in common. The variable state is created from the varying portion of the variable names. Now, instead of 7 observations on 10 variables, there are 21 observations on 5 variables.

To declare these data as a panel using xtset or tsset, an integer variable must be created to identify each unit of the panel. The encode command (see [D] **encode**) takes the string, assigns arbitrary integers 1..., and creates a variable label equal to the string variable to the variable created in its generate() option. If the variable is examined, it appears to contain the original string content, but it is numeric.

12.2.2 Reshaping the data

```
. encode state, gen(istate)
. xtset istate year, yearly
       panel variable:  istate (strongly balanced)
        time variable:  year, 1977 to 1983
                delta:  1 year
. list istate year l*, noobs sepby(istate)
```

istate	year	lfdi	lpop	len
IA	1977	4.700	14.885	1.001
IA	1978	4.990	14.886	1.128
IA	1979	5.236	14.886	1.247
IA	1980	5.375	14.885	1.545
IA	1981	5.587	14.883	1.681
IA	1982	5.598	14.876	1.742
IA	1983	5.620	14.870	1.785
MI	1977	5.513	16.032	1.179
MI	1978	5.617	16.037	1.278
MI	1979	5.841	16.042	1.401
MI	1980	5.609	16.041	1.609
MI	1981	5.945	16.036	1.758
MI	1982	5.989	16.025	1.831
MI	1983	6.047	16.018	1.890
NC	1977	3.738	14.258	0.820
NC	1978	3.761	14.263	0.944
NC	1979	3.932	14.265	1.125
NC	1980	4.007	14.266	1.504
NC	1981	3.892	14.272	1.675
NC	1982	4.078	14.274	1.726
NC	1983	4.043	14.276	1.803

If the data are in this long form, the commands

```
. drop istate
. reshape wide lfdi lpop len, i(year) j(state) string
(note: j = IA MI NC)
```

Data	long	->	wide
Number of obs.	21	->	7
Number of variables	5	->	10
j variable (3 values)	state	->	(dropped)
xij variables:			
	lfdi	->	lfdiIA lfdiMI lfdiNC
	lpop	->	lpopIA lpopMI lpopNC
	len	->	lenIA lenMI lenNC

would reshape them into the original wide form presented above. The `i()` option specifies the variable that should be used to identify the rows of the wide-format data: in this case `year`. The original `state` variable is used as the argument for option `j()`, the secondary identifier, and declared as a string. This ensures that the wide-form variable names contain the state abbreviations. Because the `istate` variable is no longer needed for this transformation, it is removed before invoking `reshape`.

12.3 The pooled model

Let y_{it} and $\{x_{1it}, x_{2it}, \ldots, x_{Kit}\}$ be, respectively, the dependent variable and K explanatory variables associated with the ith cross section at time t. The simplest possible panel model, pooled ordinary least squares, merely pools the data over time and space, treating each observation's error term ϵ_{it} as an independent and identically distributed (i.i.d.) draw from the population. The model is

$$y_{it} = \alpha + \beta_1 x_{1it} + \cdots + \beta_K x_{Kit} + \epsilon_{it}$$

in which ϵ_{it} is a (possibly heteroskedastic) disturbance term. This model is the most restrictive panel model with a single intercept (α) and K slope coefficients (β_k) that are invariant over time and across the cross sections. It is often the case in panel data that the assumption of homoskedastic disturbances is violated. In some practical situations, different variances are observed for each unit in the panel. To deal with this violation of the homoskedasticity assumption, cluster–robust standard errors are appropriate, with clustering by panel unit using the vce(cluster) option.

To fit the pooled model, the data are stacked to produce an $(NT \times 1)$ vector for the dependent variable and an $\{NT \times (K+1)\}$ matrix for the explanatory variables, where one of the columns of the matrix represents the constant term. The regress (see [R] regress) command with clustered standard errors can be used, and there is no need to tsset or xtset the data.

```
. use http://www.stata-press.com/data/eeus/phh_fix, clear
. regress lfdi lrac lwages lpop len lprox lun lump lmile, vce(cluster id)
Linear regression                               Number of obs   =        607
                                                F(8, 46)        =      72.43
                                                Prob > F        =     0.0000
                                                R-squared       =     0.7773
                                                Root MSE        =     .85824

                          (Std. Err. adjusted for 47 clusters in id)
```

| | | Robust | | | | |
lfdi	Coef.	Std. Err.	t	P>|t|	[95% Conf. Interval]	
lrac	.1571079	.2955217	0.53	0.598	-.4377465	.7519623
lwages	.7296677	.5149212	1.42	0.163	-.3068149	1.76615
lpop	-.1560636	.4670205	-0.33	0.740	-1.096127	.784
len	-1.532631	.3283146	-4.67	0.000	-2.193494	-.8717678
lprox	1.69789	.3174779	5.35	0.000	1.05884	2.33694
lun	-1.498939	.2100705	-7.14	0.000	-1.921789	-1.076089
lump	1.536789	.2347536	6.55	0.000	1.064255	2.009324
lmile	-.4941241	.2601602	-1.90	0.064	-1.017799	.0295512
_cons	1.812669	3.629154	0.50	0.620	-5.49244	9.117777

The pooled model presents an estimate of the average effect of environmental regulations on foreign direct investment in the chemical industry for each state of the continental United States. The explanatory variables in the regression include logarithms of population (lpop), average wages (lwages), union power (lun), unemployment (lunemp), the distance-weighted averages of other states' gross state product (lprox),

12.3 The pooled model

total road mileage (`lmile`), tax effort (`ltax`), land prices (`lland`), and energy prices (`len`), respectively. Contrary to expectations, the coefficient on `lrac`, the abatement costs factor, in this pooled regression is not significantly different from zero. To allow for time variation in the relationship, year indicator variables can easily be included using Stata factor-variable notation.

```
. regress lfdi lrac lwages lpop len lprox lun lump lmile i.year, vce(cluster id)
Linear regression                               Number of obs   =       607
                                                F(22, 46)       =     42.74
                                                Prob > F        =    0.0000
                                                R-squared       =    0.7936
                                                Root MSE        =    .83605

                              (Std. Err. adjusted for 47 clusters in id)
```

		Robust				
lfdi	Coef.	Std. Err.	t	P>|t|	[95% Conf.	Interval]
lrac	.152178	.2908098	0.52	0.603	-.4331918	.7375478
lwages	-.019264	1.028358	-0.02	0.985	-2.089241	2.050713
lpop	.0313179	.4867289	0.06	0.949	-.9484166	1.011052
len	-2.25172	.4232252	-5.32	0.000	-3.103629	-1.399812
lprox	1.595941	.3462451	4.61	0.000	.8989856	2.292896
lun	-1.279919	.3237878	-3.95	0.000	-1.93167	-.6281682
lump	1.206425	.2935204	4.11	0.000	.6155986	1.79725
lmile	-.6123892	.2583236	-2.37	0.022	-1.132368	-.0924107
year						
78	.2601799	.1462276	1.78	0.082	-.0341609	.5545208
79	.6960838	.2627965	2.65	0.011	.1671017	1.225066
80	1.110401	.3852128	2.88	0.006	.3350074	1.885794
81	1.658804	.5238431	3.17	0.003	.6043625	2.713246
82	1.468226	.5682314	2.58	0.013	.3244353	2.612016
83	1.475371	.6097587	2.42	0.020	.2479901	2.702751
84	1.516646	.6584715	2.30	0.026	.1912115	2.84208
85	1.4596	.7115063	2.05	0.046	.0274121	2.891788
86	1.060421	.762514	1.39	0.171	-.4744405	2.595282
87	1.198221	.7636064	1.57	0.123	-.338839	2.735281
88	1.114515	.8068302	1.38	0.174	-.5095495	2.73858
89	1.380421	.8533468	1.62	0.113	-.3372773	3.098119
90	1.557	.8778724	1.77	0.083	-.2100652	3.324066
91	1.531041	.8890998	1.72	0.092	-.2586245	3.320706
_cons	2.769844	3.746158	0.74	0.463	-4.770781	10.31047

```
. testparm i.year

 ( 1)  78.year = 0
 ( 2)  79.year = 0
 ( 3)  80.year = 0
 ( 4)  81.year = 0
 ( 5)  82.year = 0
 ( 6)  83.year = 0
 ( 7)  84.year = 0
 ( 8)  85.year = 0
 ( 9)  86.year = 0
 (10)  87.year = 0
 (11)  88.year = 0
 (12)  89.year = 0
 (13)  90.year = 0
 (14)  91.year = 0

       F( 14,    46) =    5.28
            Prob > F =  0.0000
```

The joint test for significance of time effects indicates that these should be included in the model. Their presence, however, does not change the anomalous result on the coefficient on `lrac`, pollution abatement costs. This observation prompts further investigation.

12.4 Fixed effects and random effects

Using panel data rather than a pure cross-sectional dataset or a pure time-series dataset has clear advantages in terms of the richness of the data but also introduces additional difficulties. Most important of these difficulties is that any unit-specific factor omitted from the model introduces *unobserved heterogeneity*. Heterogeneity refers to the variation across individual units in the data. Because variation related to an omitted variable cannot be measured, the term unobserved heterogeneity is used. The existence of unobserved heterogeneity is a potential source of omitted-variable bias and inconsistency in the estimates. In the current example, even if all relevant state-specific factors are included in the model, no matter how many are included, there may be other relevant differences between states that are nonquantifiable, such as the degree to which the state's residents place a high value on environmental quality. The presence of unobserved heterogeneity usually implies that the simplest panel-data estimator, the pooled ordinary least-squares model, is unreliable. The model displayed above essentially ignores the longitudinal nature of the data and merely estimates a regression over all $N \times T$ observations.

Panel-data models introduce a unit-specific intercept in the model that is essentially a catch-all for unit-specific factors. Without loss of the generality and to economize on notation, in what follows only one explanatory variable, x_{it}, will be used. The individual fixed-effects (FEs) or random-effects (REs) model can then be expressed as

$$y_{it} = \beta x_{it} + \nu_i + \epsilon_{it} \tag{12.1}$$

12.4.1 Individual FEs

in which the β parameter is still constant over time and across units and ϵ_{it} once again is a possibly heteroskedastic disturbance term. The parameter ν_i is assumed to capture the unobserved heterogeneity specific to unit i in the panel. The difference between the FEs and REs models lies in how the ν_i are interpreted.

12.4.1 Individual FEs

If the number of units in the panel is not too large, unobserved heterogeneity can be addressed by creating a set of indicators, or dummy variables, and including them as regressors. For this reason, the individual FEs model is often termed the least-squares dummy variable model. In many practical situations where N is large (in the hundreds or even thousands), the dummy variable form of the regression equation is infeasible because of computational constraints. Accordingly, an algebraic sleight of hand must be used to fit models of this kind based on the idea that each variable in a panel has two sources of variation: the *within* variation, corresponding to the variation in each unit's value over time, and the *between* variation, corresponding to the differences between units' average values.[2]

Given (12.1), it must also be true that

$$\bar{y}_i = \beta \bar{x}_i + \bar{\nu}_i + \bar{\epsilon}_i \qquad (12.2)$$

in which

$$\bar{y}_i = \frac{1}{T}\sum_{t=1}^{T} y_{it}, \qquad \bar{x}_i = \sum_{t=1}^{T} x_{it}, \qquad \bar{\epsilon}_i = \frac{1}{T}\sum_{t=1}^{T} \epsilon_{it}$$

The *between* estimator uses ordinary least squares to fit this transformed model, based on the assumption that $\bar{x}_i$ and ν_i are uncorrelated. Essentially, the between estimator averages out the within variation and uses only the information in the between variation to estimate β, the parameter of interest, in what is now a cross-sectional dataset with a single constant term. Note that any "macro" variable that is constant at each point in time over units will drop out of the between-effects model.

Subtracting (12.2) from (12.1) gives the data in deviations from the mean form given by

$$y_{it} - \bar{y}_i = \beta(x_{it} - \bar{x}_i) + (\epsilon_{it} - \bar{\epsilon}_i) \qquad (12.3)$$

in which the FEs ν_i have been eliminated. The estimates of the FEs for each of the i states are recovered from (12.2) given that $E(\bar{\epsilon}_i) = 0$. These estimates are

$$\widehat{\nu}_i = \bar{y}_i - \widehat{\beta}\bar{x}_i$$

---

2. The `xtsum` (see [XT] **xtsum**) command produces a table showing the overall standard deviation, between standard deviation, and within standard deviation for each variable.

where $\widehat{\beta}$ is the least-squares estimator obtained by estimating (12.3) by ordinary least squares.[3] This is known as the *within* estimator because only the within variation is used when estimating β. The main assumption that is required when fitting this model is that the ν_i are not assumed to have a distribution but are instead treated as fixed. For this reason, the estimator is regarded as being conditional on the sample. The *within* FEs estimator does not require a balanced panel. As long as there are at least two observations per unit, it may be applied. However, because the individual FE is essentially estimated from the observations of each unit, the precision of that effect (and the resulting slope estimates) will depend on the number of observations in that state.

Stata implements both the *between* and the *within* estimators of the FEs model with the **xtreg, be** and **xtreg, fe** (see [XT] **xtreg**) commands, respectively. The commands have a syntax similar to **regress**:

xtreg *depvar* [*indepvars*] [*if*] [*in*], **be** [*options*]

xtreg *depvar* [*indepvars*] [*if*] [*in*] [*weight*], **fe** [*options*]

As with standard regression, options include vce(robust) and vce(cluster). The vce(robust) option implies clustering by the panel ID variable.

---

3. It should also be noted that the estimates of the FEs, $\widehat{\nu}_i$, in this approach are not consistent, because of the incidental parameter problem. As the number of units, N, is increased, each additional unit adds an additional intercept to be estimated, so the precision of the estimated intercepts does not increase as $N \to \infty$.

12.4.1 Individual FEs

Fitting the individual FEs model for state-level FDI in the chemical industry gives the following output:

```
. xtreg lfdi lrac lwages lpop len lprox lun lump lmile, fe
Fixed-effects (within) regression               Number of obs      =        607
Group variable: id                              Number of groups   =         47

R-sq:                                           Obs per group:
     within  = 0.6806                                         min =          1
     between = 0.0165                                         avg =       12.9
     overall = 0.0030                                         max =         15

                                                F(8,552)           =     147.01
corr(u_i, Xb)  = -0.7181                        Prob > F           =     0.0000

------------------------------------------------------------------------------
        lfdi |      Coef.   Std. Err.      t    P>|t|     [95% Conf. Interval]
-------------+----------------------------------------------------------------
        lrac |  -.4218568   .0923929    -4.57   0.000    -.6033415   -.2403721
      lwages |  -1.206439   .6093137    -1.98   0.048    -2.403296   -.0095817
        lpop |  -2.417855    .592555    -4.08   0.000    -3.581794   -1.253917
         len |   .3534795   .1228807     2.88   0.004     .1121085    .5948506
       lprox |   2.174387   .4048215     5.37   0.000     1.379208    2.969567
         lun |  -1.000505   .2057375    -4.86   0.000    -1.404629   -.5963809
        lump |   .559056    .1153714     4.85   0.000     .3324352    .7856767
       lmile |  -1.282613   .4201928    -3.05   0.002    -2.107985   -.4572401
       _cons |   42.24652   8.425848     5.01   0.000     25.69588    58.79717
-------------+----------------------------------------------------------------
     sigma_u |  2.5228566
     sigma_e |  .39864389
         rho |  .97564014   (fraction of variance due to u_i)
------------------------------------------------------------------------------
F test that all u_i=0: F(46, 552) =  48.25              Prob > F = 0.0000
```

There are several things to note about this output.

1. Stata labels the ν_i as u_i. The actual estimates of the FEs are not reported, but the command output displays the estimate of the variance of the FEs, labeled sigma_u. Estimates of the FEs may be obtained using the postestimation command predict *newvarname*, u.

2. Stata fits a model in which the ν_i are taken as deviations from a single constant term, displayed as _cons. Testing that all ν_i are zero is equivalent in the current notation to testing that all ν_i are identical, and unobserved heterogeneity is not an issue. The output provides an F test of the null hypothesis that all the ν_i are zero. A rejection of the null implies that the pooled ordinary least-squares fit of this model yields biased and inconsistent estimates.

3. The variance of the residuals, σ_ϵ^2, is reported as sigma_e.

4. The parameter rho represents the fraction of overall variance of the disturbances due to ν_i.

5. The empirical correlation between ν_i and the regressors of -0.72 is displayed as corr(u_i, Xb). As will become apparent, this estimated correlation plays an important role in the choice between an FEs and an REs formulation of the model.

With state-level heterogeneity accounted for, the estimates now reveal that pollution abatement costs (lrac) have a statistically significant, negative effect on foreign direct investment, with an elasticity estimate of −0.42. That is, a 1% increase in abatement costs is associated with a 0.42% decrease in FDI. The large fraction of the variance due to the ν_i and the F test of the null hypothesis that all the ν_i are zero clearly reject the pooled model.

12.4.2 Two-way FE

The within transformation accounts for unobserved heterogeneity across units, in our case states. But there may also be significant time variation due to macroeconomic factors such as business cycles, federal environmental policy shifts, and the like. Because all the potential macroeconomic factors affecting this relationship are not likely to be quantified, time dummies are introduced for all but one year. This is known as the *two-way* FEs model because there are now indicators for both units and time periods.

The within transformation implies that any time-invariant factor at the state level (such as being landlocked) cannot be included in a model with individual FEs. Likewise, any macro factor that is the same for all states cannot be included in a model with time FEs. For instance, the 10-year Treasury bond rate, a benchmark for firms' cost of funds, is not state-specific, so it cannot be included in such a model. The estimates of the time effects and a test for their joint significance with testparm (see [R] test) are presented below:

12.4.2 Two-way FE

```
. xtreg lfdi lrac lwages lpop len lprox lun lump lmile i.year, fe
Fixed-effects (within) regression              Number of obs      =        607
Group variable: id                             Number of groups   =         47
R-sq:                                          Obs per group:
     within  = 0.7274                                       min =          1
     between = 0.0412                                       avg =       12.9
     overall = 0.0053                                       max =         15

                                               F(22,538)          =      65.24
corr(u_i, Xb)  = -0.6411                       Prob > F           =     0.0000
```

lfdi	Coef.	Std. Err.	t	P>|t|	[95% Conf. Interval]	
lrac	-.2794139	.0890051	-3.14	0.002	-.454254	-.1045738
lwages	-2.14084	.66339	-3.23	0.001	-3.443993	-.8376883
lpop	-1.56652	.6090244	-2.57	0.010	-2.762878	-.3701631
len	.1652627	.2057139	0.80	0.422	-.2388382	.5693635
lprox	1.299084	.4521493	2.87	0.004	.4108894	2.187278
lun	-.9459919	.2120561	-4.46	0.000	-1.362551	-.5294324
lump	.1913957	.1241526	1.54	0.124	-.0524877	.435279
lmile	-.8001443	.4093091	-1.95	0.051	-1.604184	.0038957
year						
78	.1554013	.0983753	1.58	0.115	-.0378454	.348648
79	.3871746	.131311	2.95	0.003	.1292296	.6451196
80	.5003648	.1963542	2.55	0.011	.11465	.8860797
81	1.01868	.2559155	3.98	0.000	.5159644	1.521396
82	1.141055	.2906325	3.93	0.000	.5701416	1.711969
83	1.198705	.3099136	3.87	0.000	.5899161	1.807494
84	1.161877	.3241987	3.58	0.000	.5250269	1.798728
85	1.20803	.3411276	3.54	0.000	.537925	1.878136
86	1.113008	.3411528	3.26	0.001	.4428529	1.783163
87	1.135873	.3507727	3.24	0.001	.4468213	1.824925
88	1.162595	.3613591	3.22	0.001	.4527474	1.872443
89	1.401398	.381027	3.68	0.000	.6529152	2.149881
90	1.684299	.3981127	4.23	0.000	.9022527	2.466344
91	1.868718	.4126589	4.53	0.000	1.058097	2.679338
_cons	33.09077	8.496401	3.89	0.000	16.40058	49.78096
sigma_u	2.2721991					
sigma_e	.37304716					
rho	.97375277	(fraction of variance due to u_i)				

```
F test that all u_i=0: F(46, 538) = 52.07                 Prob > F = 0.0000
```

```
. testparm i.year
 ( 1)  78.year = 0
 ( 2)  79.year = 0
 ( 3)  80.year = 0
 ( 4)  81.year = 0
 ( 5)  82.year = 0
 ( 6)  83.year = 0
 ( 7)  84.year = 0
 ( 8)  85.year = 0
 ( 9)  86.year = 0
 (10)  87.year = 0
 (11)  88.year = 0
 (12)  89.year = 0
 (13)  90.year = 0
 (14)  91.year = 0
       F( 14,   538) =     6.60
            Prob > F =    0.0000

. estimates store FE
```

The key coefficient on lrac, pollution abatement costs, now takes on a smaller value, -0.28, that is obtained by the individual FEs model. In other words, when time variation is accounted for, the coefficient is smaller, but it is still a statistically significant negative value. The test for the presence of individual FEs again strongly rejects the pooled model, and the joint test for significance of time effects indicates that these should be included in the model. Notice that the empirical correlation corr(u_i, Xb) is reported as -0.64.

12.4.3 REs

In the REs model, it is assumed that the ν_i in (12.1) are random variables drawn from a symmetric distribution with zero mean and constant variance. The REs estimator may be interpreted as a weighted average of the estimates produced by the between and within estimators. In particular, the REs estimator is equivalent to estimating

$$y_{it} - \theta \overline{y}_i = \beta (x_{it} - \theta \overline{x}_i) + \{(1-\theta)\nu_i + (\epsilon_{it} - \theta \overline{\epsilon}_i)\} \tag{12.4}$$

where the term in square brackets is regarded as a composite disturbance term and θ is a function of σ_ϵ^2 and σ_ν^2, expressing the relative importance of the two sources of random variation. The REs (ν_i) are assumed to be mean independent of the model disturbances (ϵ_{it}) and the explanatory variable (x_{it}) so that

$$E(\nu_i \epsilon_{it}) = 0, \qquad E(\nu_i x_{it}) = 0$$

Stata can fit the REs model using either a generalized least-squares method or maximum likelihood estimation.

xtreg *depvar* [*indepvars*] [*if*] [*in*], [**re** *options*]

xtreg *depvar* [*indepvars*] [*if*] [*in*] [*weight*], **mle** [*options*]

12.4.3 REs

The first option is the default option for xtreg. Given estimates of the variances $\widehat{\sigma}_\epsilon^2$ and $\widehat{\sigma}_\nu^2$, the quasidifferencing parameter θ is estimated as

$$\widehat{\theta} = 1 - \frac{\widehat{\sigma}_\epsilon}{\sqrt{\widehat{\sigma}_\epsilon^2 + T\widehat{\sigma}_\nu^2}}$$

and (12.4) is estimated on the quasidifferenced data by ordinary least squares. If maximum likelihood is required, the parameter θ is estimated along with all the other parameters of the model using an iterative optimization algorithm.

The REs estimates of the model of state-level FDI in the chemical industry are

```
. xtreg lfdi lrac lwages lpop len lprox lun lump lmile i.year, re
Random-effects GLS regression                   Number of obs     =        607
Group variable: id                              Number of groups  =         47
R-sq:                                           Obs per group:
     within  = 0.7115                                         min =          1
     between = 0.5480                                         avg =       12.9
     overall = 0.6890                                         max =         15
                                                Wald chi2(22)     =    1299.07
corr(u_i, X)  = 0 (assumed)                     Prob > chi2       =     0.0000

------------------------------------------------------------------------------
        lfdi |      Coef.   Std. Err.      z    P>|z|     [95% Conf. Interval]
-------------+----------------------------------------------------------------
        lrac |  -.155489   .0934166    -1.66   0.096    -.338582    .0276041
      lwages | -1.584249   .5169668    -3.06   0.002   -2.597486   -.5710131
        lpop |  .2321511   .3694767     0.63   0.530     -.49201    .9563122
         len | -.4504002   .2114305    -2.13   0.033   -.8647963   -.0360041
       lprox |  .9966757   .2930694     3.40   0.001    .4222701    1.571081
         lun | -.5857364   .1765871    -3.32   0.001   -.9318407   -.2396321
        lump |  .2299098   .1162936     1.98   0.048    .0019785     .457841
       lmile | -.2423064   .1769427    -1.37   0.171   -.5891078     .104495

        year |
          78 |  .1947225   .1067151     1.82   0.068   -.0144351    .4038802
          79 |  .5061586   .1382234     3.66   0.000    .2352458    .7770714
          80 |  .7594351   .2013768     3.77   0.000    .3647438    1.154126
          81 |  1.355813   .2588471     5.24   0.000    .8484825    1.863144
          82 |  1.472121   .2932677     5.02   0.000     .897327    2.046915
          83 |  1.538104    .312202     4.93   0.000    .9261992    2.150009
          84 |   1.53527   .3254973     4.72   0.000    .8973074    2.173233
          85 |  1.582343   .3410053     4.64   0.000    .9139852    2.250701
          86 |  1.421767   .3390824     4.19   0.000    .7571774    2.086356
          87 |  1.449844   .3475891     4.17   0.000    .7685823    2.131107
          88 |  1.462733    .356319     4.11   0.000    .7643607    2.161106
          89 |  1.724775   .3752557     4.60   0.000    .9892876    2.460263
          90 |   1.99864   .3907947     5.11   0.000    1.232696    2.764583
          91 |  2.152938   .4035003     5.34   0.000    1.362092    2.943784

       _cons |  .8072864   2.958714     0.27   0.785   -4.991686    6.606259
-------------+----------------------------------------------------------------
     sigma_u |  .6158525
     sigma_e | .37304716
         rho | .73157051   (fraction of variance due to u_i)
------------------------------------------------------------------------------
```

```
. testparm i.year
 ( 1)  78.year = 0
 ( 2)  79.year = 0
 ( 3)  80.year = 0
 ( 4)  81.year = 0
 ( 5)  82.year = 0
 ( 6)  83.year = 0
 ( 7)  84.year = 0
 ( 8)  85.year = 0
 ( 9)  86.year = 0
 (10)  87.year = 0
 (11)  88.year = 0
 (12)  89.year = 0
 (13)  90.year = 0
 (14)  91.year = 0

         chi2( 14) =    84.36
       Prob > chi2 =    0.0000

. estimates store RE
```

The estimate of the key coefficient on lrac is now reported as -0.16 and is barely distinguishable from zero at the 10% level of significance. Notice in the output that the application of REs implies the assumption that the correlation corr(u_i, Xb) is zero. If that assumption is not appropriate, the REs estimator will be inconsistent because the composite error term will be correlated with the regressors by construction. The validity of this assumption can be assessed by means of a Hausman test.

12.4.4 The Hausman test in a panel context

In actual empirical work, it is extremely unusual to find that the key assumption underlying the REs model is satisfied. Beyond textbook examples, it is difficult to find instances where the unobserved random effect can plausibly be uncorrelated with all observable attributes of the unit.

Because the estimator has been applied to state-level data on FDI, the state-specific random component of the error term is attributed to a draw from nature that is uncorrelated with all observable characteristics of the state's performance. This is an extreme assumption, and thus valid examples of REs are hard to find in practice. If it was possible to satisfy this assumption, the REs estimator would be preferred on the basis of efficiency because it does not consume the N degrees of freedom needed for the estimation of means by the within estimator and allows the data to produce an estimate of θ rather than setting it to 1. The REs model, subtracting $\widehat{\theta}$ of the unit's mean, allows for the inclusion of unit-specific factors, such as a state being landlocked.

A formal test for the FEs versus REs estimators of a regression model can be developed from the observation that the REs estimator is based on the fundamental assumption that the REs (ν_i) are mean independent of the explanatory variable (x_{it}) that is,

$$E(\nu_i x_{it}) = 0$$

12.4.4 The Hausman test in a panel context

which generalizes to any number of explanatory variables. If this assumption fails, the FEs model is preferred because this estimator is still consistent but inefficient. This is known as a Hausman test. Formally, the hypotheses are

$$H_0: \quad \text{REs}$$
$$H_1: \quad \text{FEs}$$

To perform the test, let $\widehat{\boldsymbol{\beta}}_{\text{FE}}$ and $\widehat{\boldsymbol{\beta}}_{\text{RE}}$ represent, respectively, the fixed and REs estimators of the slope parameters that are in common in the two parameter vectors. The general form of the test is based on the Wald statistic

$$\text{WD} = \left(\widehat{\boldsymbol{\beta}}_{\text{FE}} - \widehat{\boldsymbol{\beta}}_{\text{RE}}\right)' \left\{\text{cov}\left(\widehat{\boldsymbol{\beta}}_{\text{FE}}\right) - \text{cov}\left(\widehat{\boldsymbol{\beta}}_{\text{RE}}\right)\right\}^{-1} \left(\widehat{\boldsymbol{\beta}}_{\text{FE}} - \widehat{\boldsymbol{\beta}}_{\text{RE}}\right)$$

which is distributed asymptotically under the null hypothesis as χ^2_R, where R represents the number of explanatory variables in common. Under the null hypothesis, the REs and FEs estimators of each parameter are both consistent and should not be statistically distinguishable from each other. In this case, the Wald statistic will not be significant, and the REs estimator, which is also an efficient estimator, is preferred. Under the alternative, the REs estimator is no longer consistent, resulting in a significant Wald statistic and an indication that the REs estimates are unreliable.

This form of the Wald test is due to Hausman (1978), who showed that

$$\text{cov}\left(\widehat{\boldsymbol{\beta}}_{\text{FE}} - \widehat{\boldsymbol{\beta}}_{\text{RE}}\right) = \text{cov}\left(\widehat{\boldsymbol{\beta}}_{\text{FE}}\right) - \text{cov}\left(\widehat{\boldsymbol{\beta}}_{\text{RE}}\right)$$

We apply the Hausman test to the FEs and REs forms of our model, including time effects, after saving and naming the estimates from each form of the model with the **estimates store** (see [R] **estimates store**) command.

```
. hausman FE RE, sigmamore
                 ──── Coefficients ────
                   (b)          (B)          (b-B)       sqrt(diag(V_b-V_B))
                   FE           RE           Difference          S.E.
         lrac    -.2794139    -.155489      -.123925          .0307564
        lwages   -2.14084     -1.584249     -.5565911         .5197005
         lpop    -1.56652      .2321511     -1.798672         .5624653
          len     .1652627    -.4504002      .6156629         .0834714
        lprox    1.299084      .9966757      .3024082         .4046348
          lun    -.9459919    -.5857364     -.3602554         .154021
         lump     .1913957     .2299098     -.0385141         .0727742
        lmile    -.8001443    -.2423064     -.557838          .4162325
         year
           78     .1554013     .1947225     -.0393212         .0206957
           79     .3871746     .5061586     -.118984          .0441293
           80     .5003648     .7594351     -.2590703         .0807639
           81    1.01868      1.355813      -.3371331         .1138618
           82    1.141055     1.472121      -.331066          .1308746
           83    1.198705     1.538104      -.3393988         .14072
           84    1.161877     1.53527       -.3733929         .1496129
           85    1.20803      1.582343      -.3743129         .1606245
           86    1.113008     1.421767      -.308759          .1647086
           87    1.135873     1.449844      -.3139712         .1715076
           88    1.162595     1.462733      -.300138          .1802077
           89    1.401398     1.724775      -.3233772         .1909165
           90    1.684299     1.99864       -.3143411         .201989
           91    1.868718     2.152938      -.2842206         .2123855

                            b = consistent under Ho and Ha; obtained from xtreg
             B = inconsistent under Ha, efficient under Ho; obtained from xtreg
    Test:  Ho:  difference in coefficients not systematic
                 chi2(22) = (b-B)'[(V_b-V_B)^(-1)](b-B)
                          =    123.39
              Prob>chi2 =     0.0000
```

The Hausman test strongly rejects its null hypothesis, suggesting that the FEs estimates should be relied upon and the REs estimates should be discarded. In the context of our empirical example, the FEs model delivers economically meaningful and statistical significant effects of pollution abatement costs.

12.4.5 Correlated RE

The Hausman test produces a common finding, namely, that the assumption of REs being uncorrelated with the regressors is unlikely to hold. A method that addresses this finding is the correlated random-effects (CRE) approach of Mundlak (1978) and Wooldridge (2010), in which the REs are modeled in terms of observable variables. This device then allows the estimation of the effects of time-invariant variables like the pure REs model. Following Schunck (2013) and Schunck and Perales (2017), the CRE model can be written as

$$y_{it} = \beta_0 + \beta_w x_{it} + \beta_2 c_i + \pi \overline{x}_i + \mu_i + \epsilon_{it}$$

12.4.5 Correlated RE

in which β_w correspond to the within (FE) estimates, $\overline{x}_i$ are panel-specific means of the regressors, and π indicates the difference between within and between estimates, $(\beta_w - \beta_b)$. The μ_i denote individual REs uncorrelated with the error term (ϵ_{it}) and the other explanatory variables (x_{it}), which are modeled as functions of the panel-level group means $\overline{x}_i$. In this context, if $H_0 : \pi = 0$ cannot be rejected, a pure REs model would be appropriate. Under the alternative $H_1 : \pi \neq 0$, the data support the CRE specification and reject the hypothesis that the REs are uncorrelated with the regressors. Unlike the pure REs model, in which the REs are assumed i.i.d., the CRE model allows for other covariance structures such as clustered standard errors.

As Schunck (2013) points out, the CRE model is numerically equivalent to a hybrid model formulation given by

$$y_{it} = \beta_0 + \beta_w(x_{it} - \overline{x}_i) + \beta_2 c_i + \beta_b \overline{x}_i + \mu_i + \epsilon_{it}$$

from which both within and between estimates can be directly obtained. The hybrid model formulation might be preferred to the CRE specification because the between group estimates have a direct interpretation. While the within estimate β_w shows the effect of a variable that varies over time on the outcome for an individual panel, the between estimate β_b can be interpreted as the long-term impact of that variable.

The hybrid model can be illustrated with the PHH dataset using the community-contributed `xthybrid` command by Schunck and Perales (2017). Both within and between coefficients can be computed for the regressors with nontrivial within variation. The regressors `lpop` and `lmile` have less than 1% of their variation over time, so they are included essentially as time-invariant regressors in `xthybrid`. Because this command does not handle factor variables, the `xi` command is used to generate time effects.

```
. xi i.year
i.year            _Iyear_77-94       (naturally coded; _Iyear_77 omitted)
. xthybrid lfdi lrac lwages lpop len lprox lun lump lmile _Iyear_78-_Iyear_91,
> use(lrac lwages len lprox lun lump) clusterid(id) full test vce(cluster id)
```

Model model

Mixed-effects GLM		Number of obs	=	607
Family: Gaussian				
Link: identity				
Group variable: id		Number of groups	=	47

Obs per group:
min = 1
avg = 12.9
max = 15

Integration method: mvaghermite   Integration pts. = 7

Wald chi2(28) = 1499.94
Log pseudolikelihood = -339.7337   Prob > chi2   =   0.0000

(Std. Err. adjusted for 47 clusters in id)

lfdi	Coef.	Robust Std. Err.	z	P>|z|	[95% Conf. Interval]	
R__lpop	-1.040556	.6264296	-1.66	0.097	-2.268336	.1872231
R__lmile	-.4994911	.2556948	-1.95	0.051	-1.000644	.0016616
R___Iyear_78	.176571	.1086223	1.63	0.104	-.0363248	.3894668
R___Iyear_79	.4286151	.2172364	1.97	0.048	.0028396	.8543905
R___Iyear_80	.577793	.3183369	1.82	0.070	-.0461358	1.201722
R___Iyear_81	1.131445	.4371174	2.59	0.010	.2747105	1.988179
R___Iyear_82	1.271474	.4781895	2.66	0.008	.3342401	2.208708
R___Iyear_83	1.34558	.5121434	2.63	0.009	.341797	2.349362
R___Iyear_84	1.332694	.5417924	2.46	0.014	.2708003	2.394588
R___Iyear_85	1.397182	.5793337	2.41	0.016	.2617092	2.532656
R___Iyear_86	1.315917	.5971521	2.20	0.028	.1455209	2.486314
R___Iyear_87	1.353855	.6144552	2.20	0.028	.1495447	2.558165
R___Iyear_88	1.392488	.647497	2.15	0.032	.1234167	2.661558
R___Iyear_89	1.64819	.6779954	2.43	0.015	.3193436	2.977037
R___Iyear_90	1.948104	.7195594	2.71	0.007	.5377931	3.358414
R___Iyear_91	2.153595	.7397863	2.91	0.004	.7036401	3.603549
W__lrac	-.2618652	.1270909	-2.06	0.039	-.5109587	-.0127716
W__lwages	-2.088749	1.212446	-1.72	0.085	-4.465099	.2876007
W__len	.1715224	.4354541	0.39	0.694	-.6819519	1.024997
W__lprox	.9604426	.7812432	1.23	0.219	-.570766	2.491651
W__lun	-.8713497	.3460898	-2.52	0.012	-1.549673	-.1930262
W__lump	.1283581	.1780832	0.72	0.471	-.2206787	.4773948
B__lrac	.1719393	.5560246	0.31	0.757	-.9178489	1.261728
B__lwages	2.224552	1.314637	1.69	0.091	-.3520886	4.801193
B__len	-2.51954	.7627016	-3.30	0.001	-4.014408	-1.024673
B__lprox	2.38499	.5034469	4.74	0.000	1.398252	3.371727
B__lun	-1.878156	.346674	-5.42	0.000	-2.557625	-1.198688
B__lump	2.243784	.6424947	3.49	0.000	.9845177	3.503051
_cons	6.577805	5.599564	1.17	0.240	-4.397138	17.55275

12.4.5 Correlated RE

```
         id  |
    var(_cons)|   .4730575    .079255                      .3406477    .6569348
             |
    var(e.lfdi)|  .1341683   .0215188                       .097978    .1837263

Tests of the random effects assumption:
  _b[B__lrac] = _b[W__lrac]; p-value: 0.4429
  _b[B__lwages] = _b[W__lwages]; p-value: 0.0064
  _b[B__len] = _b[W__len]; p-value: 0.0025
  _b[B__lprox] = _b[W__lprox]; p-value: 0.0906
  _b[B__lun] = _b[W__lun]; p-value: 0.0718
  _b[B__lump] = _b[W__lump]; p-value: 0.0016
```

The `lrac` variable, measuring abatement costs, has a statistically significant short-run effect, `W__lrac`, but an insignificant long-run effect, `B__lrac`, while other regressors such as `len` and `lprox` exhibit the opposite pattern, having strong long-run effects.

To illustrate the hybrid model's utility with a time-invariant regressor, consider the variable `alwaysRep`. This variable is a state-level indicator of political affiliation in terms of U.S. Electoral College votes in the five presidential elections of 1996–1992 during the study period. During this period, 11 states voted for the Republican candidate in every election, while only one state, Minnesota, consistently voted for the Democratic candidate. This state-level variable is introduced into the hybrid model to consider whether political leanings may also be relevant in this context.

```
. use http://www.stata-press.com/data/eeus/phh_fix_aug
. xthybrid lfdi lrac lwages lpop len lprox lun lump lmile alwaysRep
> _Iyear_78-_Iyear_91, use(lrac lwages len lprox lun lump) clusterid(id)
> full keep vce(cluster id)
```

Model model

Mixed-effects GLM		Number of obs	=	607
Family: Gaussian				
Link: identity				
Group variable: id		Number of groups	=	47
		Obs per group:		
		min	=	1
		avg	=	12.9
		max	=	15
Integration method: mvaghermite		Integration pts.	=	7
		Wald chi2(29)	=	1557.66
Log pseudolikelihood = -338.11729		Prob > chi2	=	0.0000

(Std. Err. adjusted for 47 clusters in id)

lfdi	Coef.	Robust Std. Err.	z	P>|z|	[95% Conf. Interval]	
R__lpop	-1.108564	.6419574	-1.73	0.084	-2.366777	.1496494
R__lmile	-.4510817	.2575429	-1.75	0.080	-.9558565	.053693
R__alwaysRep	-.6219524	.3000185	-2.07	0.038	-1.209978	-.033927
R___Iyear_78	.1790418	.1078024	1.66	0.097	-.0322472	.3903307
R___Iyear_79	.4344028	.2161351	2.01	0.044	.0107857	.8580199
R___Iyear_80	.5885483	.3170395	1.86	0.063	-.0328376	1.209934
R___Iyear_81	1.147315	.4360071	2.63	0.009	.2927564	2.001873
R___Iyear_82	1.289873	.4765821	2.71	0.007	.3557896	2.223957
R___Iyear_83	1.364538	.510289	2.67	0.007	.3643896	2.364686
R___Iyear_84	1.351491	.5401442	2.50	0.012	.292828	2.410154
R___Iyear_85	1.416919	.5775629	2.45	0.014	.2849167	2.548922
R___Iyear_86	1.335029	.5951015	2.24	0.025	.1686515	2.501406
R___Iyear_87	1.373894	.6125052	2.24	0.025	.1734057	2.574382
R___Iyear_88	1.411825	.645407	2.19	0.029	.146851	2.6768
R___Iyear_89	1.668185	.6759962	2.47	0.014	.3432568	2.993113
R___Iyear_90	1.969852	.7174109	2.75	0.006	.5637528	3.375952
R___Iyear_91	2.176037	.7370899	2.95	0.003	.7313674	3.620707
W__lrac	-.260221	.1261985	-2.06	0.039	-.5075656	-.0128764
W__lwages	-2.165752	1.211731	-1.79	0.074	-4.540702	.2091983
W__len	.1671645	.4356673	0.38	0.701	-.6867277	1.021057
W__lprox	.999871	.7853105	1.27	0.203	-.5393093	2.539051
W__lun	-.8726905	.3446954	-2.53	0.011	-1.548281	-.1970998
W__lump	.1333424	.1789195	0.75	0.456	-.2173334	.4840183
B__lrac	-.0720838	.5419268	-0.13	0.894	-1.134241	.9900732
B__lwages	2.041622	1.247429	1.64	0.102	-.4032945	4.486538
B__len	-3.298026	.8720232	-3.78	0.000	-5.00716	-1.588892
B__lprox	2.360298	.4970549	4.75	0.000	1.386088	3.334507
B__lun	-1.896446	.3470484	-5.46	0.000	-2.576648	-1.216244
B__lump	1.899699	.6127483	3.10	0.002	.6987346	3.100664
_cons	9.753441	5.957664	1.64	0.102	-1.923366	21.43025

id				
var(_cons)	.439762	.0836087	.3029599	.6383374
var(e.lfdi)	.1341891	.0215137	.0980053	.1837321

```
Please remember to remove any variables beginning with the prefix B__, W__ or
> R__ from the data before executing xthybrid again
```

The results indicate that the time-invariant regressor, `alwaysRep`, has a statistically significant negative effect on `lfdi`.

12.5 Dynamic panel-data models

The panel-data models discussed thus far specify a relationship between the current value of the dependent variable y_{it} and current and lagged values of the explanatory variables, x_{it}. This sort of specification may not be adequate to model many economic, social, and physical processes, because it implicitly assumes that when the data are observed at time t, they are in equilibrium. Depending on the frequency at which observations are made, that may be an inappropriate assumption. For instance, in long-term environmental projects such as cleanup efforts of waterways or brownfield sites, observed variables are likely to be on a trajectory toward an equilibrium in terms of some partial adjustment process. Consequently, it is reasonable to assume that the level of y_{it} depends on its own lag ($y_{i\,t-1}$) and possibly higher-order lags, as recognized previously in the specification of dynamic models in chapter 4.

In the context of panel data, a problem immediately arises. If a single unit's level of y_{it} is modeled as determined by $y_{i\,t-1}$ and a set of exogenous factors x_{it}, an ordinary least-squares regression would still yield unbiased and consistent estimates if $y_{i\,t-1}$ is considered as a predetermined variable.[4] Given this result, a straightforward extension of the earlier approach to panel-data modeling could be expressed as

$$y_{it} = \nu_i + \beta x_{it} + \rho y_{i\,t-1} + \epsilon_{it}$$

where ϵ_{it} is the disturbance term assumed to be independent over time, $E(\epsilon_{it}\epsilon_{is}) = 0$, $t \neq s$. The parameter ρ controls the strength of the dynamics. For the case where $-1 < \rho < 1$, from the discussion in chapter 4, the dynamics are stationary, with the effects of shocks on the dependent variable dissipating over time.

To remove the panel-specific intercept ν_i, the within transformation is applied, as described in section 12.4.1. The centered lagged dependent variable $(y_{i\,t-1} - \bar{y}_{i,t-1})$ is correlated by construction with the centered disturbance term, $(\epsilon_{it} - \bar{\epsilon}_{it})$, thus violating the conditions needed for least squares to yield consistent estimates. Nickell (1981) shows that $\hat{\rho}$ is biased and that for reasonably large values of T, the approximate size of the bias is

$$\plim_{N \to \infty} (\hat{\rho} - \rho) \simeq -\frac{1+\rho}{T-1}$$

---

[4]. An exception arises in the case where the error process is autocorrelated.

Nickell bias, as it is known, appears in both the coefficient of the lagged dependent variable and the coefficients of exogenous regressors, rendering the estimates unreliable.[5]

One solution to this problem involves taking first differences of the original model. Consider a model containing a lagged dependent variable and a single regressor X:

$$y_{it} = \nu_i + \rho y_{i\,t-1} + \beta x_{it} + \epsilon_{it}$$

The first-difference transformation removes the individual effect

$$\Delta y_{it} = \rho \Delta y_{i\,t-1} + \beta \Delta x_{it} + \Delta \epsilon_{it}$$

There is correlation between the differenced lagged dependent variable and the disturbance process, but with the individual effects removed, an instrumental-variables estimator is available.[6] Instruments can be constructed for the lagged dependent variable from the second and third lags of y_{it}, in the form of either differences or lagged levels. If ϵ_{it} is i.i.d., those lags of y_{it} will be highly correlated with the lagged dependent variable (and its difference) but uncorrelated with the composite error process. This approach is the Anderson–Hsiao (Anderson and Hsiao 1981) estimator implemented by the Stata command xtivreg, fd (see [XT] **xtivreg**).

In the empirical literature, the most commonly used approach to the dynamic panel-data (DPD) model is based on the work of Holtz-Eakin, Newey, and Rosen (1988) and Arellano and Bond (1991), which uses a generalized method of moments (GMM) estimator to construct more efficient estimates of the DPD model. Arellano and Bover (1995) and Blundell and Bond (1998) recognized that lagged levels are often rather poor instruments for first-differenced variables, especially if the variables are close to a random walk. Their modification of the estimator includes lagged levels as well as lagged differences and is known as the system GMM estimator. An excellent alternative to Stata's official commands[7] is David Roodman's **xtabond2**, which is well documented in Roodman (2009). The **xtabond2** command provides flexibility and several additional features not available in official Stata commands. A detailed discussion of these approaches to estimation of the DPD model is beyond the scope of this introductory text. An exercise for this chapter provides an example of its use.

In summary, the widespread availability of longitudinal data for environmental measures has made it possible to take advantage of a broad set of empirical methodologies for their analysis. Although there are many advantages to longitudinal data, there are also several issues that arise in this context: unobserved heterogeneity, the modeling of dynamics, and the like. These issues require careful consideration of the appropriate techniques to be used.

---

5. Note also that this bias is not caused by an autocorrelated error process ϵ. The bias arises even if the error process is i.i.d. If the error process is autocorrelated, the problem is even more severe given the difficulty of deriving a consistent estimate of the autoregressive parameters in that context.
6. The former contains $y_{i\,t-1}$ and the latter contains $\epsilon_{i\,t-1}$.
7. The official commands include xtabond, xtdpd, and xtdpdsys (see [XT] **xtabond**, [XT] **xtdpd**, and [XT] **xtdpdsys**).

Exercises

1. The vector autoregressive (VAR) model introduced in chapter 4 was extended to a panel setting by Holtz-Eakin, Newey, and Rosen (1988), and a Stata implementation of a panel VAR estimator, pvar, was provided by Abrigo and Love (2016). Their implementation considers a homogenous panel VAR of order p with panel-specific FEs. The coefficient matrices of the VAR for lagged values of the set of y_{jt} variables and any exogenous variables included as X_t are assumed to be common across panel units. The inclusion of lagged dependent variables in a FEs context creates the problem of Nickell bias, so an instrumental-variables GMM approach is used to produce consistent parameter estimates. The panel-specific FEs can be removed by first differencing (as in the DPD approach) or via the forward orthogonal deviation transformation, which minimizes data loss. Fitting the panel VAR as a system of equations may produce efficiency gains relative to estimating the individual equations separately. As discussed in the case of the seemingly unrelated regression estimator, estimating the system also allows for cross-equation hypothesis testing. Just as in VARs applied to a single unit, all the features of VAR estimation are available in the panel VAR setting. Impulse–response functions, forecast-error vector decompositions, and multiperiod forecasts can be computed with the pvar suite of commands.

   phh.dta contains the PHH data for 48 states of the United States.

   a. Use the pvar command to fit a panel VAR(2) model for the first differences of the variables len, lump, and lfdi are jointly modeled using the first differences of lpop, lwages, lprox, lun, and lrac as exogenous variables. Is there statistical evidence to support the choice of two lags?
   b. Check the estimated panel VAR for dynamic stability using the pvarstable command. Comment on the results.
   c. Examine the Granger causality relationships embedded in the panel VAR using the pvargranger command.
   d. Examine the effect on foreign direct investment of an orthogonalized one standard deviation shock to energy prices. Outline the assumptions you have made to allow identification of the orthogonalized shocks.
   e. Compute the variance decomposition of the panel VAR. Do the results accord with your assumptions?

2. The pollution haven model can be viewed as a DPD by expressing FDI as a function of its lagged value, as well as controls lrac, lwages, and lpop. This exercise relies upon the features of xtabond2 as described in Roodman (2009).

   a. Fit the model of lfdi using the Anderson–Hsiao first-difference estimator, instrumenting the lagged value of lfdi with its second and third lags, with phh.dta. Include those states that have at least 10 years of data on lfdi. Comment on the results.
   b. Fit this model of lfdi using the Arellano–Bond two-step difference GMM estimator, using the xtabond2 command, to the data for those states that

have at least 10 years of data on lfdi. Use the first 5 lags of L.lfdi as instruments for L.lfdi. Comment on the results.

c. Compute the long-run (or steady-state) multiplier for abatement costs using the point and interval estimates from exercise 2b, using nlcom (see [R] **nlcom**).

d. Fit this model of lfdi using the Blundell–Bond two-step system GMM estimator, using the xtabond2 command, to the data for those states that have at least 10 years of data on lfdi. Use the first 5 lags of L.lfdi as instruments for L.lfdi. Comment on the results.

e. Compute the long-run (or steady-state) multiplier for abatement costs using the point and interval estimates from exercise 2b, using nlcom.

13 Spatial models

Environmental data are often characterized by the fact that they are measured at specific locations whose coordinates are typically known. With the increasing availability of environmental monitoring of sea levels, temperatures, and other climactic conditions at several stations, spatial econometric techniques that account for the spatiotemporal relationships related to the underlying geography may be particularly advantageous to the applied researcher. Indeed, in many cases the spatial dimension of the data may be the most important aspect of the problem. Standard econometric models would start by assuming that all observations i, where i represents a particular location, are independent of observations made at other locations. By contrast, spatial models start by assuming that the values observed at location i depend on the values measured at neighboring locations. This assumption gives rise to the need to develop econometric models capable of dealing with spatial dependence and spatial heterogeneity in the data.

Spatial data also give rise to the problem of simultaneity, in that variables measured at i depend on variables measured at j, and the reverse is also true. Trying to capture this simultaneity can potentially lead to problems because the number of parameters in the model can quickly overwhelm the available sample size. The solution to this difficulty is to impose a structure on the spatial interactions. One of the key tools in spatial econometrics is the spatial weighting matrix. Typically, the spatial weighting matrix is defined in terms of the inverse distance between the locations of observations i and j, so that each element of the matrix is of the form $1/d_{ij}$ where d_{ij} is the absolute distance between i and j. The concept of distance can also be taken to mean an economic or political distance so that the elements of the weighting matrix relate to differences in socioeconomic characteristics.

This chapter discusses simple models for fitting spatial models using environmental data and serves as an introduction to the topic. The analysis of spatial data lends itself to the use of matrix notation, so this will be one of the instances in the book when matrix and vector notation will be used. The data that will be used here are cross-sectional data to provide a tractable introduction to the subject although Stata is capable of handling longitudinal spatial models. A good introductory text is by LeSage and Pace (2009) and a more advanced discussion can be found in Anselin (1988).

13.1 Regulatory compliance

In recent years, there has been tremendous growth in the number of international agreements on environmental issues such as ozone depletion, pollution, and the overexploitation of land and marine mammals, birds and fish. The evidence on whether these agreements are effective is an interesting empirical question, as is the related issue of why a country would participate in international agreements when they can enjoy the benefits of a free ride. One interesting line of argument is that strategic interaction between governments is an important consideration when deciding whether to join an international agreement. For example, in testing the pollution haven hypothesis using state-level United States data discussed in chapter 12, states with a more strict regulatory environment may have a positive pull effect on neighboring states' decisions on abatement (Fredriksson and Millimet 2002). Similarly, Murdoch, Sandler, and Vijverberg (2003) provide evidence of a spatial element to cooperation by signatories of the Helsinki Protocol on Sulphur Emissions.

Davies and Naughton (2014) provide empirical evidence to support the hypothesis that a county's decision to participate in an international agreement on the environment is positively related to the number of agreements its neighbors have signed. However, this evidence relates only to the participation decision and does not address the spatial element of enforcement effort. In a recent study, Borsky and Raschky (2015) examine the compliance of signatory countries with the voluntary obligations of Article 7 of the 1995 United Nations Code of Conduct (CoC) for Responsible Fisheries. The idea is to measure the effect of spatial influences, such as geographical and political distance between the signatory nations, on their compliance effort to determine the extent of intergovernmental interaction on compliance behavior. From a theoretical perspective, there are good reasons to expect a spatial element to compliance. The degree of (non)compliance is more easily observed for countries that are geographically proximate, the scale of economic transactions and political relations decrease with distance, and the treaty itself is a common good that has a strong local impact.

The dataset used in this chapter is based on that used by Pitcher et al. (2009) who evaluated the compliance performance of all 53 signatory countries using a set of 44 different score variables.[1] The dependent variables are two index variables—`behavior` and `intention`—that range from 0 to 10, with higher scores indicating better compliance. The behavior indicator is constructed from score variables that have an immediate effect on fishing stocks (fleet capacity, fishing methods, action to restore depleted stocks). It turns out that the indices of `behavior` and `intention` are largely similar, so attention here is focused on `behavior`.

The spatial distribution of the `behavior` measure of compliance is illustrated on the choropleth map in figure 13.1, in which the shading is proportional to the `behavior` measure. Dark shaded areas indicate good compliance, and lighter shaded areas indicate intermediate and poor compliance. The areas with no shading are those for which there are no data. There certainly seems to be casual empirical evidence to support the notion that there are spatial differences in the distribution of these compliance indicators. For example, there is distinct clustering of compliance in the countries of the European Union.

---

[1]. The data were provided by Stefan Borsky of the University of Southern Denmark and Paul Raschky of Monash University.

13.1 Regulatory compliance

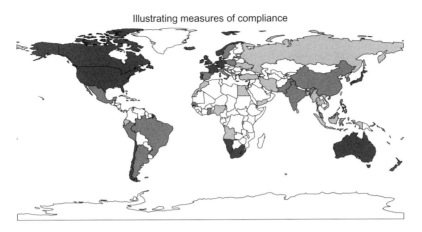

Figure 13.1. Choropleth map of the fishing compliance indicator `behavior` for 53 signatory countries to the United Nations CoC on responsible fishing. The dark shaded areas indicate good compliance, and lighter shaded areas indicate intermediate and poor compliance. The areas with no shading are those for which there are no data.

As a check on this claim, the summary statistics of the `behavior` measure for European Union and non-European Union members are as follows:

```
. use http://www.stata-press.com/data/eeus/spatialfish_small
. by eu_dum, sort: summarize behavior

-> eu_dum = 0
    Variable |       Obs        Mean    Std. Dev.       Min        Max
-------------+--------------------------------------------------------
    behavior |        39    3.322955    1.866542    .8650794   7.206349

-> eu_dum = 1
    Variable |       Obs        Mean    Std. Dev.       Min        Max
-------------+--------------------------------------------------------
    behavior |        12    4.053571    .608403     2.833333   4.753968

. ttest behavior, by(eu_dum)
Two-sample t test with equal variances
------------------------------------------------------------------------------
   Group |     Obs        Mean    Std. Err.   Std. Dev.   [95% Conf. Interval]
---------+--------------------------------------------------------------------
       0 |      39    3.322955    .2988859    1.866542    2.717892    3.928018
       1 |      12    4.053571    .1756308    .608403     3.667011    4.440132
---------+--------------------------------------------------------------------
combined |      51    3.494865    .2354481    1.681436    3.021953    3.967776
---------+--------------------------------------------------------------------
    diff |             -.7306166   .5508982               -1.837688    .3764548
------------------------------------------------------------------------------
    diff = mean(0) - mean(1)                                  t =  -1.3262
Ho: diff = 0                                 degrees of freedom =       49

    Ha: diff < 0                 Ha: diff != 0                 Ha: diff > 0
 Pr(T < t) = 0.0955         Pr(|T| > |t|) = 0.1909          Pr(T > t) = 0.9045
```

The mean of `behavior` for the European Union countries is higher than that of non-European Union countries, indicating better compliance. However, the size of the estimated standard errors mean that the difference may not be statistically significant. This observation is verified by the formal t test of equality of the means. Given that the dataset is not confined to European Union members, however, there is sufficient evidence to suggest that there is a spatial element in the issue of compliance with Article 7 of the United Nations CoC.

13.2 The spatial weighting matrix

A central concept in spatial models is the spatial weighting matrix. Consider a simple case in which what happens at one location affects all other locations,

$$y_i = f(y_j), \quad \text{for all } j \neq i$$

If, in addition, we are entitled to assume a linear relationship, then

$$\begin{bmatrix} y_1 \\ y_2 \\ \vdots \\ y_n \end{bmatrix} = \begin{bmatrix} 0 & \pi_{12} & \cdots & \pi_{1n} \\ \pi_{21} & 0 & \cdots & \pi_{2n} \\ \vdots & \vdots & \vdots & \vdots \\ \pi_{n1} & \pi_{n2} & \cdots & 0 \end{bmatrix} \begin{bmatrix} y_1 \\ y_2 \\ \vdots \\ y_n \end{bmatrix} + \begin{bmatrix} u_1 \\ u_2 \\ \vdots \\ u_n \end{bmatrix}$$

in which all the diagonal elements are zero and the $n(n-1)$ off-diagonal elements π_{ij} are unknown. Of course, trying to estimate $n(n-1)$ parameters from n data points is not possible, so additional structure must be imposed to make the problem tractable.

13.2.1 Specification

To identify this model, the spatial approach is to impose restrictions on the model of the form

$$\begin{bmatrix} 0 & \pi_{12} & \cdots & \pi_{1n} \\ \pi_{21} & 0 & \cdots & \pi_{2n} \\ \vdots & \vdots & \vdots & \vdots \\ \pi_{n1} & \pi_{n2} & \cdots & 0 \end{bmatrix} = \phi \begin{bmatrix} 0 & w_{12} & w_{13} & \cdots & w_{1n} \\ w_{21} & 0 & w_{23} & \cdots & w_{2n} \\ w_{31} & w_{32} & 0 & \cdots & w_{3n} \\ \vdots & \vdots & \vdots & \vdots & \vdots \\ w_{n1} & w_{n2} & w_{n3} & \cdots & 0 \end{bmatrix} = \phi \mathbf{W}$$

where ϕ is now the only parameter to be estimated and the elements w_{ij} of the spatial weighting matrix $\mathbf{W}$ are usually determined by means of an exogenous distance metric. Note that a fundamental feature of $\mathbf{W}$ is that there are zeros on the main diagonal, denoting zero distance.

13.2.1 Specification

This specification of the spatial dependence has two important consequences for the nature of the model.

1. The spatial dependency parameter, ϕ, is a common scale factor that governs the average strength of spatial dependence in the model. If ϕ is large in absolute value, the spatial dependence in the model is strong. On the other hand, if $\phi = 0$, then all spatial structure disappears.

2. The spatial weighting matrix, $\mathbf{W}$, governs the relative strength between the spatial units because
$$\frac{\phi w_{ij}}{\phi w_{kl}} = \frac{w_{ij}}{w_{kl}}$$

So, in general, the specification task involves placing restrictions on the nature of the spatial variation in the data with all the associated questions of the efficacy and appropriateness of the chosen specification. The spatial weighting matrix is usually constructed in one of two ways.

Distance weights

Stata uses two different metrics to compute distance. If the coordinates of the two points are Cartesian and given by (x_1, y_1) and (x_2, y_2), the familiar Euclidean distance can be expressed as
$$d_{12} = \sqrt{(x_2 - x_1)^2 + (y_2 - y_1)^2}$$

Often, however, the coordinates are in latitude (t) and longitude (n) given in degrees. Let the two coordinate pairs be (t_1, n_1) and (t_2, n_2), where the measurements have been converted into radians. Then, the distance is given by
$$d_{12} = r \operatorname{invhav} \{r \operatorname{hav}(t_2 - t_1) + \cos t_1 \cos t_2 \operatorname{hav}(n_2 - n_1)\}$$
$$\operatorname{hav}(x) = \frac{1 - \cos x}{2}$$
$$\operatorname{invhav}(x) = 2 \operatorname{asin}(\sqrt{x})$$

where r is the radius of the Earth.

Let d_{ij} be the distance between two units i and j. Then, two common specifications are

$$w_{ij} = d_{ij}^{-\alpha} \qquad \text{(Inverse distance)}$$
$$w_{ij} = \exp(-\alpha d_{ij}) \qquad \text{(Exponential distance)}$$

where α is positive and usually chosen as 1 or 2. The inverse distance spatial weighting matrix is created in Stata using the spmatrix create idistance (see [SP] **spmatrix create**) command.

Contiguity weights

Contiguity weights simply indicate whether spatial units share a boundary or not:

$$w_{ij} = \begin{cases} 1 & \text{if } i \text{ and } j \text{ share a boundary} \\ 0 & \text{otherwise} \end{cases}$$

Rook contiguity requires that the whole boundary is shared, as shown in the following illustration:

Note that the gray shaded block has only four contiguous neighbors.

A less strict definition of contiguity is queen contiguity, which extends rook contiguity to the sharing of a vertex alone as in the illustration.

In this instance, the gray shaded cell has eight neighbors.

13.2.2 Construction

Contiguity matrices may be created using Stata's `spmatrix create contiguity` command. However, to determine boundaries, contiguity matrices are usually constructed from information contained in a shapefile, a computer file that stores a map. If you have a shapefile, Stata can determine neighbors and also the distances between the centroids of the places so that both distance and contiguity matrices can be created. The focus in this chapter is on inverse distance weighting matrices computed using coordinates of latitude and longitude.

To aid interpretation of the results and also to make estimation easier, the spatial weighting matrix is usually normalized. Traditionally, the rows of the matrix are scaled to sum to 1:

$$\widetilde{w}_{ij} = \frac{w_{ij}}{\sum_{j=1}^{n} w_{ij}}, \quad \text{so that} \sum_{j=1}^{n} \widetilde{w}_{ij} = 1$$

13.2.2 Construction

Note that each row-normalized weight can then be interpreted as the fraction of all spatial influence on unit i attributable to unit j.

The default in Stata is scalar normalization, in which $\mathbf{W}$ is scaled by a suitable scalar

$$\widetilde{w}_{ij} = \alpha \mathbf{W}$$

One choice for α is the inverse of the maximum element of $\mathbf{W}$, but a better choice and one adopted by Stata is the inverse of the maximal eigenvalue, $\lambda_{\max}$ of $\mathbf{W}$

$$\widetilde{w}_{ij} = \frac{w_{ij}}{\lambda_{\max}}$$

To construct the spatial weighting matrix for the compliance data, the data were augmented by the latitude and longitude readings of the main cities of all the countries present in the dataset. This information was obtained from the CEPII–GeoDist database[2] (Mayer and Zignago 2011). Note that because the data are in a Stata .dta file and there is no shapefile, it is not possible to construct a contiguity matrix. However, the spatial weights can easily be constructed based on distance.

The data are cross-sectional data, and the variable countrycode contains a three character country code to identify the cross-sectional units. The variables lat and lon are the coordinates of latitude and longitude, respectively, for each country's leading city. The following steps are now required:

1. Encode the string variable countrycode into a numeric identifier id, and check that this variable uniquely identifies the cross section.
2. Set the data as a spatial dataset using the spset (see [SP] **spset**) command.
3. Set the coordinate units to be latitude and longitude using the spset option coordsys(latlong, kilometers).
4. Create the spatial weighting matrix using **spmatrix create**, and normalize it using the maximal eigenvalue by specifying the option normalize(spectral). This use of the spectral option is not strictly necessary, because it is the default option.

The code to implement these steps is as follows:

```
encode countrycode, gen(id)
assert id != .
bysort id: assert _N==1
spset id, coord(lat lon) coordsys(latlong, kilometers)
spmatrix create idistance WDIST, normalize(spectral)
```

The command **spmatrix create** by default normalizes the weighting matrix it creates by dividing the entries by the absolute value of the largest eigenvalue of the matrix;

---
2. The data were downloaded from http://www.cepii.fr/CEPII/en/bdd_modele/bdd_modele.asp.

this is the normalize(spectral) option. The other commonly used normalization is normalize(row), which scales each row of the matrix by its row sum so that each row sums to 1.

Borsky and Raschky (2015) construct a custom version of a spatial weighting matrix for 53 countries used in their study of compliance with Article 7 of the United Nations CoC on responsible fishing. The main assumption is that the strength of interaction between the signatory countries is inversely related to the distance between two countries because of direct trade links and because the interactions between official representatives decreases with distance. They construct an inverse distance matrix where the weights are defined as

$$w_{ij} = \exp(-3d_{ij})$$

where d_{ij} is the distance between the most populated cities in country i and country j and the matrix is row normalized. The exponential operator gives greater weight to closer countries than would be the case with a simple inverse distance matrix. Note that although there is no thresholding in this definition, the operation of the exponential will mean that for larger distances, the weights are zero for all practical purposes. The resulting weighting matrix is plotted in figure 13.2. Note that row normalizing a symmetric matrix produces an asymmetric weighting matrix except in very special cases (Drukker et al. 2013). The first thing to note is that the matrix has zeros on the diagonal as required. The matrix is also quite sparse, indicating that the exponential function does provide a strong level of natural thresholding.

13.2.2 Construction

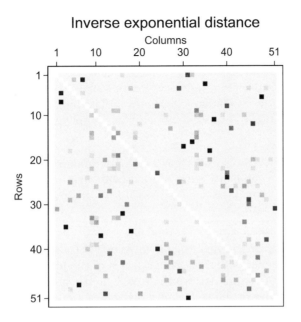

Figure 13.2. Inverse exponential distance spatial weighting matrix for the 53 countries used in the study of regulatory compliance with the United Nations CoC on responsible fishing.

To give an idea of the actual size of the entries in the matrix, the (10×10) top left corner is given by

	1	2	3	4	5	6	7	8	9	10
1	0.000	0.000	0.000	0.000	0.000	0.000	0.000	0.000	0.000	0.000
2	0.000	0.000	0.000	0.000	0.156	0.000	0.842	0.000	0.000	0.000
3	0.000	0.000	0.000	0.000	0.000	0.000	0.000	0.000	0.000	0.000
4	0.000	0.000	0.000	0.000	0.000	0.000	0.000	0.001	0.000	0.000
5	0.000	0.932	0.000	0.000	0.000	0.000	0.063	0.000	0.000	0.000
6	0.000	0.000	0.000	0.000	0.000	0.000	0.000	0.000	0.000	0.000
7	0.000	0.970	0.000	0.000	0.012	0.000	0.000	0.000	0.000	0.000
8	0.000	0.000	0.000	0.001	0.000	0.000	0.000	0.000	0.000	0.000
9	0.000	0.000	0.000	0.001	0.000	0.000	0.000	0.000	0.000	0.108
10	0.000	0.000	0.000	0.001	0.000	0.000	0.000	0.000	0.142	0.000

The dark gray cells match the black cells from figure 13.2, and the light gray match the grey cells. Cells that are shaded black in figure 13.2 have weights close to 1. To give an idea of how the values listed here relate to those in the entire matrix, the column maximum values range from a low of 0.004 to the highest recorded entry of 0.999.

13.3 Exploratory data analysis

The compliance effort on the responsible fishing CoC is influenced by the related costs and benefits. Domestic costs are related to the loss in voters' utility due to compliance with the CoC and the costs of enforcing it on the fishing grounds, given that fishing is by nature a geographically dispersed industry. International costs stem from the sanctions or loss of trust from other signatory governments. The full set of variables used in the analysis and the direction of expected influence on compliance is shown in table 13.1.

Table 13.1. Summary of the cross-sectional variables from 51 countries used to capture spatial interactions in compliance with international environmental agreements

Variable	Description	Sign
overall	overall measure of compliance	
behavior	summary of quantifiable measures of compliance	
intention	summary of intangible measures of compliance	
gdp	logarithm of per capita gross domestic product	+
cost	average distance from fishing zone	−
gov	measure of good governance/political stability	+
compet	degree of competition in fishing zone	−
bio	biodiversity index measuring overall health of the environment	+
export	logarithm of value of fish exports relative to gross domestic product	−
eezcatch	fraction of catch in country's exclusion zone	−
treaty	total involvement in 443 other international agreements	+

Prior to undertaking any spatial analysis, Stata requires that the data be spset. In the case of the compliance example, the data are cross-sectional data for 51 countries with a cross-sectional identifier called countrycode, which is a string variable. The code to load the spatial weighting matrix WDIST, which was previously created and saved, is

```
. spmatrix use WDIST using wdist.stswm, replace
```

To detect spatial correlation in the data, the most well-known and commonly used visualization tool is Moran's scatterplot. The plot is a scatter of the dependent variable, **y**, against the so-called spatial lag of **y**, which is computed as **Wy**. A positive correlation will indicate that nearby values y_i and y_j tend to take similar values and a negative correlation will indicate different values. A Moran's scatterplot of behavior against its spatial lag is given in figure 13.3, in which there appears to be a strong positive spatial correlation present in behavior. This positive relationship is clearly illustrated by the line of best fit through the scatter.

13.3 Exploratory data analysis

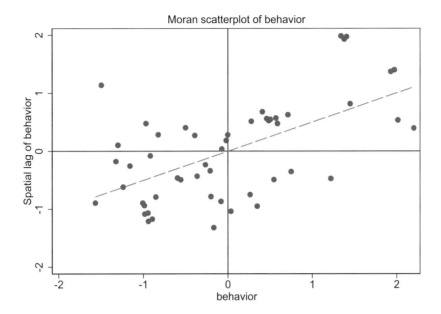

Figure 13.3. Moran's scatterplot of the `behavior` measure against its spatial lag for the 53 countries used in the study of regulatory compliance with the United Nations CoC on responsible fishing. The dashed line represents the line of best fit.

A formal test of spatial dependence between N cross-sectional units may be computed using the residuals of a linear regression model, say, $\hat{\mathbf{u}}$, and an $(N \times N)$ spatial weighting matrix $\mathbf{W}$ with elements w_{ij}. Moran's I (MI) statistic is computed as

$$\text{MI} = \frac{N}{S_0} \frac{\hat{\mathbf{u}}'\mathbf{W}\hat{\mathbf{u}}}{\hat{\mathbf{u}}'\hat{\mathbf{u}}}, \qquad S_0 = \sum_{i=1}^{N}\sum_{j=1}^{N} w_{ij}$$

The MI statistic is distributed as $N(0,1)$, but Stata reports it as MI^2, which has a χ_1^2 distribution. The MI statistic is computed using the `regress postestimation` (see [R] **regress postestimation**) command `estat moran` and also passing the name of the spatial weighting matrix. Note that even though the command for a linear regression is being used, the data must have been `spset` first. In the case of `behavior` with the control variables in table 13.1 listed in the global macro `controls`, the Moran test is implemented as follows:

```
. rename distance cost
. global controls "gdp cost gov competition bio export eezcatch treaty"
. regress behavior $controls

      Source │       SS           df       MS      Number of obs   =        51
─────────────┼──────────────────────────────────    F(8, 42)        =     12.45
       Model │  99.4243508         8  12.4280439    Prob > F        =    0.0000
    Residual │   41.936929        42  .998498309    R-squared       =    0.7033
─────────────┼──────────────────────────────────    Adj R-squared   =    0.6468
       Total │   141.36128        50   2.8272256    Root MSE        =    .99925

────────────┬──────────────────────────────────────────────────────────────────
    behavior│      Coef.   Std. Err.      t    P>|t|     [95% Conf. Interval]
────────────┼──────────────────────────────────────────────────────────────────
         gdp│   .5002757   .2133115     2.35   0.024     .0697958    .9307556
        cost│   .5039543   .2079324     2.42   0.020     .0843297    .9235788
         gov│   .7604702   .3128219     2.43   0.019     .1291701    1.39177
 competition│    2.27132   5.599502     0.41   0.687    -9.028933   13.57157
         bio│   .0150848   .0101667     1.48   0.145    -.0054324    .035602
      export│   .1318825   .1104565     1.19   0.239    -.0910277    .3547927
     eezcatch│  1.107973   .5969949     1.86   0.070    -.0968114   2.312758
      treaty│   .0013664   .0040879     0.33   0.740    -.0068834    .0096161
       _cons│  -1.82646    1.584475    -1.15   0.256    -5.024059    1.371139
────────────┴──────────────────────────────────────────────────────────────────

. encode countrycode, gen(id)
. spset id
  Sp dataset spatialfish_small.dta
              data:   cross sectional
    spatial-unit id:  _ID (equal to id)
       coordinates:   none
   linked shapefile:  none
. estat moran, errorlag(WDIST)
Moran test for spatial dependence
        Ho: error is i.i.d.
        Errorlags: WDIST
          chi2(1)     =     4.41
          Prob > chi2 =     0.0358
```

The MI statistic is significant at the 5% level, indicating that the null hypothesis of no spatial autocorrelation is rejected.

13.4 Spatial models

The simplest possible model that allows for all possible spatial interactions is the spatial autoregressive model (Whittle 1954; Ord 1975) given by

$$y_i = \phi \sum_{j=1}^{N} w_{ij} y_j + u_i$$

13.4 Spatial models

where the constant term is omitted without loss of generality, $\mathbf{W}$ is a spatial weighting matrix, and u_i is a disturbance term. The first term on the right-hand side of this equation is known as the spatial lag. Spatial models are particularly amenable to manipulation using the tools of matrix algebra, so this is one area in the book where matrix notation will be relied upon. In matrix form, the spatial autoregressive model becomes

$$\mathbf{y} = \phi \mathbf{W} \mathbf{y} + \mathbf{u}$$

By contrast, the most general spatial model allows for explanatory variables both with and without spatial effects and also for a spatial dimension to the error structure. The model is given by

$$\mathbf{y} = \phi \mathbf{W} \mathbf{y} + \mathbf{X} \boldsymbol{\beta} + \mathbf{W} \mathbf{X} \boldsymbol{\gamma} + \mathbf{u}$$
$$\mathbf{u} = \rho \mathbf{W} \mathbf{u} + \mathbf{v}$$
$$\mathbf{v} \sim N(0, \sigma_v^2 \mathbf{I}_N) \tag{13.1}$$

in which $\mathbf{X}$ is a matrix of explanatory variables, $\mathbf{v}$ is a disturbance term, and $\mathbf{I}_N$ is a $(N \times N)$ identity matrix. Note that there is no theoretical necessity for all the instances of the spatial weight matrix in (13.1) to be the same matrix. Although Stata allows up to three different weight matrices to be specified, in most practical applications it is difficult to justify the use of different matrices.

The two most frequently encountered models in practice are the spatial lag model and the spatial error model.

Spatial lag model

The spatial lag model is given by

$$\mathbf{y} = \phi \mathbf{W} \mathbf{y} + \mathbf{X} \boldsymbol{\beta} + \mathbf{u} \tag{13.2}$$

which imposes the restrictions $\boldsymbol{\gamma} = 0$ and $\rho = 0$ on the general model in (13.1). The reduced form of the model is

$$\mathbf{y} = (I_n - \phi \mathbf{W})^{-1} \mathbf{X} \boldsymbol{\beta} + (I_n - \phi \mathbf{W})^{-1} \mathbf{u} \tag{13.3}$$

if the matrix inverse $(I_n - \phi \mathbf{W})^{-1}$ exists. It turns out the fundamental condition for this inverse to exist is

$$|\phi| < 1/\lambda_{\max}$$

where $\lambda_{\max}$ is the largest eigenvalue of the spatial weighting matrix $\mathbf{W}$. The usefulness of the normalization of the matrix $\mathbf{W}$ to have a maximal eigenvalue of 1 is now apparent because the invertibility of the matrix is simply the usual condition that the first-order autoregressive parameter of the process be strictly less than 1.

Estimation of (13.2) by ordinary least squares will result in biased and inconsistent parameter estimates. The inconsistency of the parameter estimates stems from the violation of the exogeneity assumption, $E\{(\mathbf{Wy})\mathbf{u}'\} \neq 0$. Because endogeneity is a pervasive problem in spatial models, it is worth tracing out this result. Note that in this model the weighting matrix $\mathbf{W}$ is taken as fixed, and conditioning is with respect to $\mathbf{X}$. To establish results that are based on taking expectations, conditional expectations with respect to $\mathbf{X}$ are appropriate, and the unconditional expectations are then calculated using the law of iterated expectations.[3]

From the reduced form in (13.3), it follows that

$$\mathbf{Wy} = \mathbf{W}(I_n - \phi\mathbf{W})^{-1}\mathbf{X}\boldsymbol{\beta} + \mathbf{W}(I_n - \phi\mathbf{W})^{-1}\mathbf{u} \tag{13.4}$$

Using this form, the expectation $E\{(\mathbf{Wy})\mathbf{u}'\}$ can be obtained as follows. Concentrating on the first term on the right-hand side of (13.4), it follows that

$$E\left\{\mathbf{W}(I_n - \phi\mathbf{W})^{-1}\mathbf{X}\boldsymbol{\beta}\mathbf{u}'\right\} = E\left[E_x\left\{\mathbf{W}(I_n - \phi\mathbf{W})^{-1}\mathbf{X}\boldsymbol{\beta}\mathbf{u}'\right\}\right]$$
$$= \mathbf{W}(I_n - \phi\mathbf{W})^{-1}\mathbf{X}\,E\left\{E_x\left(\mathbf{u}'\right)\right\} = 0$$

Similarly, from the second term on the right-hand side of (13.4), it follows that

$$E\left\{\mathbf{W}(I_n - \phi\mathbf{W})^{-1}\mathbf{u}\mathbf{u}'\right\} = \mathbf{W}(I_n - \phi\mathbf{W})^{-1}\,E\left\{E_x\left(\mathbf{u}\mathbf{u}'\right)\right\}$$
$$= \mathbf{W}(I_n - \phi\mathbf{W})^{-1}\sigma^2 I_n \neq 0$$

Therefore, it follows that

$$E\{(\mathbf{Wy})\mathbf{u}'\} = \mathbf{W}(I_n - \phi\mathbf{W})^{-1}\sigma^2 I_n \neq 0$$

which illustrates the violation of the exogeneity assumption in the spatial lag model.

Spatial error model

The spatial error model is given by

$$\mathbf{y} = \mathbf{X}\boldsymbol{\beta} + \mathbf{u}$$
$$\mathbf{u} = \rho\mathbf{W}\mathbf{u} + \mathbf{v} \tag{13.5}$$

which imposes the restrictions $\phi = 0$ and $\boldsymbol{\gamma} = \mathbf{0}$ on the general model in (13.1). Equation (13.5) has the interpretation of a linear regression model, where each disturbance is regressed on all the other disturbances apart from itself.

---

3. Expressed in simple terms, the law of iterated expectations states
$$E(Y) = E\{E(Y|\mathbf{X})\}$$
or in slightly simpler notation
$$E(Y) = E\{E_x(Y)\}$$
where E_x denotes that the expectation is taken conditional on $\mathbf{X}$.

13.5 Fitting spatial models by maximum likelihood

The reduced form for this model is

$$y = \mathbf{X}\boldsymbol{\beta} + (I_n - \rho\mathbf{W})^{-1}\mathbf{v}$$

Similar to the case of the spatial lag model, the condition for the existence of $(I_n - \rho\mathbf{W})^{-1}$ is

$$|\rho| < 1/\lambda_{\max}$$

If the model in (13.5) is fit by ordinary least squares, ignoring the spatial error structure, the coefficient estimates $\boldsymbol{\beta}$ are inefficient but consistent because of the heteroskedastic error term. No estimate of ρ is available.

13.5 Fitting spatial models by maximum likelihood

To fit spatial autoregressive models in Stata, the spregress (see [SP] **spregress**) command is issued after the data have been spset and a weighting matrix $\mathbf{W}$ has been created. The treatment here will focus exclusively on maximum likelihood estimation, for which the option ml must be specified. There are several important options to spregress that define the model. The option dvarlag(W) specifies that the spatial lag of the dependent variable, ivarlag(W: varlist) is the spatial lag of the independent variables, while errorlag(W) is the lagged error term. $\mathbf{W}$ appears as the argument here, but as noted previously, Stata can in fact handle up to three different weighting matrices.

Spatial lag model

From the structure of the model in (13.2), using the notation $\mathbf{v} = (I_n - \phi\mathbf{W})^{-1}\mathbf{u}$, taking $\mathbf{W}$ as fixed, and conditioning with respect to $\mathbf{X}$ gives the relationships

$$\mathbf{v} = \mathbf{y} - (I_n - \phi\mathbf{W})^{-1}\mathbf{X}\boldsymbol{\beta}$$
$$E_\mathbf{x}(\mathbf{v}) = 0$$
$$E_\mathbf{x}(\mathbf{v}\mathbf{v}') = E\left[\left\{(I_n - \phi\mathbf{W})^{-1}\mathbf{u}\right\}\left\{(I_n - \phi\mathbf{W})^{-1}\mathbf{u}\right\}'\right]$$
$$= (I_n - \phi\mathbf{W})^{-1}(I_n - \phi\mathbf{W}')^{-1}E\{(\mathbf{u}\mathbf{u}')\}$$
$$= \boldsymbol{\Omega}\sigma^2$$

Using the assumption that the $\mathbf{v}$ is conditionally normally distributed, the log-likelihood function of the spatial lag model is given by

$$\log L\left(\phi, \boldsymbol{\beta}, \sigma^2 | \mathbf{X}\right) = -\frac{N}{2}\log(2\pi) - \frac{N}{2}\log(\sigma^2) - \frac{1}{2}\log|\boldsymbol{\Omega}| - \frac{1}{2}\mathbf{v}'\boldsymbol{\Omega}^{-1}\mathbf{v}$$

which is maximized, taking the spatial weighting matrix $\mathbf{W}$ as given, to yield the maximum likelihood estimates of $\widehat{\phi}$, $\widehat{\boldsymbol{\beta}}$, and $\widehat{\sigma}^2$.

The spatial lag model for the `behavior` measure of compliance with the explanatory variables once again given in the global variable controls can be fit by maximum likelihood in Stata.

```
. spregress behavior $controls, noconstant dvarlag(WDIST) ml nolog
  (51 observations)
  (51 observations (places) used)
  (weighting matrix defines 51 places)

Spatial autoregressive model                    Number of obs    =        51
Maximum likelihood estimates                    Wald chi2(9)     =    905.73
                                                Prob > chi2      =    0.0000
Log likelihood = -66.949648                     Pseudo R2        =    0.6767
```

behavior	Coef.	Std. Err.	z	P>|z|	[95% Conf. Interval]	
behavior						
gdp	.247196	.1262418	1.96	0.050	-.0002335	.4946254
cost	.378055	.184613	2.05	0.041	.0162203	.7398898
gov	.963907	.2210252	4.36	0.000	.5307055	1.397109
competition	-.7707438	5.372734	-0.14	0.886	-11.30111	9.759621
bio	.0132426	.008901	1.49	0.137	-.0042032	.0306883
export	.1026307	.1013215	1.01	0.311	-.0959557	.3012171
eezcatch	.829639	.5217085	1.59	0.112	-.192891	1.852169
treaty	.000831	.0036928	0.23	0.822	-.0064067	.0080687
WDIST						
behavior	.1726258	.1079708	1.60	0.110	-.038993	.3842445
var(e.behav~r)	.7983139	.1586306			.5407966	1.178456

```
Wald test of spatial terms:          chi2(1) = 2.56        Prob > chi2 = 0.1099
```

In this simple specification, there is no spatial lag on the explanatory variables. The coefficient labeled [WDIST] `behavior` is the estimate of $\widehat{\phi}$, while the parameter labeled `var(e.behavior)` is the estimate of $\widehat{\sigma}^2$. The estimates of β are all prefixed by [`behavior`]. The results of the estimation indicate that many of the explanatory variables' coefficients are in fact not significantly different from zero. More concerning, the estimate of the strength of the spatial relationships $\widehat{\phi}$ is not significant at the 5% level but almost significant at the 10% level. Given the clear evidence in favor of spatial patterns in both **y** and in the residuals of the simple linear regression, it can be conjectured that the insignificance of $\widehat{\phi}$ is probably due to the inclusion of too many insignificant controls. The estimation results suggest that `gpd`, `cost`, and `gov` are the control variables that have most explanatory power, so it is reasonable to refit the model with only these variables included as explanatory variables.

13.5 Fitting spatial models by maximum likelihood

```
. spregress behavior gdp cost gov, noconstant dvarlag(WDIST) ml nolog
  (51 observations)
  (51 observations (places) used)
  (weighting matrix defines 51 places)
Spatial autoregressive model              Number of obs    =        51
Maximum likelihood estimates              Wald chi2(4)     =    742.68
                                          Prob > chi2      =    0.0000
Log likelihood = -71.905942               Pseudo R2        =    0.5836
```

behavior	Coef.	Std. Err.	z	P>|z|	[95% Conf. Interval]	
behavior						
gdp	.3094564	.0468352	6.61	0.000	.2176611	.4012517
cost	.3125211	.1843971	1.69	0.090	-.0488905	.6739327
gov	.8441503	.1610212	5.24	0.000	.5285546	1.159746
WDIST						
behavior	.2193839	.1027047	2.14	0.033	.0180864	.4206813
var(e.behav~r)	.9617128	.191441			.651033	1.420652

```
Wald test of spatial terms:         chi2(1) = 4.56       Prob > chi2 = 0.0327
```

There is now clear evidence of a significant spatial effect with the χ_1^2 statistic of the significance of the spatial element having a p-value of 0.033, thus allowing rejection at the 5% level of the null hypothesis that the spatial effect is zero. These results indicate that despite the positive and statistically significant influences of income and government regulation, the conduct of neighboring states has a positive effect on the intention to abide by the CoC of responsible fishing. The coefficient on the cost of compliance that is expected to be negatively related to the intention to comply is not statistically significant. Although this seems counterintuitive, it is a better outcome than a positive and significant coefficient.

Spatial error model

From the structure of the model in (13.5), it may be deduced that

$$\mathbf{u} = \mathbf{y} - \mathbf{X}\boldsymbol{\beta}$$
$$E_x(\mathbf{u}) = 0$$
$$E_x(\mathbf{uu}') = E\left[\left\{(I_n - \rho\mathbf{W})^{-1}\mathbf{v}\right\}\left\{(I_n - \rho\mathbf{W})^{-1}\mathbf{v}\right\}'\right]$$
$$= (I_n - \rho\mathbf{W})^{-1}(I_n - \rho\mathbf{W}')^{-1}E\{(\mathbf{vv}')\}$$
$$= \boldsymbol{\Omega}\sigma_v^2$$

Once again, assuming that the disturbances $\mathbf{u}$ are conditionally normally distributed, the log-likelihood function for the spatial error model is given by

$$\log L\left(\rho, \boldsymbol{\beta}, \sigma_v^2 | \mathbf{X}\right) = -\frac{N}{2}\log(2\pi) - \frac{N}{2}\log(\sigma_v^2) - \frac{1}{2}\log|\boldsymbol{\Omega}| - \frac{1}{2}\mathbf{u}'\boldsymbol{\Omega}^{-1}\mathbf{u}$$

which is maximized, taking the spatial weighting matrix **W** as given, to yield the maximum likelihood estimates of $\widehat{\rho}$, $\widehat{\beta}$, and $\widehat{\sigma}_v^2$.

Taking `gdp` and `gov` to be the only control variables, the spatial error model yields the following results:

```
. spregress behavior gdp gov, noconstant errorlag(WDIST) ml nolog
  (51 observations)
  (51 observations (places) used)
  (weighting matrix defines 51 places)

Spatial autoregressive model                    Number of obs     =         51
Maximum likelihood estimates                    Wald chi2(2)      =     273.68
                                                Prob > chi2       =     0.0000
Log likelihood = -68.842695                     Pseudo R2         =     0.5894
```

behavior	Coef.	Std. Err.	z	P>|z|	[95% Conf. Interval]
behavior					
gdp	.4358691	.0281671	15.47	0.000	.3806626 .4910757
gov	.987286	.1770535	5.58	0.000	.6402676 1.334305
WDIST					
e.behavior	.4624008	.1060727	4.36	0.000	.2545021 .6702995
var(e.behav~r)	.7857121	.1611531			.5256276 1.174488

```
Wald test of spatial terms:         chi2(1) = 19.00      Prob > chi2 = 0.0000
```

The coefficient labeled [WDIST] `e.behavior` is the estimate of $\widehat{\rho}$, while the parameter labeled `var(e.behavior)` is the estimate of $\widehat{\sigma}_v^2$. The estimate of the strength of the spatial relationship $\widehat{\rho}$ is significant and indicates a strong spatial influence even when accounting for differences in country income and the regulatory environment.

13.6 Estimating spillover effects

The dependence structure in a spatial regression model allows us to retrieve information on spatial spillover effects. A change in a variable in country i will have a direct effect on the dependent variable of country i as well as an indirect effect on the dependent variable of all other countries due to spatial dependence. The magnitude of the spatial spillover will depend upon the spatial weighting matrix **W**, which summarizes the degree of spatial connectivity; the parameter ϕ, which represents the strength of spatial dependence; and the parameter vector β. Spatial spillover effects implied by a fitted spatial autoregressive model are computed in Stata using the `spregress postestimation` (see [SP] **spregress postestimation**) command `estat impact`. Note that the scaling properties of the spectral normalization and the spatial lag coefficient estimates implies that the estimates of the direct and indirect effects should be scale invariant if `spregress` is used with the option `ml`.

13.6 Estimating spillover effects

Consider the spatial lag model in (13.2). Solving for **y** and taking conditional expectations with **W** regarded as fixed gives

$$E_x(\mathbf{y}) = (I_n - \phi \mathbf{W})^{-1} \mathbf{X}\boldsymbol{\beta}$$

It follows that the impact on the dependent variable **y** of changing an independent variable **X** will have a spatial element to it. Specifically, changing x_i, which is the ith element of **X**, will affect not only y_i (the direct impact effect) but also y_j for all $j \neq i$ (the indirect impact or spillover effects).

For the spatial lag model, let the direct impact of individual x_i on y_i be denoted

$$\frac{\partial E_x(y_i)}{\partial x_i}$$

The direct impact of the variable x_i on **y** is then given by the average of these individual impacts over the cross section of N units:

$$\frac{1}{N} \sum_{i=1}^{N} \frac{\partial E_x(y_i)}{\partial x_i}$$

Similarly, the indirect impact or spillover effects of the variable x_i are given by

$$\frac{1}{N(N-1)} \sum_{i=1}^{N} \sum_{j=1, j \neq i}^{N} \frac{\partial E_x(y_i)}{\partial x_j}$$

The total impact is therefore

$$\frac{1}{N^2} \sum_{i=1}^{N} \sum_{j=1}^{N} \frac{\partial E_x(y_i)}{\partial x_j}$$

By default, Stata uses the delta method to calculate the estimated variance of these impacts. This variance is conditional on the values of the independent variables and so does not account for the sampling variance of the independent variables in the model.[4]

---

4. Using `estat impact` with option `vce(unconditional)` uses a generalized method of moments estimation strategy to compute the unconditional variance of the impacts and thus accounts for the sampling variance of the independent variables in the model.

The spatial lag estimated with `gdp` and `gov` as explanatory variables produces the following table of spillover effects:

```
. quietly spregress behavior gdp gov, noconstant dvarlag(WDIST) ml nolog
. estat impact
progress    : 50% 100%
Average impacts                                 Number of obs   =         51
```

	dy/dx	Delta-Method Std. Err.	z	P>|z|	[95% Conf. Interval]	
direct						
gdp	.3297616	.0413151	7.98	0.000	.2487854	.4107378
gov	.7725569	.1596756	4.84	0.000	.4595985	1.085515
indirect						
gdp	.0939939	.0373236	2.52	0.012	.020841	.1671468
gov	.2202065	.1231012	1.79	0.074	-.0210673	.4614804
total						
gdp	.4237556	.0224345	18.89	0.000	.3797847	.4677264
gov	.9927635	.2394485	4.15	0.000	.5234531	1.462074

In this particular application, the indirect effects are all statistically significant at the 10% level so that the total effect of changing either `gdp` or `gov` differs from the direct effects expressed by the coefficients of these variables. Also note that the use of the `margins` command will only yield estimates of the total effect and not provide a breakdown into direct and indirect effects, so `estat impact` is preferable when analyzing spatial lag models.

```
. margins, dydx(*)
Average marginal effects                        Number of obs   =         51
Model VCE    : OIM

Expression   : Reduced-form mean, predict()
dy/dx w.r.t. : gdp gov
```

	dy/dx	Delta-method Std. Err.	z	P>|z|	[95% Conf. Interval]	
gdp	.4237556	.0224345	18.89	0.000	.3797847	.4677264
gov	.9927635	.2394485	4.15	0.000	.5234531	1.462074

By contrast with the spatial error model in (13.5), solving for **y** and taking expectations gives

$$E_x(\mathbf{y}) = \mathbf{X}\boldsymbol{\beta}$$

indicating that only direct effects can be deduced from the model. Consequently, the use of `estat impact` after the spatial error model provides no additional information in terms of computing the total effect.

```
. quietly spregress intention gdp gov, noconstant errorlag(WDIST) ml nolog
. estat impact
progress   : 50% 100%
Average impacts                                      Number of obs    =     51

                      Delta-Method
              dy/dx   Std. Err.      z     P>|z|     [95% Conf. Interval]

direct
      gdp    .5252884  .0263674    19.92   0.000     .4736092    .5769675
      gov    1.062088  .1888259     5.62   0.000     .6919957    1.43218

indirect
      gdp           0  (omitted)
      gov           0  (omitted)

total
      gdp    .5252884  .0263674    19.92   0.000     .4736092    .5769675
      gov    1.062088  .1888259     5.62   0.000     .6919957    1.43218
```

13.7 Model selection

The research question examined in the example running through this chapter is whether there is strategic interaction between governments when they decide on the level of compliance to a treaty. Significant interaction requires that the parameter ϕ in the spatial lag model or the parameter ρ in the spatial error model be statistically significant. At this stage, however, no guidance has been offered as to which model represents the correct specification in which to test for strategic spatial interaction and examine the nature of the spatial spillover effects. To this end, some statistical tests can be performed on the model and judgment exercised as to the correct type of spatial model to choose.

Broadly speaking, there are two approaches to the problem of specification in the context of spatial models (Florax, Folmer, and Rey 2003).

1. The "classical" or *simple-to-general* approach, in which a simple nonspatial model is fit and the errors tested for omitted spatial effects. If the null hypothesis is rejected, then the appropriate spatial model is chosen based on the level of significance of the tests of the spatial effects. For example, if the strength of the rejection of $H_0: \phi = 0$ in the spatial lag model is stronger in terms of its *p*-value than the rejection of the null hypothesis $H_0: \rho = 0$ in the spatial error model, then the former is chosen. In nonspatial applications, this approach has been termed "excessive presimplification with inadequate diagnostic testing" (Maddala 1992, 494).

2. The *general-to-specific* methodology (Hendry 1979, 1995; Gilbert 1986) requires that the most general model be fit and then sequentially tested down by means of a series of tests until a parsimonious model is found that is admissible by the data and that encompasses rival models. This approach is labeled "intended overparameterization with data-based simplification" (Maddala 1992, 494).

While Stata is able to fit the general model in (13.1), which includes a spatial lag of the dependent variable, a spatial lag on the independent variables, and a spatially lagged error term, there are problems in implementing the general-to-specific methodology in a spatial context partly because automated software that implements the Hendry methodology, such as *Autometrics* (Doornik 2009), did not originally include a spatial capability and partly because original spatial software, such as SpaceStat (Anselin 1995), did not include any automated specification search. The main conclusion reached by Florax, Folmer, and Rey (2003) is that under simulation, the specific-to-general approach outperforms the general-to-specific methodology in terms of identifying the true model, although it should be borne in mind that this conclusion applies only to the limited simulation study reported in the article and cannot be endorsed as a general rule.

Ultimately, the choice of the specification of the spatial model will require experimentation and the use of judgment. Fitting the general spatial model in (13.1) for the fisheries data and using gpd and gov as explanatory variables yields the following results:

```
. spregress behavior gdp gov, noconstant dvarlag(WDIST) ivarlag(WDIST: gdp gov)
> errorlag(WDIST) ml nolog
  (51 observations)
  (51 observations (places) used)
  (weighting matrix defines 51 places)

Spatial autoregressive model                   Number of obs    =         51
Maximum likelihood estimates                   Wald chi2(5)     =    1189.04
                                               Prob > chi2      =     0.0000
Log likelihood = -67.043041                    Pseudo R2        =     0.6200
```

behavior	Coef.	Std. Err.	z	P>|z|	[95% Conf.	Interval]
behavior						
gdp	.4707441	.1632586	2.88	0.004	.1507632	.790725
gov	1.077039	.198848	5.42	0.000	.6873043	1.466774
WDIST						
gdp	-.2491177	.1778269	-1.40	0.161	-.597652	.0994166
gov	-.8554532	.3034955	-2.82	0.005	-1.450293	-.2606129
behavior	.50612	.2371679	2.13	0.033	.0412794	.9709606
e.behavior	-.0850594	.3677346	-0.23	0.817	-.8058061	.6356872
var(e.behav~r)	.7128869	.1824608			.4316768	1.177288

```
Wald test of spatial terms:         chi2(4) = 28.22    Prob > chi2 = 0.0000
```

13.7 Model selection

Based on these results, one might first be tempted to eliminate the spatial error model from consideration because the *p*-value of the null hypothesis $H_0: \rho = 0$ is 0.817, the largest of the *p*-values of any of the coefficients in the output table. This decision would then lead to the simple spatial lag model as the preferred specification, with `gpd` and `gov` as explanatory variables with no spatial lag. Fitting this model and computing the associated information criteria using `estat ic` yields

```
. spregress behavior gdp gov, noconstant dvarlag(WDIST) ml nolog
  (51 observations)
  (51 observations (places) used)
  (weighting matrix defines 51 places)

Spatial autoregressive model                    Number of obs   =         51
Maximum likelihood estimates                    Wald chi2(3)    =     703.44
                                                Prob > chi2     =     0.0000
Log likelihood = -73.310148                     Pseudo R2       =     0.5536
```

	Coef.	Std. Err.	z	P>|z|	[95% Conf. Interval]	
behavior						
gdp	.3212043	.0472238	6.80	0.000	.2286474	.4137611
gov	.752509	.1554751	4.84	0.000	.4477835	1.057235
WDIST						
behavior	.2420057	.1031516	2.35	0.019	.0398323	.4441792
var(e.behav~r)	1.011372	.2015943			.6842945	1.494785

```
Wald test of spatial terms:         chi2(1) = 5.50         Prob > chi2 = 0.0190
. estat ic
Akaike's information criterion and Bayesian information criterion
```

Model	N	ll(null)	ll(model)	df	AIC	BIC
.	51	.	-73.31015	4	154.6203	162.3476

Note: BIC uses N = number of observations. See [R] BIC note.

These results are perfectly acceptable, and the spatial effect is strong. However, consider the spatial error model using the same basic setup for the explanatory variables, yielding

```
. spregress behavior gdp gov, noconstant dvarlag(WDIST) ml nolog
  (51 observations)
  (51 observations (places) used)
  (weighting matrix defines 51 places)
Spatial autoregressive model              Number of obs    =         51
Maximum likelihood estimates              Wald chi2(3)     =     703.44
                                          Prob > chi2      =     0.0000
Log likelihood = -73.310148               Pseudo R2        =     0.5536
```

behavior	Coef.	Std. Err.	z	P>|z|	[95% Conf. Interval]	
behavior						
gdp	.3212043	.0472238	6.80	0.000	.2286474	.4137611
gov	.752509	.1554751	4.84	0.000	.4477835	1.057235
WDIST						
behavior	.2420057	.1031516	2.35	0.019	.0398323	.4441792
var(e.behav~r)	1.011372	.2015943			.6842945	1.494785

```
Wald test of spatial terms:          chi2(1) = 5.50        Prob > chi2 = 0.0190
. estat ic
Akaike's information criterion and Bayesian information criterion
```

Model	N	ll(null)	ll(model)	df	AIC	BIC
.	51	.	-73.31015	4	154.6203	162.3476

Note: BIC uses N = number of observations. See [R] BIC note.

Both in terms of the strength of the rejection of the test for zero spatial effect and in terms of both the Akaike information criterion and the Bayesian information criterion, the spatial error model is preferable, whereas fitting the general model first leads to the conclusion that it should be dropped in favor of the spatial autoregressive model. This simple demonstration should not be taken as support for a specific-to-general method always being preferred. It does, however, emphasize the point that the final model reached in a general-to-specific setting is path dependent. If correctly applied, the methodology requires searching many paths. The endpoints of these paths should be evaluated against one another. This is a complex search procedure and is best undertaken using software specially designed for that purpose.

With the increasing availability of detailed data on spatial units (counties, provinces, census tracts, etc.) and geocode information for these units, it has become feasible to account for spatial relationships that may play an important role in modeling many outcomes. As has been noted in the COVID-19 pandemic, viruses do not respect political boundaries. Although different jurisdictions may have their own regulations to combat the spread of infection, spatial effects are likely to be important. It is feasible to evaluate the importance of spatial effects and incorporate them where they are relevant in settings of several thousand spatial units such as U.S. counties. Coupled with mapping capabilities, spatial modeling can provide useful tools for the environmental researcher.

13.7 Model selection

Exercises

1. Inverse distance is not a perfect weighting mechanism in some cases. In the cases of the code of conduct on responsible fishing, there may be instances where contiguous states are politically distant. This means that any noncompliance can be observed, but the lack of political cooperation means that any response to noncompliance is likely to be ineffectual. Borsky and Raschky (2015) suggest using a weighting matrix that adjusts the geographical distance by trying to account for political ties in the form of mutual involvement in other international environmental agreements. The weights are

$$w_{ij} = \frac{1}{T} \sum_{t=1}^{T} \frac{g_{ijt}}{d_{ij}}$$

where $g = 1$ if country i and j signed treaty t and $g = 0$ otherwise. The weighting matrix is saved in the file wpol.stwm, and the main data file is saved as spatialfish_small.dta.

   a. Use the official suite of Stata commands relating to spmatrix to create a standard inverse distance matrix for the code of conduct data. Explore the performance of the various spatial models proposed in the chapter using this weighting matrix. What do you conclude?

   b. The Stata command grmat allows weighting matrices to be graphed. Graph the political distance weighting matrix saved as wpol.spmat. Comment on the differences, if any, between this matrix and the custom-made inverse exponential weighting matrix used in the chapter.

   c. Fit separate spatial lag models for the behavior and intention dependent variables using the political distance matrix. Comment on the results.

   d. Estimate the spillover effects for these two models, and interpret the results.

   e. Fit separate spatial error models for the behavior and intention dependent variables using the political distance matrix. Comment on the results and indicate which specification you prefer: the spatial lag model or the spatial error model.

2. This exercise uses daily data on confirmed cases of COVID-19 in 48 U.S. states.

   a. Use usastates.dta to create an inverse distance spatial weighting matrix. In this dataset, the _CX and _CY variables contain the longitude and latitude of each state's centroid, respectively.

   b. confirmed_adj_st_pol.dta contains state-level data on confirmed cases of the coronavirus for April 23, 2020 in confirmed_adj_0. Test this variable for spatial effects by running a linear regression with only a constant term, followed by estat moran using the spatial distance matrix you prepared.

   c. Fit a spatial lag model for confirmed_adj_0 using the command spregress, with popestimate2019, predomRep and their interaction as explanatory variables. The variable predomRep is an indicator of whether a state's electoral votes were predominantly won by the Republican presidential candidate over several election cycles. Interpret the findings using `estat impact` and the evidence for spatial effects.

d. `confirmed_adj_st_long.dta` contains state-level data on confirmed cases for April 23, 2020 and the preceding 14 calendar days, reflecting the period during which the coronavirus is thought to be transmissible. It is daily panel data in the long format. Create the spatial distance matrix using a single day's data by applying the `if` condition `if past==0` because the geography does not differ from day to day.

e. Fit a panel spatial lag model for confirmed cases `conf` based on six lags of that variable, stored as `l1conf`, `l2conf`, `l3conf`, `l4conf`, `l5conf`, and `l6conf` as spxtregress (see [SP] **spxtregress**) does not support lag operators. Use the `re` random-effects specification because there are time-invariant regressors. Interpret the evidence for autoregressive effects and spatial effects.

14 Discrete dependent variables

In most of the models previously discussed in this book, the dependent variable is assumed to be a continuous random variable. There are several situations where this assumption is inappropriate, and alternative classes of models must be specified. This often calls for a model of discrete choice in which the response variable is restricted to a Boolean or binary choice, indicating that a particular course of action was or was not selected. In other models, the dependent variable may take only integer values such as the ordered values on a Likert scale.[1] Alternatively, the dependent variable may appear to be a continuous variable with several responses at a threshold value. For instance, the response to the question of how much you are willing to pay for environmental conservation will be bounded by zero, making these kinds of dependent variables unsuitable for modeling by means of linear regression methods.

There are two important types of discrete dependent variables that will not be covered in this chapter. These are categorical data, in which the dependent variable falls into one of several mutually exclusive but unordered categories, and count data, where the dependent variable takes only nonnegative values. In both of these cases, just as with the other limited dependent variable models, linear regression is not an appropriate estimation technique. Interested readers are referred to Cameron and Trivedi (2010, 2013) for excellent treatments of discrete dependent variable models.

The discrete time-series models that will be discussed here are logit, probit, ordered probit, and Tobit regression models. This choice should not be interpreted as suggesting that these are the only important models for discrete dependent variables but rather that the ideas and methods in these models are broadly compatible and lend themselves to being modeled in one chapter. From a purely practical perspective, obtaining one dataset that facilitates the discussion of a broader set of discrete dependent variable models is difficult.

14.1 Humpback whales

Each year between April and November, eastern humpback whales migrate north for about 10,000 kilometers from their feeding grounds in Antarctic waters to subtropical waters where they mate and give birth. Australia's eastern coastline comes alive with the spectacular acrobatic displays of humpback whales. Before commercial whaling, an

---

[1]. A Likert scale is a rating scale used in surveys that is aimed at ascertaining how people feel about something. Typically, responders are asked to indicate their agreement or disagreement with a statement typically in five or seven points.

estimated population of around 40,000 humpback whales migrated along the east coast of Australia.

Shortly after European colonization, whaling and the export of whale products became Australia's first primary industry. Australian and New Zealand whalers of the early 19th century hunted from small boats, towing their catch back for processing at shore stations. The development of harpoon guns, explosive harpoons, and steam-driven whaling ships later that century made large-scale commercial whaling so efficient that many whale species were overexploited in the 20th century and came very close to extinction. It is believed that up to 95% of the east coast population of humpbacks was killed in the decade from 1952 to 1962. By 1963, when whaling ended, there may have only been 500 whales left. This disastrous period in the history of the eastern humpback whale is graphically illustrated in figure 14.1, where the annual catch for Australian and New Zealand is plotted, representing a grand total of 14399 eastern humpbacks; see Jackson et al. (2008).

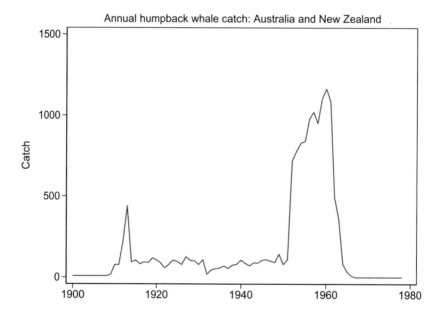

Figure 14.1. Time-series plot of the estimated annual catch of humpback whales for Australia and New Zealand

14.2 The data

The International Whaling Commission banned humpback whaling in the Southern Hemisphere in 1963, and a series of research and monitoring programs has allowed accurate estimates to be made of the growth rate and size of the humpback whale population that migrates along the east coast of Australia. Left alone, the remnants of the population staged a miraculous recovery, increasing in number by around 10% each year. In 2013, the total population estimate was around 19,000 and is currently estimated to be back to about 90% of the prewhaling population.

The recovery of the humpback population has contributed significantly to the rapid growth of Australia's whale-watching industry. During their annual migration, humpbacks attract thousands of visitors to coastal towns along the subtropical east coast of Australia.

14.2 The data

The dataset used in this chapter is the result of a survey conducted in the Queensland town of Hervey Bay by a group of academics seeking to measure the willingness to pay for whale conservation efforts.[2] The survey was conducted in 2000 and had 701 respondents who were surveyed right after returning from a whale-watching trip in Hervey Bay.

The discrete dependent variables for the econometric methods demonstrated in this chapter are constructed from the answers to the following three questions in the survey.

---

1. Would you be willing to have your take home income reduced by $2 per week for the next 10 years to protect and conserve whales that come to breed in Australian waters?

    Yes ☐ No ☐

2. Following your visit to Hervey Bay, are you willing to pay

    More ☐ Less ☐ Same ☐

    for humpback whale protection and conservation as before your visit?

3. To protect and conserve humpback whales that come to Australia to breed, what is the **maximum** amount you would be willing to pay per week for the next 10 years?

    Aus $ .... per week

---

[2]. The data used in this chapter were kindly made available by Professor Clevo Wilson of the Queensland University of Technology.

The answers to these questions give rise to different kinds of dependent variables and fundamental differences in the kinds of econometric models that are appropriate in each case. The answer to the first question creates a binary dependent variable (PayConserve$_k$), the answer to the second defines an ordered dependent variable (wtp$_k$), and the answer to the third should be considered as a censored dependent variable (Max_wtp$_k$). Each of these is now examined in more detail.

Using the survey data, the results from the first 20 respondents are as follows:

```
. use http://www.stata-press.com/data/eeus/whalesdata
. list PayConserve wtp Max_wtp in 1/20
```

	PayCon~e	wtp	Max_wtp
1.	1	Less	0
2.	0	More	0
3.	0	More	0
4.	1	More	0
5.	0	More	0
6.	1	Same	1
7.	1	Less	2
8.	0	More	0
9.	1	Same	5
10.	0	More	0
11.	0	More	0
12.	1	Same	5
13.	1	Same	2
14.	0	More	0
15.	1	Same	2
16.	1	Same	4
17.	0	More	0
18.	1	Same	10
19.	0	More	0
20.	1	More	0

The binary dependent variable, PayConserve$_k$, is created by applying the rule

$$\text{PayConserve}_k = \begin{cases} 1 & \text{yes} \\ 0 & \text{no} \end{cases}$$

The linear probability model and the binomial logit or probit models are appropriate for this kind of dependent variable.

14.2 The data

The second column of data is an example of an ordered dependent variable. Here the question about willingness to pay is phrased slightly differently. After the whale-watching trip, the respondents were asked to say whether they would now be willing to pay less, the same, or more for whale conservation. The ordered data are created by applying the rule

$$\mathtt{wtp}_k = \begin{cases} 1 & \text{willing to pay less than before} \\ 2 & \text{willing to pay the same as before} \\ 3 & \text{willing to pay more than before} \end{cases}$$

These data are obviously more informative than the simple binary variable, and the appropriate econometric models are the ordered probit or logit models.

In the case of the censored model, respondents were asked to indicate how much per week over a 10-year period they were willing to contribute to whale conservation. Let $\widetilde{y}_t$ now represent the amount offered, a variable that is censored at zero, so the censored dependent variable y_t is created as follows:

$$\mathtt{Max\_wtp}_k = \begin{cases} \widetilde{y}_t & \widetilde{y}_t > 0 \\ 0 & \widetilde{y}_t \leq 0 \end{cases}$$

with the results recorded in the column headed censored. This model is a mixture of the full information model where $\widetilde{y}_t > 0$ and the binary model where $\widetilde{y}_t < 0$. This is the censored regression or Tobit model.

In terms of explanatory variables used in the study, the `describe` command produces the following output:

```
. describe
Contains data from http://www.stata-press.com/data/eeus/whalesdata.dta
  obs:           701
 vars:            15                          23 Oct 2018 14:30

              storage   display    value
variable name   type    format     label      variable label

Country        str15    %15s
SeeWhales      byte     %8.0g                 Have you seen humpback whales at
                                              Hervey Bay before
Age            byte     %8.0g                 Q3.1A Age of respondent
Gender         byte     %8.0g      GENDER     Q3.1G Gender of respondent
Education      byte     %8.0g      EDUCATIO   Q3.2 Highest qualification of
                                              respondent
Income_AUD     byte     %8.0g      INCOME_A   Q3.4.1 Household income in
                                              Australian dollars
Income         float    %9.0g                 Mid point
AdultWhales    byte     %8.0g                 Q4.1How many adult humpback whales
                                              did you see
YoungWhales    byte     %8.0g      YOUNGWHA   Q4.2 Did you see young humpback
                                              whales
Max_wtp        double   %10.0g                Q7.4 Maximum WTP/ week to conserve
                                              whales
wtp            byte     %8.0g      WTP        Q7.5 Following WW trip are you
                                              willing to pay for protection
PayConserve    byte     %9.0g                 Willingness to pay to conserve
                                              whales (1=Yes; 0=No)
foreign        byte     %9.0g
highered       byte     %9.0g
scaledWtp      float    %9.0g

Sorted by:
```

The distribution of income, measured in Australian dollars, of the respondents reveals the following:

```
. tabulate Income_AUD
   Q3.4.1 Household |
         income in  |
        Australian  |
           dollars  |    Freq.     Percent        Cum.

 Less than $20,000  |      138       20.47       20.47
  $20,001 - $30,000 |       87       12.91       33.38
  $30,001 - $40,000 |      116       17.21       50.59
  $40,001 - $50,000 |       62        9.20       59.79
  $50,001 - $60,000 |       80       11.87       71.66
  $60,001 - $70,000 |       51        7.57       79.23
  More than $70,000 |      140       20.77      100.00

             Total |      674      100.00
```

14.2 The data

The distribution of income reflects the age distribution in figure 14.2. The large number of respondents earning less than $40,000 follows directly from the spike in the age distribution in the early to mid 20s. The large number of respondents above $70,000 is perhaps indicative that the survey prematurely "top-coded" the income scale. This is partly due to the fact that there were many visitors from the United States and the United Kingdom. In September 2000, the height of the whale-watching season, the Australian dollar was weak against these currencies, resulting in a bracket creep for these respondents.

About 70% of the respondents were Australian, and 60% had completed secondary education and had some sort of tertiary education. The age distribution of the respondents is shown in figure 14.2. The distribution is fairly typical of the age of visitors to Queensland with a relatively higher proportion than would be expected in the early to mid-20s (backpackers) and in the late 50s and early 60s (the gray nomads).

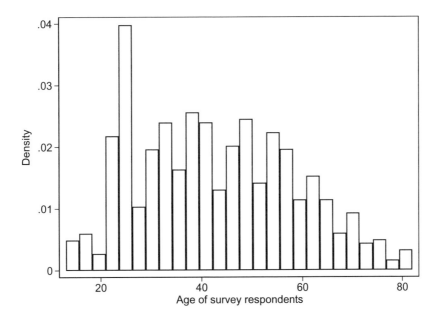

Figure 14.2. Histogram of age of the 701 respondents to the Hervey Bay whale conservation questionnaire

Finally, note that the `foreign` variable takes the value 1 if the respondent's nationality is not Australian and 0 otherwise. Also note the `highered` variable is a simplified version of `Education` that takes the value 1 if the respondent has education above the secondary level.

14.3 Binary dependent variables

There are two approaches to modeling a binary dependent variable. The first simply ignores the fact that the dependent variable is binary and uses linear regression to fit the proposed model. This approach is known as the linear probability model. The second way of dealing with binary dependent variables is to use a nonlinear link function to ensure that the predictions of the model fall between 0 and 1. This approach gives rise to the binomial logit and probit models depending on the choice of link function.

14.3.1 Linear probability model

The linear probability model is the multiple regression model

$$y_k = \beta_0 + \beta_1 x_{1k} + \beta_2 x_{2k} + \cdots + \beta_q x_{qk} + u_k$$
$$= \mathbf{x}_k' \boldsymbol{\beta} + u_k \qquad (14.1)$$

where $\boldsymbol{\beta} = [\beta_1 \ \ldots \ \beta_q]'$, $x_k = [x_{1k} \ \ldots \ x_{qk}]'$, y_k is a binary dependent variable and u_k is a disturbance term. In this specification, the binary nature of the dependent variable is simply ignored, and the regression coefficients are estimated by ordinary least squares.

Because y_k can only take two values (0 and 1) it follows that

$$E(y_t|\mathbf{x}_k) = 1 \times P_k + 0 \times (1 - P_k) = P_k$$

in which $P_k = \Pr(y_k = 1)$ and $1 - P_k = \Pr(y_k = 0)$. From (14.1),

$$E(y_k|\mathbf{x}_k) = \mathbf{x}_k' \boldsymbol{\beta}$$

and it follows that

$$\Pr(y_k = 1|\mathbf{x}_k) = \mathbf{x}_k' \boldsymbol{\beta}$$

This implies that the regression coefficient β_j has the interpretation that it measures the change in the probability that $y_k = 1$ associated with a unit change in x_j, holding all the other regressors constant. In other words, the coefficient estimates in the linear probability model, notwithstanding the fact that the dependent variable is a binary response variable, have the same interpretation as marginal effects as in any linear regression model. This ease of interpretation of the coefficients and the fact that the ordinary least-squares estimator remains a consistent estimator makes the linear probability model a common way of handling the issue of binary dependent variables.

The properties of the disturbance term (u_k) are slightly different from those normally associated with a linear regression.

1. For any k, the disturbance term u_k can take only one of two values:

$$y_k = 1 \to u_k = 1 - \mathbf{x}_k' \boldsymbol{\beta} \quad \text{with probability} \quad P_k$$
$$y_k = 0 \to u_k = -\mathbf{x}_k' \boldsymbol{\beta} \quad \text{with probability} \quad 1 - P_k$$

14.3.1 Linear probability model

2. The mean and variance of the error term u_k are

$$E(u_k|\mathbf{x}_k) = (1 - \mathbf{x}_k'\boldsymbol{\beta}) \times P_k + (-\mathbf{x}_k'\boldsymbol{\beta}) \times (1 - P_k)$$
$$E(u_k^2|\mathbf{x}_k) = (1 - \mathbf{x}_k'\boldsymbol{\beta})^2 \times P_k + (-x_t'\boldsymbol{\beta})^2 \times (1 - P_k)$$

Because var(u_k) depends on $\mathbf{x}_k$, the disturbances are heteroskedastic, so robust standard errors should be used when making inference using the linear probability model.

In addition to the unusual properties of the disturbances, the predictions for y_k obtained from the linear probability model cannot be guaranteed to be interpretable as probabilities. The predictions $\mathbf{x}_k'\widehat{\boldsymbol{\beta}}$ can potentially take on values greater than 1 or less than 0 so that the predicted values for y_k are unbounded.

Consider a linear probability model used to model the binary variable PayConserve$_k$, generated as the answer to the first question in the survey on whale protection and conservation. The willingness to pay for conservation is hypothesized to be a function of nationality, age, gender, education, and income as well as the number of adult whales spotted on the current trip and whether any young whales were seen. The results are as follows:

```
. use http://www.stata-press.com/data/eeus/whalesdata, clear
. regress PayConserve i.foreign i.highered Age i.Gender Income_AUD AdultWhales
> i.YoungWhales, vce(robust)
Linear regression                               Number of obs     =        668
                                                F(7, 660)         =       4.17
                                                Prob > F          =     0.0002
                                                R-squared         =     0.0419
                                                Root MSE          =     .46333
```

| | | Robust | | | | |
PayConserve	Coef.	Std. Err.	t	P>|t|	[95% Conf. Interval]
1.foreign	-.1865396	.0446778	-4.18	0.000	-.2742672	-.0988119
1.highered	.0864871	.04021	2.15	0.032	.0075322	.1654421
Age	-.0046948	.0012805	-3.67	0.000	-.0072092	-.0021804
Gender						
Male	-.0068869	.0380012	-0.18	0.856	-.0815046	.0677309
Income_AUD	-.0110343	.0090025	-1.23	0.221	-.0287114	.0066428
AdultWhales	.0069031	.004496	1.54	0.125	-.001925	.0157312
YoungWhales						
Yes	.0122331	.0415212	0.29	0.768	-.0692965	.0937627
_cons	.8645838	.0823153	10.50	0.000	.7029523	1.026215

The results indicate that nationality, age, and education are the major drivers influencing willingness to pay. Not surprisingly, foreign nationals appear less likely to sponsor protection and conservation of whales found in Australian waters. The negative coefficient on Age strongly indicates the altruism of youth and the success of conservation awareness programs among the younger members of the population. The positive coefficient on `highered` is also expected. Interestingly, `Gender` and `Income_AUD` do not appear to be statistically significant and neither are the variables summarizing the success of the recent whale-watching trip. A surprising result is the fact that the coefficient on `YoungWhales`, the variable that captures whether any young whales were seen on the trip, does not seem to have a significant effect on willingness to pay. Indeed, the success or otherwise of the current trip in terms of whale spotting does not seem to play a major role in influencing willingness to pay.

14.3.2 Binomial logit and probit models

A probability distribution for outcomes that take on only two values is known as the Bernoulli distribution

$$f(1) = \pi \qquad f(0) = 1 - \pi$$

where the event occurs with probability π and fails to occur with probability $1 - \pi$. Given K observations on the binary variable ($y_k = \{y_1, \ldots y_K\}$) that are assumed to be independent and identically distributed, the likelihood function is

$$L = \prod_{k=1}^{K} \pi^{y_k}(1-\pi)^{1-y_k}$$

and the log-likelihood function is

$$\log L = \sum_{k=1}^{K} \{y_k \log(\pi) + (1-y_k)\log(1-\pi)\}$$

This version of the Bernoulli distribution is too restrictive as it stands because every instance y_k has constant probability π of taking the value 1. In a heterogeneous sample of respondents, this assumption is inappropriate. Ideally, π would vary for each individual, although letting each observation have its own π_k would imply that the model would not be identified.

To both reduce the number of parameters and add substantive explanatory variables, the compromise is to express π_k as a function of explanatory variables. Although the combination of explanatory variables in linear, it is related to π_k via a link function that preserves the integrity of π_k as a probability, namely, $\pi_k = G(\mathbf{x}'_k \boldsymbol{\beta})$. The function $G(\cdot)$ is chosen to ensure that $0 \le \pi_k \le 1$. The log-likelihood function is then

$$\log L = \sum_{t=1}^{T} [y_t \log\{G(\mathbf{x}'_k \boldsymbol{\beta})\} + (1-y_t)\log\{1 - G(\mathbf{x}'_k \boldsymbol{\beta})\}]$$

There are several possible functions, $G(\cdot)$, that are appropriate for use in this context, but two functions are usually used in the econometric literature.

14.3.2 Binomial logit and probit models

1. The cumulative standard normal distribution function is given by

$$\Pr(y_k = 1|x_k) = \int_{-\infty}^{\mathbf{x}'_k\boldsymbol{\beta}} \frac{1}{\sqrt{2\pi}} \exp\left(-\frac{(y_t - \mathbf{x}'_k\boldsymbol{\beta})^2}{2}\right) = \Phi(\mathbf{x}'_k\boldsymbol{\beta})$$

The log-likelihood function is

$$\log L(\boldsymbol{\beta}) = \sum_{t=1}^{K} [y_k \log \{\Phi(\mathbf{x}'_k\boldsymbol{\beta})\} + (1 - y_k) \log \{1 - \Phi(\mathbf{x}'_k\boldsymbol{\beta})\}]$$

$$= \sum_{t=1}^{T} [y_k \log \{\Phi(\mathbf{x}'_k\boldsymbol{\beta})\} + (1 - y_k) \log \{\Phi(-\mathbf{x}'_k\boldsymbol{\beta})\}]$$

where the second line follows by symmetry. This model is known as the binomial probit model.

2. The cumulative logistic distribution function is given by

$$\Pr(y_k = 1|\mathbf{x}_k) = \frac{e^{\mathbf{x}'_k\boldsymbol{\beta}}}{1 + e^{\mathbf{x}'_k\boldsymbol{\beta}}} = \frac{1}{1 + e^{-\mathbf{x}'_k\boldsymbol{\beta}}} = \Lambda(\mathbf{x}'_k\boldsymbol{\beta})$$

The log-likelihood function of the model is then

$$\log L = \sum_{t=1}^{K} [y_k \log \{\Lambda(\mathbf{x}'_k\boldsymbol{\beta})\} + (1 - y_t) \log \{1 - \Lambda(\mathbf{x}'_k\boldsymbol{\beta})\}]$$

This model is known as the binomial logit model.

In both of these instances, the log-likelihood function is nonlinear, and there is no analytical solution for $\boldsymbol{\beta}$. However, it turns out that the numerical optimisation of the function is straightforward.

Applying the probit model to the PayConserve$_k$ binary dependent variable and maintaining the same set of explanatory variables gives

```
. probit PayConserve i.foreign i.highered Age Income_AUD AdultWhales
> i.YoungWhales

Iteration 0:   log likelihood = -424.02685
Iteration 1:   log likelihood =  -409.9695
Iteration 2:   log likelihood = -409.94912
Iteration 3:   log likelihood = -409.94912

Probit regression                               Number of obs   =        668
                                                LR chi2(6)      =      28.16
                                                Prob > chi2     =     0.0001
Log likelihood = -409.94912                     Pseudo R2       =     0.0332
```

PayConserve	Coef.	Std. Err.	z	P>|z|	[95% Conf. Interval]	
1.foreign	-.5199672	.123693	-4.20	0.000	-.762401	-.2775334
1.highered	.241608	.1127846	2.14	0.032	.0205542	.4626618
Age	-.0132854	.00359	-3.70	0.000	-.0203218	-.0062491
Income_AUD	-.0305215	.0242875	-1.26	0.209	-.0781242	.0170812
AdultWhales	.0199928	.0140895	1.42	0.156	-.0076221	.0476077
YoungWhales						
Yes	.0335357	.1170163	0.29	0.774	-.1958121	.2628835
_cons	.9931701	.235131	4.22	0.000	.5323218	1.454018

The iteration log confirms that the numerical optimization of the log-likelihood function for the probit model is particularly straightforward. Note that while the signs and the statistical significance of the explanatory variables are the same as those obtained for the linear probability model, the actual coefficient estimates $\widehat{\beta}_j$ of the coefficient on the explanatory variable x_j do not match those obtained in the linear probability model. This occurs because the coefficients of the probit model do not express the change in the probability but rather the change in the linear combination $\mathbf{x}'_k\boldsymbol{\beta}$.

14.3.2 Binomial logit and probit models

The same conclusions hold for the logit model, and the numerical optimization is again straightforward because the likelihood function can be shown to be a concave function.

```
. logit PayConserve i.foreign i.highered Age Income_AUD AdultWhales
> i.YoungWhales

Iteration 0:   log likelihood = -424.02685
Iteration 1:   log likelihood = -409.98527
Iteration 2:   log likelihood = -409.91236
Iteration 3:   log likelihood = -409.91235

Logistic regression                               Number of obs   =        668
                                                  LR chi2(6)      =      28.23
                                                  Prob > chi2     =     0.0001
Log likelihood = -409.91235                       Pseudo R2       =     0.0333
```

PayConserve	Coef.	Std. Err.	z	P>|z|	[95% Conf. Interval]
1.foreign	-.8578591	.2044841	-4.20	0.000	-1.258641 -.4570778
1.highered	.3971133	.1868998	2.12	0.034	.0307964 .7634301
Age	-.0219375	.0059704	-3.67	0.000	-.0336393 -.0102358
Income_AUD	-.0522243	.0403744	-1.29	0.196	-.1313567 .0269081
AdultWhales	.0332282	.0232544	1.43	0.153	-.0123495 .0788059
YoungWhales					
Yes	.0602844	.1923266	0.31	0.754	-.3166688 .4372376
_cons	1.630999	.3920357	4.16	0.000	.8626228 2.399374

Note that in the probit model, the cumulative standard normal distribution adopts the normalizing restriction that the standard deviation of the distribution, σ, is equal to 1. In the case of the logistic regression model, the distribution of $\Lambda(\cdot)$ adopts the normalizing assumption $\sigma = \sqrt{\pi^2/3}$, the exact form of which arises from the properties of this distribution. For this reason, the coefficients in a logit model will be larger by that scale factor than those of a corresponding probit model.

Consider now the maximum likelihood estimate $\widehat{\beta}_j$ of the coefficient on the explanatory variable x_j, which as just shown is not an estimate of the marginal effect of x_j on $\Pr(y_k = 1|\mathbf{x}_k)$. This quantity is given by

$$\frac{\partial \Pr(y_k = 1|\mathbf{x}_k)}{\partial x_j} = g\left(\mathbf{x}_k'\widehat{\boldsymbol{\beta}}\right)\widehat{\beta}_j = \widehat{m}_j$$

where $g(\cdot)$ is the probability density function, or derivative, of the cumulative distribution function $G(\cdot)$. Because $g(\cdot)$ depends on the values of all the independent variables, $\mathbf{x}_k$, the marginal effect must be computed at interesting values of these explanatory variables. There are three choices facing the econometrician:

1. choose a single value for the $\mathbf{x}_k$ that is of particular importance to the problem;
2. choose the sample averages of the $\mathbf{x}_k$ and perform the computation at these values; or

3. evaluate the marginal effect of $\mathbf{x}_k$ at each of the k observations, and compute the sample average of the individual marginal effects—the so-called average marginal effects (AMEs).

The Stata command `margins` computes the average of the marginal effects (see case 3) as the default. The option `atmeans` is a convenient way of implementing case 2, while the `at()` option can be used to specify a particular value of some or all variables at which to compute the marginal effects (see case 1). In most cases, the AMEs are to be preferred because they account for multivariate distribution of the explanatory variables.

The standard errors of the marginal effect $(\widehat{m}_j)$ are computed using the delta method

$$\operatorname{var}(\widehat{m}_j) = \left(\frac{\partial g\left(\mathbf{x}_k'\widehat{\boldsymbol{\beta}}\right)\widehat{\beta}_j}{\partial \widehat{\beta}_j} \right)^2 \operatorname{var}\left(\widehat{\beta}_j\right) \tag{14.2}$$

where $\operatorname{var}(\widehat{\beta}_j)$ is already available from the estimation.

Consider the binomial probit model where the $\Phi(\cdot)$ is used as the link function. Using the multiplication rule and the fact that for the standard normal distribution

$$\frac{\partial \phi(z)}{\partial z} = -z\phi(z)$$

the partial derivative on the right-hand side of (14.2) becomes

$$\frac{\partial \phi\left(\mathbf{x}_k'\widehat{\boldsymbol{\beta}}\right)\widehat{\beta}_j}{\partial \widehat{\beta}_j} = \phi\left(\mathbf{x}_k'\widehat{\boldsymbol{\beta}}\right) + \widehat{\beta}_j \frac{\partial \phi\left(\mathbf{x}_k'\widehat{\boldsymbol{\beta}}\right)}{\partial \left(\mathbf{x}_k'\widehat{\boldsymbol{\beta}}\right)} x_j$$

$$= \phi\left(\mathbf{x}_k'\widehat{\boldsymbol{\beta}}\right) - \phi\left(\mathbf{x}_k'\widehat{\boldsymbol{\beta}}\right)\widehat{\beta}_j^2 x_j^2$$

$$= \phi\left(\mathbf{x}_k'\widehat{\boldsymbol{\beta}}\right)\left(1 - \widehat{\beta}_j^2 x_j^2\right)$$

The variance of the marginal effect therefore takes the simple form

$$\operatorname{var}(\widehat{m}_j) = \phi\left(\mathbf{x}_k'\widehat{\boldsymbol{\beta}}\right)^2 \left(1 - \widehat{\beta}_j^2 x_j^2\right)^2 \operatorname{var}\left(\widehat{\beta}_j\right)$$

14.3.2 Binomial logit and probit models

The AMEs are now computed for both the probit and the logit models and reported for the variables `foreign`, `highered`, and `Income_AUD`.

```
. quietly probit PayConserve i.foreign i.highered Age Income_AUD AdultWhales
. margins, dydx(foreign highered Income_AUD)
Average marginal effects                          Number of obs     =        668
Model VCE    : OIM

Expression   : Pr(PayConserve), predict()
dy/dx w.r.t. : 1.foreign 1.highered Income_AUD
```

	dy/dx	Delta-method Std. Err.	z	P>|z|	[95% Conf. Interval]
1.foreign	-.1891066	.0445283	-4.25	0.000	-.2763805 -.1018326
1.highered	.0852746	.039963	2.13	0.033	.0069485 .1636007
Income_AUD	-.0103929	.0084119	-1.24	0.217	-.02688 .0060942

Note: dy/dx for factor levels is the discrete change from the base level.

```
. quietly logit PayConserve i.foreign i.highered Age Income_AUD AdultWhales
> i.YoungWhales
. margins, dydx(foreign highered Income_AUD)
Average marginal effects                          Number of obs     =        668
Model VCE    : OIM

Expression   : Pr(PayConserve), predict()
dy/dx w.r.t. : 1.foreign 1.highered Income_AUD
```

	dy/dx	Delta-method Std. Err.	z	P>|z|	[95% Conf. Interval]
1.foreign	-.1902019	.0449094	-4.24	0.000	-.2782227 -.1021811
1.highered	.0853276	.040247	2.12	0.034	.0064449 .1642103
Income_AUD	-.0110734	.0085232	-1.30	0.194	-.0277786 .0056318

Note: dy/dx for factor levels is the discrete change from the base level.

It is now clear that despite differences in the coefficient estimates due to the differences in scaling, the marginal effects obtained from the binomial logit model are almost identical to those for the binomial probit model. Furthermore, the estimates of the marginal effects now line up with the results obtained from the linear probability model. This has led some econometricians to argue for using the linear probability model; see, for example, Angrist and Pischke (2009, 107). This result, however, is not a general one, and the estimates of the marginal effects yielded by the linear probability model cannot always be relied upon, especially near the 0–1 boundaries of the probability measure.

The choice between probit and logit models is largely a matter of taste. Because of the reliance of the assumption of normality in econometrics, econometricians usually use the probit model. In disciplines such as public health, the binomial logit model is more popular because the results of logistic regression can be evaluated as odds ratios, given the closed form of the logistic distribution. By using the `or` option, the odds ratios, which are merely exponentials of the estimated logit coefficients, are displayed.

```
. logit, or
Logistic regression                          Number of obs   =        668
                                             LR chi2(6)      =      28.23
                                             Prob > chi2     =     0.0001
Log likelihood = -409.91235                  Pseudo R2       =     0.0333
```

PayConserve	Odds Ratio	Std. Err.	z	P>|z|	[95% Conf. Interval]	
1.foreign	.424069	.0867153	-4.20	0.000	.2840399	.6331311
1.highered	1.487524	.278018	2.12	0.034	1.031276	2.145623
Age	.9783014	.0058408	-3.67	0.000	.9669202	.9898164
Income_AUD	.949116	.03832	-1.29	0.196	.876905	1.027273
AdultWhales	1.033786	.02404	1.43	0.153	.9877264	1.081994
YoungWhales						
Yes	1.062139	.2042775	0.31	0.754	.728572	1.548424
_cons	5.108974	2.0029	4.16	0.000	2.369367	11.01628

Note: _cons estimates baseline odds.

An odds ratio of 1.0 indicates that the probability of $y_k = 1$ is not affected by that particular x_j. If the odds ratio is significantly less than 1.0, say, 0.43, then a one-unit change in x_j reduces the probability that $y_k = 1$ by 57%. If the odds ratio is significantly greater than 1.0, say, 1.49, then a one-unit change in x_j increases the probability that $y_k = 1$ by 49%. In some disciplines, the odds ratios in logistic regression are commonly used to evaluate the importance of the explanatory variables rather than the marginal effects presented above.

Because of the binary nature of the outcome variable in logit or probit models, there is no summary statistic such as R^2 applicable as a measure of goodness of fit. A glance at the estimation results presented previously will show that Stata reports a Pseudo R2 figure in the estimation output. This statistic is also known as McFadden's R^2. Let ll(1) be the value of the log-likelihood function of the fitted model, and let ll(0) be the log-likelihood value of a model that includes only an intercept as the predictor [so that every individual $\Pr(y_k = 1)$ is constant for all k]. Then,

$$\text{Pseudo } R^2 = 1 - \frac{\text{ll}(1)}{\text{ll}(0)}$$

If the model has no predictive ability, then although ll(1) will always be larger than the likelihood of the restricted model [ll(0)] it will not be much greater. Consequently, the ratio ll(1)/ll(0) will be close to 1, and the pseudo-R^2 will be close to zero as required. It turns out that R^2 measures in these models can be quite a contentious subject, and many alternatives have been suggested.[3]

Once fit, a probit or logit model provides predicted probabilities associated with the state of the binary dependent variable. At the very least, a good model should do relatively well in classifying the binary outcomes correctly. The standard procedure is to classify predictions using a cutoff probability of 0.5 so that $\widehat{y}_k = 1$ when the

---

[3]. See, for example, the package fitstat (Long and Freese 2000) from the Statistical Software Components Archive (type search fitstat).

14.3.2   Binomial logit and probit models

predicted probability obtained from the model is greater than 0.5. Stata computes this classification table using the postestimation command `estat classification`.

```
. estat classification
Logistic model for PayConserve
                        True
Classified  |     D          ~D       |    Total
------------+-------------------------+---------
     +      |    431         202      |     633
     -      |     16          19      |      35
------------+-------------------------+---------
   Total    |    447         221      |     668

Classified + if predicted Pr(D) >= .5
True D defined as PayConserve != 0
-----------------------------------------------------
Sensitivity                     Pr( +| D)      96.42%
Specificity                     Pr( -|~D)       8.60%
Positive predictive value       Pr( D| +)      68.09%
Negative predictive value       Pr(~D| -)      54.29%
-----------------------------------------------------
False + rate for true ~D        Pr( +|~D)      91.40%
False - rate for true  D        Pr( -| D)       3.58%
False + rate for classified +   Pr(~D| +)      31.91%
False - rate for classified -   Pr( D| -)      45.71%
-----------------------------------------------------
Correctly classified                           67.37%
-----------------------------------------------------
```

The probability of correctly predicting $y_k = 1$ using the model is known as the model's sensitivity. On the other hand, the probability of correctly predicting the alternative, $y_k = 0$, is known as the specificity of the model. The correct classifications are on the main diagonal in the output table, while the off-diagonal observations are misclassified by the model. In this particular example, based on the estimates of the logit model with a cutoff probability of 0.5, the model's ability to predict correctly that a survey respondent is willing to pay for the conservation of whales is very good—a 96% sensitivity. On the other hand, the model's ability to classify those who are not willing to pay is not particularly good, because the specificity is only 8.6%. This flip side of this result is that there is a false positive prediction 91.4% of the time.

Ideally, we would like to maximize both sensitivity and specificity. For a given model, lowering the probability cutoff point is one way of improving the sensitivity of the model, but this improvement comes with a cost because specificity is necessarily reduced. Consequently, there is a tradeoff between sensitivity and specificity when manipulating the probability cutoff. The postestimation command `lsens` illustrates the tradeoff between sensitivity and specificity as a function of probability cutoff. Figure 14.3 illustrates the tradeoff between sensitivity and specificity based on the parameter estimates of the logit model. The plot illustrates quite clearly that using a probability cutoff of 0.5 gives increased sensitivity at the cost of sensitivity and that the optimal balance between these factors in this particular model may require the use of a cutoff probability significantly greater than 0.5 and closer to 0.7.

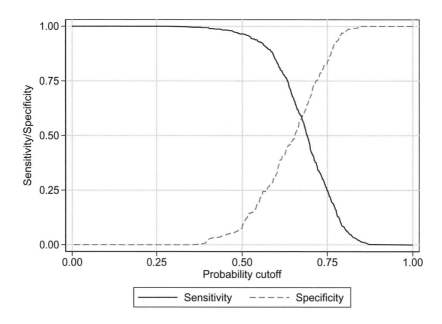

Figure 14.3. A plot of sensitivity versus specificity of the logit model of willingness to pay for whale conservation based on respondents to the Hervey Bay whale conservation questionnaire.

14.4 Ordered dependent variables

The answer to the second willingness-to-pay question in the survey gives rise to an ordered polychotomous variable, wtp_k. Note that the recode (see [D] **recode**) command is applied to the wtp variable in order to ensure that its values are in ascending order of willingness to pay rather than in the order given in the questionnaire. The recoded variable is named wtp2. This variable now takes three values, namely, $y_k = 1$ if the respondent is willing to pay less than before the current visit, $y_k = 2$ if the willingness to pay is unchanged, and $y_k = 3$ if the respondent is willing to pay more after the current visit.

14.4 Ordered dependent variables

```
. use http://www.stata-press.com/data/eeus/whalesdata, clear
. recode wtp (2=1 "Less") (3=2 "Same") (1=3 "More"), gen(wtp2) label(wtp2)
(701 differences between wtp and wtp2)
. tabulate wtp2
```

RECODE of wtp (Q7.5 Following WW trip are you willing to pay for protection)	Freq.	Percent	Cum.
Less	280	39.94	39.94
Same	149	21.26	61.20
More	272	38.80	100.00
Total	701	100.00	

The appropriate model to use in this instance is known as the ordered probit model. Note that the linear probability model should not be applied to ordered data, because the ordering of the outcome values is merely transitive. That is, an individual with $y = 2$ is not willing to pay twice as much as an individual with $y = 1$.

The ordered probit and ordered logit models are generalizations of the binomial probit and binomial logit models, where zero is the natural threshold between $y_i = 0$ and $y_i = 1$. With m multiple ordered alternatives,

$$\Pr(y_k = i) = \Pr\left(\kappa_{i-1} < \mathbf{x}_k'\boldsymbol{\beta} + u < \kappa_i\right)$$

where $\kappa_0 = -\infty$ and $\kappa_m = +\infty$. In this case of three ordered outcomes, the probabilities associated with each outcome are

$$\Phi_{1k} = \Pr(y_k = 1) = \Phi(\kappa_1 - \mathbf{x}_k'\boldsymbol{\beta}) - \Phi(\kappa_0 - \mathbf{x}_k'\boldsymbol{\beta})$$
$$\Phi_{2k} = \Pr(y_k = 2) = \Phi(\kappa_2 - \mathbf{x}_k'\boldsymbol{\beta}) - \Phi(\kappa_1 - \mathbf{x}_k'\boldsymbol{\beta})$$
$$\Phi_{3k} = \Pr(y_k = 3) = \Phi(\kappa_3 - \mathbf{x}_k'\boldsymbol{\beta}) - \Phi(\kappa_2 - \mathbf{x}_k'\boldsymbol{\beta})$$

noting that for any cumulative probability density function, $\Phi(-\infty) = 0$ and $\Phi(+\infty) = 1$. As before, $\mathbf{x}_k$ is the set of explanatory variables with parameter vector $\boldsymbol{\beta}$ and $\kappa_1 < \kappa_2$ are parameters, known as cut points, to be estimated. These are the threshold values separating the outcomes. The ordering restriction on the κ parameters is needed to ensure that the probabilities of each outcome are positive. Any number of ordered outcomes can be modeled.

In our case, define three indicator variables $I_{jk} = 1$ if $y_k = j$ for $j = 1, 2, 3$. The log-likelihood function for a sample of size K is

$$\ln L = \sum_{k=1}^{K} \left(I_{1k} \log \Phi_{1k} + I_{2k} \log \Phi_{2k} + I_{3k} \log \Phi_{3k}\right)$$

The parameters are estimated by a numerical optimization algorithm. In Stata, the ordered probit model is fit using the oprobit (see [R] **oprobit**) command. An ordered logit model is also available using ologit (see [R] ologit), but for brevity only the ordered probit model will be discussed here.

Fitting an ordered probit model for the ordered variable wtp2 using the same explanatory variables as for the probit model fit earlier produces the following results:

```
. oprobit wtp2 i.foreign i.highered Age Income_AUD AdultWhales i.YoungWhales
Iteration 0:   log likelihood = -710.68065
Iteration 1:   log likelihood = -657.23015
Iteration 2:   log likelihood = -657.06783
Iteration 3:   log likelihood = -657.06783

Ordered probit regression                         Number of obs   =        668
                                                  LR chi2(6)      =     107.23
                                                  Prob > chi2     =     0.0000
Log likelihood = -657.06783                       Pseudo R2       =     0.0754

------------------------------------------------------------------------------
        wtp2 |      Coef.   Std. Err.      z    P>|z|     [95% Conf. Interval]
-------------+----------------------------------------------------------------
   1.foreign |   1.126976   .1187037     9.49   0.000     .8943211    1.359631
  1.highered |  -.2538755   .1015795    -2.50   0.012    -.4529676   -.0547834
         Age |   .0186367   .0033439     5.57   0.000     .0120827    .0251906
  Income_AUD |    .001079   .0220938     0.05   0.961     -.042224    .0443821
 AdultWhales |  -.0218444   .0125243    -1.74   0.081    -.0463916    .0027028
             |
 YoungWhales |
         Yes |  -.0646411    .105857    -0.61   0.541    -.272117    .1428349
-------------+----------------------------------------------------------------
       /cut1 |   .4846884   .2137083                      .0658278    .9035489
       /cut2 |   1.091049   .2157384                      .6682098    1.513889
------------------------------------------------------------------------------
```

These estimation results from the ordered probit model are best interpreted using margins. Before doing so, it is worth examining how the marginal effects are computed for each value of the ordered dependent variable, noting that for any cumulative probability density function, $\Phi(-\infty) = 0$ and $\Phi(+\infty) = 1$.

1. Case 1: $y_k = 1$

$$\frac{\partial \Pr(y_k = 1 | \mathbf{x}_k)}{\partial x_j} = -\beta_j \underbrace{\phi(\kappa_1 - \mathbf{x}'_k \boldsymbol{\beta})}_{>0}$$

Given $\beta_j > 0$, the probability that y_k takes the smallest possible value will decrease if x_j increases, redistributing the probability mass toward higher values.

2. Case 2: $y_k = 2$

$$\frac{\partial \Pr(y_k = 2 | \mathbf{x}_k)}{\partial x_j} = \{\phi(\kappa_2 - \mathbf{x}'_k \boldsymbol{\beta}) - \phi(\kappa_1 - \mathbf{x}'_k \boldsymbol{\beta})\} \beta_j$$

The relation that $\kappa_2 > \kappa_1$ implies that if $\beta_j > 0$, the probability that $y_k = 2$ will increase may rise or fall as the probability mass is redistributed among the three outcomes.

14.4 Ordered dependent variables

3. Case 3: $y_k = 3$

$$\frac{\partial \Pr(y_k = 3|\mathbf{x}_k)}{\partial x_j} = \beta_j \underbrace{\phi(\kappa_2 - \mathbf{x}'_k\boldsymbol{\beta})}_{>0}$$

The change in the probability of being in the highest bin will rise if $\beta_j > 0$, and vice versa, as the probability mass is redistributed toward (or away from) the third bin.

The sum of the marginal effects over the three alternatives must be zero because a change in an explanatory factor can only reallocate the probabilities over the possible outcomes. Once again, the standard errors of the marginal effects may be computed using the delta method as for the probit model.

Using the `margins` command gives the following output for the ordered probit model:

```
. margins, dydx(*)
Average marginal effects                        Number of obs   =        668
Model VCE    : OIM

dy/dx w.r.t. : 1.foreign 1.highered Age Income_AUD AdultWhales 1.YoungWhales
1._predict   : Pr(wtp2==1), predict(pr outcome(1))
2._predict   : Pr(wtp2==2), predict(pr outcome(2))
3._predict   : Pr(wtp2==3), predict(pr outcome(3))
```

	dy/dx	Delta-method Std. Err.	z	P>|z|	[95% Conf. Interval]	
0.foreign	(base outcome)					
1.foreign _predict						
1	-.3608081	.0301199	-11.98	0.000	-.4198421	-.3017742
2	-.0445936	.0120158	-3.71	0.000	-.0681442	-.0210431
3	.4054018	.03717	10.91	0.000	.33255	.4782536
0.highered	(base outcome)					
1.highered _predict						
1	.0871967	.0343326	2.54	0.011	.019906	.1544874
2	-.0002135	.0019746	-0.11	0.914	-.0040836	.0036566
3	-.0869832	.0345681	-2.52	0.012	-.1547355	-.019231
Age _predict						
1	-.0064544	.0010937	-5.90	0.000	-.0085979	-.0043108
2	.0000797	.0001438	0.55	0.579	-.0002021	.0003616
3	.0063746	.0010876	5.86	0.000	.0042429	.0085063
Income_AUD _predict						
1	-.0003737	.0076518	-0.05	0.961	-.015371	.0146236
2	4.62e-06	.0000953	0.05	0.961	-.0001821	.0001913
3	.0003691	.0075569	0.05	0.961	-.0144422	.0151804

AdultWhales _predict						
1	.0075653	.0043131	1.75	0.079	-.0008883	.0160189
2	-.0000934	.0001753	-0.53	0.594	-.0004371	.0002502
3	-.0074718	.0042649	-1.75	0.080	-.0158308	.0008872
0.YoungWhales	(base outcome)					
1.YoungWhales _predict						
1	.0223328	.036453	0.61	0.540	-.0491137	.0937794
2	-.0001146	.0005093	-0.22	0.822	-.0011128	.0008837
3	-.0222183	.0365461	-0.61	0.543	-.0938472	.0494107

Note: dy/dx for factor levels is the discrete change from the base level.

Looking at the results for `foreign`, recall that this variable takes the value 0 if the respondent is Australian. There are three cases to consider.

1. If the prediction is $y_k = 1$ and the respondent is expected to pay less than before, changing `foreign` from 0 to 1 will decrease the probability that y_k remains in the lowest category.

2. If the prediction is $y_k = 2$, changing `foreign` from 0 to 1 will decrease the probability that y_k remains in the second category.

3. If the prediction is that $y_k = 3$, changing `foreign` from 0 to 1 substantially increases the probability that y_k remains in the top bracket.

In every instance, the marginal effect is statistically significant. These results are at odds with the simple probit model fit earlier and perhaps a little counterintuitive because one might not expect foreign nationals to be willing to pay for Australian conservation projects. Perhaps the framing of the relevant questions in the survey provide a partial explanation of the results. The `PayConserve`$_k$ variable used in the binary dependent variable example asks respondents to commit to having their take-home income reduced for the next 10 years, while the `wtp`$_k$ variable does not ask for this kind of long-term commitment. Consequently, foreign nationals could have felt more comfortable answering the willingness-to-pay question.

The variable `highered` takes the value 1 if the respondent has any further education beyond secondary school, be that a degree, diploma, or trade certificate. The main result here is that if the prediction is $y_k = 1$ and the respondent's willingness to pay is predicted to be in the lowest category, then changing `highered` from 0 to 1 increases the probability that y_k remains in the lowest category. By contrast, if the prediction is for $y_k = 2, 3$, then this change decreases the probability that the prediction remains in the current category, although for the case $y_k = 2$ the effect is insignificant. This result is interesting because one might expect that higher education and conservation awareness would be positively related. Perhaps the rather broad distinction between secondary school and all other education is simply too broad to capture any meaningful results in this instance.

14.5 Censored dependent variables

On the contrary, the `age` variable follows the opposite pattern to that of `highered` where, from the perspective of funding whale protection and conservation, older individuals appear to be more generous in their attitude to funding conservation efforts. Rather surprisingly, although the pattern for the `income` variable is similar in the sense that increasing income decreases the probability of remaining in the lowest category and increases the probability of remaining in the second and third categories, the effects are not significant.

Note that `estat classification` is not available after an ordered probit estimation.

14.5 Censored dependent variables

The answer to the third question in the survey gives rise to a censored dependent variable, Max_wtp_k. Censoring occurs when a response variable is set to an arbitrary value above or below a certain value: the censoring point. A censored response variable should be considered as being generated by a mixture of distributions: the binary choice of being willing to contribute to conservation or not and the continuous response of how much to pay conditional on being willing to contribute. So while all the demographic information on a set of individuals is available, their responses are set at the censoring point.

Although it would appear that the variable Max_wtp_k could be used as the dependent variable in a regression, it should not be used in that manner, becuase it is generated by a censored distribution. Figure 14.4 is a scatterplot of the censored variable Max_wtp_k that illustrates the problem. The large number of censored observations renders any attempt to fit a linear model impossible.

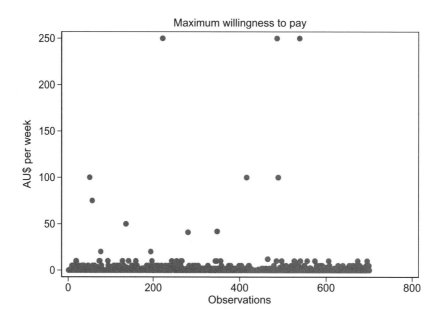

Figure 14.4. Scatterplot of maximum willingness to pay for whale conservation. Survey data for 701 individuals taken in Hervey Bay, Queensland, Australia, in 2002.

A solution to this problem was first proposed by Tobin (1958) as the censored regression model. It has subsequently became known as "Tobin's probit" or the tobit model.[4] The model can conveniently be expressed in terms of a latent variable, y_k^*, as follows:

$$y_k^* = \mathbf{x}_k' \boldsymbol{\beta} + u_k \tag{14.3}$$

$$y_k = \begin{cases} y_k^* & y_k^* > 0 \\ 0 & y_k^* \leq 0 \end{cases} \tag{14.4}$$

The model combines aspects of the binomial probit for the distinction of $y_k = 0$ versus $y_k > 0$ and the regression model for $(y_k | y_k > 0)$. Of course, it is possible to collapse all positive observations on y_k and treat this as a binomial probit estimation problem, but that would discard the information on the dollar amounts that respondents would be willing to pay. Likewise, all the $y_k = 0$ observations could be discarded, but this would leave a truncated distribution with the various problems that this creates (see [R] **truncreg**).

---

4. The term "censored regression" is now more commonly used for a generalization of the Tobit model in which the censoring values may vary from observation to observation. See the documentation for Stata's `intreg` command (see [R] **intreg**).

14.5 Censored dependent variables

To account for all the information in y_k properly, the model must be fit by a full maximum-likelihood method that combines the probit and regression components in the log-likelihood function. The distribution of y_k is given by

$$f(y_k|\mathbf{x}_k) = \left(\frac{1}{\sigma}\right)^{d_t} \phi(\mathbf{x}'_k\boldsymbol{\beta})^{d_t} \{1 - \Phi(\mathbf{x}'_k\boldsymbol{\beta})\}^{1-d_t}$$

where $\phi(\cdot)$ and $\Phi(\cdot)$ are, as usual, the standard normal and cumulative standard normal distributions, respectively, and d_t is the indicator variable

$$d_t = \begin{cases} 1 & y^*_t > 0 \\ 0 & y^*_t \leq 0 \end{cases}$$

The log-likelihood function for a sample of K observations is

$$\ln L = \sum_{t=1}^{K} [-d_t \ln \sigma + d_t \ln \phi(\mathbf{x}'_k\boldsymbol{\beta}) + (1-d_t) \ln \{1 - \Phi(\mathbf{x}'_k\boldsymbol{\beta})\}]$$

Tobit models are fit in Stata using the **tobit** command (see [R] **tobit**). The development here has focused on the case of censoring at zero, but tobit models may be defined with a threshold other than zero. Censoring from below may be specified at any point on the y scale with the ll(#) option for left-censoring. Similarly, the standard tobit formulation may use an upper threshold (censoring from above, or right censoring) using the ul(#) option to specify the upper limit. This form of censoring, also known as top coding, will occur with a variable that takes on a value of "$x or more". For instance, the value could be the answer to a question about income, where the respondent is asked to indicate whether their income was greater than $200,000 last year in lieu of the exact amount. Stata's **tobit** also supports the two-limit tobit model, where observations on y are censored from both left and right by specifying both the ll(#) and ul(#) options.

Fitting the tobit model for the survey data with Max_wtp$_k$ as the censored dependent variable gives an interesting result:

```
. tobit Max_wtp i.foreign i.highered Age Income_AUD AdultWhales i.YoungWhales,
> ll(0)
Refining starting values:
Grid node 0:   log likelihood = -1710.6209
Fitting full model:
Iteration 0:   log likelihood = -1710.6209
Iteration 1:   log likelihood = -1555.6881
Iteration 2:   log likelihood = -1521.6018
Iteration 3:   log likelihood = -1519.8718
Iteration 4:   log likelihood = -1519.8596
Iteration 5:   log likelihood = -1519.8596
```

```
Tobit regression                                Number of obs   =        668
                                                Uncensored      =        280
Limits: lower = 0                               Left-censored   =        388
        upper = +inf                            Right-censored  =          0

                                                LR chi2(6)      =      69.14
                                                Prob > chi2     =     0.0000
Log likelihood = -1519.8596                     Pseudo R2       =     0.0222
```

Max_wtp	Coef.	Std. Err.	t	P>|t|	[95% Conf.	Interval]
1.foreign	-15.52233	3.622621	-4.28	0.000	-22.63555	-8.409123
1.highered	6.508825	3.025971	2.15	0.032	.5671678	12.45048
Age	-.5041619	.1020998	-4.94	0.000	-.7046404	-.3036834
Income_AUD	.4125914	.6606609	0.62	0.533	-.884652	1.709835
AdultWhales	1.944485	.3541514	5.49	0.000	1.24909	2.639881
YoungWhales						
Yes	4.203855	3.193804	1.32	0.189	-2.067351	10.47506
_cons	-9.727677	6.236153	-1.56	0.119	-21.9727	2.517347
var(e.Max_wtp)	858.4412	75.31582			722.592	1019.83

What is noticeably different in the results of the tobit model is the statistical significance of the coefficient on the number of adult whales seen on the trip. This variable is a fundamental influence on the maximum willingness to contribute to protection and conservation. It is clear that seeing more whales is positive in terms of its effect on maximum willingness to pay. Interestingly, the income variable is less significant than in the previous models.

Even in the case of a single censoring point, predictions from the tobit model are quite complex because one may want to calculate the regression-like xb with predict but could also compute the predicted probability that $[y|\mathbf{x}_k]$ falls within a particular interval (that may be open-ended on left or right). This may be specified with the pr(a,b) option, where arguments a, b specify the limits of the interval; the missing-value code (.) is taken to mean infinity (of either sign). Another predict option, e(a,b), calculates the expectation $E(y_k) = E(\mathbf{x}'_k\boldsymbol{\beta} + u_k)$ conditional on $[y_k|\mathbf{x}_k]$ being in the a, b interval. Finally, the ystar(a,b) option computes the prediction from (14.3) and (14.4). This prediction is a censored prediction where the threshold is accounted for.

The marginal effects of the tobit model are also quite complex. The estimated coefficients are the marginal effects of a change in x_j on y_k^*, the unobservable latent variable, given by

$$\frac{\partial E(y_k^*|\mathbf{x}_k)}{\partial x_j} = \beta_j$$

If instead the effect on the observable y_k is evaluated, it follows that

$$\frac{\partial E(y_k|\mathbf{x}_k)}{\partial x_j} = \beta_j \times \Pr(a < y_k^* < b)$$

14.5 Censored dependent variables

where a, b are defined as above for `predict`. For instance, for left-censoring at zero, $a = 0, b = +\infty$. Because that probability is at most unity (and will be reduced by a larger proportion of censored observations), the marginal effect of x_j is attenuated from the reported coefficient toward zero.

The `margins` command with the option `predict(pr(0,.))` indicating left-censoring at zero for the survey data gives

```
. quietly tobit Max_wtp i.foreign i.highered Age Income_AUD AdultWhales
> i.YoungWhales, ll(0)
. margins, predict(pr(0,.))
Predictive margins                              Number of obs    =        668
Model VCE    : OIM
Expression   : Pr(Max_wtp>0), predict(pr(0,.))
```

	Margin	Delta-method Std. Err.	z	P>|z|	[95% Conf. Interval]	
_cons	.3134524	.0155316	20.18	0.000	.283011	.3438938

```
. margins, dydx(*) predict(ystar(0,.))
Average marginal effects                        Number of obs    =        668
Model VCE    : OIM
Expression   : E(Max_wtp*|Max_wtp>0), predict(ystar(0,.))
dy/dx w.r.t. : 1.foreign 1.highered Age Income_AUD AdultWhales 1.YoungWhales
```

	dy/dx	Delta-method Std. Err.	z	P>|z|	[95% Conf. Interval]	
1.foreign	-4.32585	.9016468	-4.80	0.000	-6.093046	-2.558655
1.highered	1.984137	.8990005	2.21	0.027	.2221287	3.746146
Age	-.1580308	.0325104	-4.86	0.000	-.22175	-.0943116
Income_AUD	.1293278	.2071726	0.62	0.532	-.2767231	.5353786
AdultWhales	.6095037	.1151707	5.29	0.000	.3837732	.8352342
YoungWhales						
Yes	1.268852	.9291687	1.37	0.172	-.5522852	3.089989

Note: dy/dx for factor levels is the discrete change from the base level.

The first set of margins results indicates that the model predicts that the probability of being uncensored ($\text{Max\_wtp}_k > 0$) is 0.313. The second set of margins results provides the AMEs of increasing each explanatory variable, conditional on being uncensored. An increase in an explanatory variable with a positive coefficient will imply that a left-censored individual is less likely to be censored. The predicted probability of a nonzero value will increase. For a noncensored individual, an increase in x_j will imply that $E(y_k|y_k > 0)$ will increase. In the case above, if an individual is foreign and willing to pay a positive amount, that amount is \$4.87 less than that of an Australian citizen. Likewise, an individual with higher education is willing to pay \$2.04 more than an individual with a high school degree. Although still statistically insignificant at the conventional levels, the influence of `YoungWhales` increases in this approach relative to the other models considered in the chapter.

Exercises

1. Although the linear probability model enjoys some popularity because of its simplicity, it has also been criticized as not being able to yield reliable estimates of the true marginal effects. Consider the model

$$y_k^* = \beta_0 + \beta_1 x_k + u_k$$

$$y_k = \begin{cases} = 1 & y_k^* > 0 \\ = 0 & y_k^* \leq 0 \end{cases}$$

with $\beta_0 = -1$, $\beta_1 = 2$, $x_k \sim \text{uniform}(0, 1)$ and $u_k \sim N(0, 1)$.

   a. Run 1,000 repetitions of the following Monte Carlo experiment:
      i. Simulate data from the model with the number of observations K set to 200.
      ii. Estimate a linear regression of y_k on x_k (linear probability model), and save the estimate of β_1 (marginal effect of x_k) and its standard error.
      iii. Fit a binomial probit model, and save the marginal effect of x_k and its standard error computed at the mean of x_k (use the post option of margins [see [R] margins]).

   b. Compute the means of the saved marginal effects and standard errors. Graph the empirical distributions of the coefficient estimates and the estimated standard errors, and comment on the results.

   c. Repeat exercise 1a and exercise 1b with $K = 500$. Comment on the results.

2. Although the ratification of international environmental agreements on pollution are crucial for solving global pollution problems, relatively little is known about the factors that contribute to ratification (or the absence of it). Arguably, a government will be more likely to ratify an international environmental agreement the stronger the environmental lobby is in that country, an effect that is reinforced if the government is more corruptible. A subset of their data panel set on 170 countries for the years 1998 to 2002 is found in corruption.dta. The main variables of interest are as follows:

14.5 Censored dependent variables

Variable	Description
ratify	1 if country ratified the Kyoto protocol, 0 otherwise
seller	former socialist eastern European country (low abatement costs)
icc	1 if country is member of International Chamber of Commerce, 0 otherwise
nngo	number of national environmental nongovernmental organizations that are members of the World Conservation Union
annex1	1 if OECD or former eastern European country, 0 otherwise
pop	population in 100 millions
island	ratio coastline/area
gdppc	gross domestic product per capita
co2pc	per capita CO_2 emissions
democracy	Freedom House democracy ratings (1–3 increasing in democracy)
industry	share of the labor force employed in the industrial sector
integrity	international country risk guide measurement of corruption/integrity
fuel	share of fuel exports as a percentage of total merchandise exports

  a. Compute summary statistics for the variables in the dataset, and compare your results with table 2 of Fredriksson, Neumayer, and Ujhelyi (2007).
  b. Fit linear probability models in which ratify is regressed against the following explanatory variables:
     i. annex1, island, pop, gdppc, democracy, nngo, industry, and integrity1;
     ii. the variables in exercise 2bi together with year dummy variables; and
     iii. the variables in exercise 2bii together with interaction effects between nngo and integrity1 and also industry and integrity1.

  In each case, interpret the results and test for the statistical significance of the interaction terms both individually and as a group.

  c. Fit binary logit models in which ratify is modeled using the following explanatory variables:
     i. annex1, island, pop, gdppc, democracy, nngo, industry, and integrity1;
     ii. the variables in exercise 2ci together with year dummy variables; and
     iii. the variables in exercise 2cii together with interaction effects between nngo and integrity1 and also industry and integrity1.

  In each case, compute the marginal effects and interpret the results. Test for the statistical significance of the interaction terms both individually and as a group.

d. For the model in exercise 2ciii, estimate the log odds-ratio for each of the variables, and interpret the results.
e. Fit a binary logit model for `ratify` with `annex1`, `island`, `pop`, `gdppc`, and `democracy` together with year dummies and interaction terms between `nngo` and `integrity1` and `fuel` and `integrity1`. Interpret the results. Test for the significance of the interaction between `fuel` and `integrity1`. What do you conclude?

15 Fractional integration

There is a class of stochastic processes known as long-memory processes that is easily confused with nonstationary integrated processes. Long-memory processes occur in hydrology, astronomy, biology, computer networks, chemistry, agriculture, geophysics, economics, and finance. In simple terms, long-memory processes have autocorrelations that do not decay exponentially like those of stationary processes (Granger and Joyeux 1980; Hosking 1981). Using the terminology of chapter 7, a long-memory series exhibits too much long-range dependence to be classified as $I(0)$ but is also not $I(1)$; in other words, long-memory processes represent a bridge between those presenting infinite memory (nonstationary) and short memory (stationary). Time series that display long memory are known as fractionally integrated or $I(d)$ variables, where d is no longer an integer and falls in the interval $-1/2 \leq d \leq 1$ but excludes 0.

The literature on long memory and, in particular, on fractional integration is fairly well developed in econometrics and particularly financial econometrics: see, for example, the survey by Baillie (1996). In comparison, the econometrics of fractional integration in the context of environmental econometrics is relatively underdeveloped. Exceptions to this rule are examinations of the long-memory properties of the level of the Nile over a millennium (Hurst 1951), sea level pressure (Percival, Overland, and Morfjeld 2001; Stephenson, Pravan, and Bojariu 2000), global temperature (Bloomfield 1992), rainfall (Motanari, Rosso, and Taqqu 1996), and wind power (Haslett and Raftery 1989). There is no doubt that many environmental time series exhibit persistence, but there is still debate on the degree of this persistence. Consequently, the relaxation of the concepts of integration and cointegration to include noninteger orders is a potentially useful tool in the environmental econometrics toolkit.

15.1 Mean sea levels and global temperature

As noted in chapter 9, the thermal expansion of oceans and melting of land-based ice implies that global warming likely contributed to the sea level rise observed during the 20th century. The structural time-series model used to estimate the relationship between these two variables provided statistical support for the impact of global temperature on the increase in the mean sea level. However, this empirical approach ignores the possibility of spurious inference that is likely to be drawn when one nonstationary process is regressed on another. The effects of spurious regression may still be present if the data possess long memory. Figure 15.1 plots monthly time-series data for the period January 1880 to December 2009.

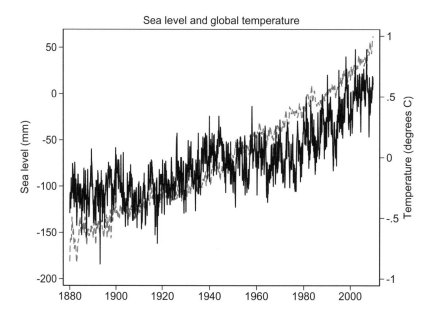

Figure 15.1. Time-series plots of mean global sea level and temperature. The data are monthly for the period January 1880 to December 2009.

There is enough evidence in this plot to suggest that there are strong temporal effects in both series, which warrants further investigation. The analysis of the relationship between mean sea level and global temperature now proceeds in three steps, which follows quite closely the modeling strategies discussed in chapter 7. First, it must be established whether the time series exhibit long memory. This requires that the definition of long memory be outlined and methods for detecting long memory be examined. Second, if there is evidence of long memory, the parameter governing the long-range dependence must be estimated. Third, if it can also be established that both series exhibit identical long-memory characteristics, then this fact can be used to build a model to estimate the long-run effect of global temperature on mean sea level, an approach adopted by Ventosa-Santaulària, Heres, and Martínez-Hernández (2014).

15.2 Autocorrelations and long memory

Once the restriction that the differencing parameter d in a time-series context takes only integer values is relaxed, the autoregressive fractional integration moving-average [ARFIMA(p, d, q)] class of model is introduced. The ARFIMA(p, d, q) model is defined as

$$\Delta^d y_t = \phi_1 \Delta^d y_{t-1} + \cdots + \phi_p \Delta^d y_{t-p} + v_t + \psi_1 v_{t-1} + \cdots + \psi_1 v_{t-q} \tag{15.1}$$

15.2 Autocorrelations and long memory

where d is no longer required to be an integer. The notion of fractional integration requires a slight change to the difference operator, which is now defined by the binomial expansion

$$\Delta^d y_t = y_t - dy_{t-1} + \frac{d(d-1)}{2!} y_{t-2} - \frac{d(d-1)(d-2)}{3!} y_{t-3} + \cdots$$

For integer values $d = 1$ and $d = 2$, we have immediately that

$$\Delta y_t = y_t - y_{t-1} \qquad \Delta^2 y_t = y_t - 2y_{t-1} + y_{t-2}$$

It turns out that the value of d and the notion of long memory in a time series as summarized by its autocorrelations are intimately related (Hosking 1981).

1. For $d = 0$, the process is $I(0)$ and exhibits short memory corresponding to stationary and invertible autoregressive moving-average (ARMA) modeling. The autocorrelations exhibit exponential decay.
2. For $d = 1$, the process is $I(1)$. In theory, the autocorrelations are not defined, because the variance of the process is infinite.
3. If $d \in (0, 1)$, the autocorrelations take longer to disappear than in the $I(0)$ case and decay hyperbolically to zero in contrast with the faster decay of a stationary ARMA process.

Figure 15.2 illustrates the autocorrelation functions of the levels of the variables y_{it} that are generated by the stochastic processes

$$\Delta y_{1t} = 0.5 y_{1\,t-1} + u_{1t} \qquad \text{(Stationary AR(1) model)}$$
$$\Delta^d y_{2t} = u_{2t}, \quad d = 0.5, \qquad \text{(Fractionally integrated model)}$$

with $u_{it} \sim N(0, 1)$ in all cases.

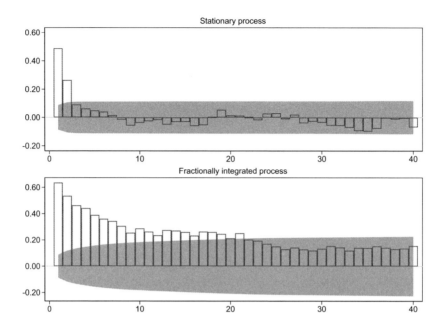

Figure 15.2. The autocorrelation function of simulated $I(0)$ and $I(d)$ processes with $T = 500$

As required by theory, the autocorrelations of the stationary first-order autoregressive [AR(1)] process diminish exponentially and approach zero after a few lags. Any stationary ARMA process therefore exhibits short memory. The fractionally integrated model has autocorrelations that exhibit hyperbolic decay, where the autocorrelations die out faster than the integrated process at short horizons, but the speed of decay slows down at longer lags.

The autocorrelation function for mean sea level and mean global temperature after detrending using a simple linear trend are shown in figure 15.3. The autocorrelations for both series exhibit a similar pattern to that of the fractionally integrated series in figure 15.2, namely, hyperbolic decay.

15.3 Testing for long memory

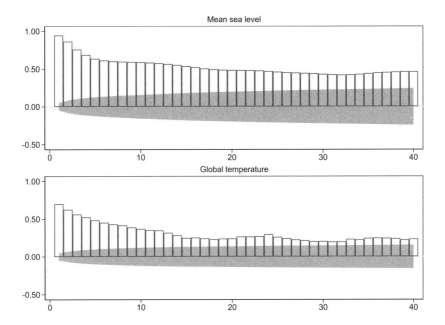

Figure 15.3. The autocorrelation functions of mean sea level and global temperature. The data are monthly for the period January 1880 to December 2009.

15.3 Testing for long memory

The classical test for long memory is the range over scale, or R/S statistic, devised by Hurst (1951) and Mandelbrot (1972). The test statistic is based on the range of partial sums of deviations of a time series from its mean, rescaled by the standard deviation of the series. For a sample of T values $\{x_1, x_2, \ldots x_T\}$, the statistic is computed as

$$Q_T = \frac{1}{s_T} \left\{ \max_{1 \leq k \leq T} \sum_{j=1}^{k} (x_j - \bar{x}_T) - \min_{1 \leq k \leq T} \sum_{j=1}^{k} (x_j - \bar{x}_T) \right\}$$

where s_T is the maximum likelihood estimator of the standard deviation of x. The first term in the square bracket is the maximum of the partial sums of the first k deviations of x_j from the full-sample mean, which is a nonnegative quantity. The second bracketed term is the corresponding minimum, which is a nonpositive quantity. The difference of these two quantities is thus nonnegative, so $Q_T > 0$. Empirical studies have demonstrated that the R/S statistic has the ability to detect long-range dependence in the data.

Like many other estimators of long-range dependence, the R/S statistic has been shown to be excessively sensitive to short-range dependence, or short-memory, features of the data. Lo (1991) shows that a sizable AR(1) component in the data generating process will seriously bias the R/S statistic. He modifies the R/S statistic to account for the effect of short-range dependence by applying a Newey–West correction (see the discussion concerning the computation of standard errors in chapter 2) to derive a consistent estimate of the long-range variance of the time series.

The `lomodrs` command (Baum and Rööm 2000) performs the modified rescaled range test for long-range dependence of a time series (Baum and Rööm 2001). For `maxlag` > 0, the denominator of the statistic is computed as the Newey–West estimate of the long-run variance of the series. If `maxlag` is set to zero, the test performed is the classical Hurst–Mandelbrot rescaled-range statistic. Critical values for the test are taken from Lo (1991, table II).

```
. use http://www.stata-press.com/data/eeus/fci
. lomodrs Sea, maxlag(0)

Hurst-Mandelbrot Classical R/S test for Sea

Critical values for H0: Sea is not long-range dependent

90%: [ 0.861, 1.747 ]
95%: [ 0.809, 1.862 ]
99%: [ 0.721, 2.098 ]

Test statistic:       4.97     N = 130
. lomodrs Temp, maxlag(0)

Hurst-Mandelbrot Classical R/S test for Temp

Critical values for H0: Temp is not long-range dependent

90%: [ 0.861, 1.747 ]
95%: [ 0.809, 1.862 ]
99%: [ 0.721, 2.098 ]

Test statistic:       4.23     N = 130
. lomodrs Sea

Lo Modified R/S test for Sea

Critical values for H0: Sea is not long-range dependent

90%: [ 0.861, 1.747 ]
95%: [ 0.809, 1.862 ]
99%: [ 0.721, 2.098 ]

Test statistic:       2.61  (99 lags via Andrews criterion)   N = 31
```

15.3 Testing for long memory

```
. lomodrs Temp

Lo Modified R/S test for Temp

Critical values for H0: Temp is not long-range dependent

90%: [ 0.861, 1.747 ]
95%: [ 0.809, 1.862 ]
99%: [ 0.721, 2.098 ]

Test statistic:    1.11   (32 lags via Andrews criterion)   N = 98
```

The classical R/S test rejects the null hypothesis of no long-range dependence for both series at the 1% level of significance. The statistics returned by Lo's modified R/S test are much smaller but still reject the null hypothesis in favor of long-range dependence at the 5% level.

The Kwiatkowski, Phillips, Schmidt, and Shin (KPSS) test, discussed in section 6.4.1, can also be considered a test for long memory in that a rejection of its null hypothesis of trend stationarity implies that the series is not $I(0)$, but $I(d), d > 0$. In that sense, it does not necessarily imply that the series is a unit-root process, $d = 1$, but merely that $d > 0$.

```
. kpss Sea, qs

KPSS test for Sea

Maxlag = 4 chosen by Schwert criterion
Autocovariances weighted by Quadratic Spectral kernel

Critical values for H0: Sea is trend stationary

10%: 0.119   5% : 0.146   2.5%: 0.176   1% : 0.216

Lag order      Test statistic
     0              1.8
     1              1.49
     2              .818
     3              .591
     4              .469
```

```
. kpss Temp, qs

KPSS test for Temp

Maxlag = 4 chosen by Schwert criterion
Autocovariances weighted by Quadratic Spectral kernel

Critical values for H0: Temp is trend stationary

10%: 0.119   5% : 0.146   2.5%: 0.176   1% : 0.216

  Lag order    Test statistic
      0              .863
      1              .74
      2              .442
      3              .335
      4              .274
```

The KPSS test rejects trend stationarity, $d = 0$, at any level of confidence. Both tests therefore suggest that mean sea level and mean global temperature exhibit long-range dependence, or long memory.

15.4 Estimating d in the frequency domain

One approach to providing consistent and asymptotically normal estimates of the fractional differencing parameter, d, without fully specifying the ARMA components of the model requires shifting from the time domain to the frequency domain. Once an estimate of d is available, an ARMA model is then fit to the transformed series, $\Delta^{\widehat{d}} y_t$, to obtain consistent estimators of the remaining model parameters.

Up to this point, all the techniques discussed in the book are concerned with how a stochastic process changes over time. In the frequency domain, the behavior of the process is decomposed into contributions from a range of frequencies. At a given frequency ω_j, the link between the time domain and the frequency domain is given by the discrete Fourier transform

$$S_x(\omega_j) = \frac{1}{\sqrt{2\pi T}} \sum_{t=0}^{T} x_t e^{-i\omega_j t}$$

where $S_x(\omega_j)$ is known as the spectral density[1] of the series x_t at frequency ω_j. In essence, the value of $S_x(\omega_j)$ gives an indication of the importance of frequency component ω_j in the time series x_t. The squared modulus of the discrete Fourier transform given by

$$I_x(\omega_j) = \frac{|S_y(\omega_j)|^2}{T}$$

---

1. Stata has the capacity to estimate the spectral density of a stationary time series, but rather than using a discrete Fourier transform, it is reconstructed using the parameters of a previously fitted time-series model.

15.4 Estimating d in the frequency domain

is known as the periodogram, and it shows the power of the signal at each frequency. The periodogram is also important in construction of the frequency-domain estimators of d.

By convention, it is usual to consider only positive frequencies at which to evaluate the spectral density (Hamilton 1994, 160–161). A periodic function that completes one full cycle (2π radians) over the length of a sample of size T is given by $2\pi/T$, which is known as the fundamental frequency. A set of frequencies, ω_j, given by

$$\omega_j = \frac{2\pi j}{T}, \qquad \text{for } j = 0, 1, 2, \ldots, T/2$$

trace out a set of frequencies defined on the interval $[0, \pi]$, known as the Fourier frequencies, which are used in this context.

Consider the situation in which $\Delta^d y_t \equiv v_t$. The relationship between the spectral density of y_t and that of v_t is given by (Brockwell and Davis 1991, theorem 13.2.2)

$$S_y(\omega) = \left|1 - e^{-i\omega}\right|^{-2d} S_v(\omega)$$

which, after evaluating at frequency ω_j and taking logarithms, becomes

$$\log S_y(\omega_j) = -d \log \left|1 - e^{-i\omega_j}\right|^2 + \log S_v(\omega_j)$$

As it stands, this expression is not amenable to estimation, because $\log S_v(\omega_j)$ is unobservable, but Geweke and Porter-Hudak (1983) manipulate it into a form that allows the fractional difference parameter d to be estimated. Specifically, they show that in the vicinity of the zero frequency, d can be estimated from the linear regression equation

$$\log \left\{I_y(\omega_j)\right\} = \beta_0 + \beta_1 \log \left|1 - e^{-i\omega_j}\right|^2 + \epsilon_j \qquad (15.2)$$

with $d = -\beta_1$ and $j = 1, \ldots, M$, and $I_y(\omega_j)$ is the periodogram of y_t at ω_j. Note that if too few frequencies are used to estimate d, small-sample problems are encountered. If too many are included, medium and high frequencies may contaminate the estimate. The choice $T^{-0.5}$ is often used, although it is also common practice to compute the test for a range of indices for T ranging from -0.40 to -0.75. The command gphudak (Baum and Wiggins 2000) implements the Geweke–Porter-Hudak (GPH) estimator and performs a test of the null hypothesis $H_0 : d = 0$.

```
. use http://www.stata-press.com/data/eeus/fractionalcointegration
. tsset datevec
        time variable:  datevec, 1880m1 to 2009m10
                delta:  1 month
. generate trend = _n
. regress Sea trend
  (output omitted)
. predict double sea, res
```

```
. regress Temp trend
  (output omitted)
. predict double temp, res
. gphudak sea, powers(0.4(0.1)0.7)
GPH estimate of fractional differencing parameter
-----------------------------------------------------------------------
                                              Asy.
 Power   Ords   Est d   StdErr  t(H0: d=0) P>|t|  StdErr  z(H0: d=0) P>|z|
-----------------------------------------------------------------------
  .4      19   .745241  .09296   8.0168   0.000   .194    3.8418   0.000
  .5      40   .61581   .0884    6.9660   0.000   .1194   5.1574   0.000
  .6      83   .572206  .0711    8.0481   0.000   .07768  7.3663   0.000
  .7     172   .471417  .04833   9.7536   0.000   .05206  9.0560   0.000
-----------------------------------------------------------------------

. gphudak temp, powers(0.4(0.1)0.7)
GPH estimate of fractional differencing parameter
-----------------------------------------------------------------------
                                              Asy.
 Power   Ords   Est d   StdErr  t(H0: d=0) P>|t|  StdErr  z(H0: d=0) P>|z|
-----------------------------------------------------------------------
  .4      19   .477422  .1953    2.4450   0.026   .194    2.4612   0.014
  .5      40   .312223  .1307    2.3886   0.022   .1194   2.6149   0.009
  .6      83   .386991  .08057   4.8031   0.000   .07768  4.9819   0.000
  .7     172   .460974  .05253   8.7761   0.000   .05206  8.8554   0.000
-----------------------------------------------------------------------
```

Over a range of choices for the number of frequencies to include in the test regression, here ranging from $T^{-0.4}$ to $T^{-0.7}$, the point estimate of d for the detrended variables differs for powers of T but converges to ≈ 0.46 at $T^{-0.7}$. Furthermore, the GPH test clearly rejects the null hypothesis of $d = 0$. Note that two estimates of the d coefficient's standard error are commonly used: the usual ordinary least-squares regression standard error, which gives rise to standard t tests, and an asymptotic standard error based upon the theoretical variance of the log-periodogram of $(\pi^2)/6$. The test statistic of the null hypothesis of $\widetilde{d} = 0$ based upon this asymptotic standard error has a standard normal distribution under the null hypothesis.

There are two important refinements of the GPH estimator:

1. Phillips (1999, 2007) modifies the GPH estimator by using an exact representation of the discrete-time Fourier transform for a fractionally integrated process. The modified transform and periodogram are given by

$$\widetilde{S}_y(\omega_j) = S_y(\omega_j) + \frac{e^{i\omega_j}}{e^{i\omega_j} - 1}\left(\frac{y_{T-1} - y_0}{\sqrt{2\pi}}\right) \quad (15.3)$$

$$\widetilde{I}_y(\omega_j) = \frac{|\widetilde{S}_y(\omega_j)|^2}{T}$$

where the second term on the right-hand side of (15.3) is a correction term for the fractional integration. The test regression is then

$$\log\left\{\widetilde{I}_y(\omega_j)\right\} = \beta_0 + \beta_1 \log\left|1 - e^{-i\omega_j}\right|^2 + \epsilon_j$$

15.4 Estimating d in the frequency domain

The asymptotic variance of the modified log periodogram is given by $\pi^2/24$. If the standard error used on this theoretical variance is used to construct the t test of $H_0: d = 0$, then the statistic has a standard normal distribution under the null hypothesis. The `modlpr` command (Baum and Wiggins 2000) computes the modified estimator of d. The command also provides a test of the null hypothesis of nonstationarity, $H_0: d = 1$.

2. Robinson (1995) uses a variant of the GPH estimator making use of the fact that for stationary Gaussian time series

$$\lim_{\omega \to 0} |1 - e^{-i\omega}|^{-2d} \approx |\omega|^{-2d}\{1 + o(1)\}$$

This means that the term on the right-hand side of the regression (15.2) becomes $-2\log \omega_j$. Although the GPH estimator is a consistent estimator of d for a more general class of time-series models than fractionally differenced time series, the simpler form for the regressor used by Robinson does not affect the asymptotic properties of $\widehat{d}$ in fractionally integrated models (Andrews and Guggenberger 2003). Robinson also trims out small values of j from the regression and allows for averaging over adjacent values of j in the log-periodogram. He argues that these modifications offer modest asymptotic efficiency gains over the GPH estimator. The `roblpr` command (Baum and Wiggins 2000) computes the Robinson modified log-periodogram estimate of d.

Using Phillips' modified GPH estimator, d, can be clearly distinguished from both 0 and 1 for each of the time series. This indicates that neither series is $I(0)$ nor $I(1)$ and that long-memory behavior is an appropriate conclusion. The point estimates of d are remarkably similar to those returned for the detrended version of the GPH estimator, suggesting that both series are covariance stationary.

```
. modlpr Sea, powers(0.4(0.1)0.7)
Modified LPR estimate of fractional differencing parameter for Sea
------------------------------------------------------------------
Power   Ords    Est d     Std Err    t(H0: d=0)   P>|t|    z(H0: d=1)   P>|z|
------------------------------------------------------------------
 .4      18    .9218841   .2361814    3.9033      0.001    -0.5168      0.605
 .5      39    .6135026   .1324401    4.6323      0.000    -3.7639      0.000
 .6      82    .5449442   .0819157    6.6525      0.000    -6.4258      0.000
 .7     171    .4766134   .0526453    9.0533      0.000   -10.6728      0.000
------------------------------------------------------------------

. modlpr Temp, powers(0.4(0.1)0.7)
Modified LPR estimate of fractional differencing parameter for Temp
------------------------------------------------------------------
Power   Ords    Est d     Std Err    t(H0: d=0)   P>|t|    z(H0: d=1)   P>|z|
------------------------------------------------------------------
 .4      18    .4631596   .2764676    1.6753      0.111    -3.5517      0.000
 .5      39    .2535564   .1412067    1.7956      0.080    -7.2692      0.000
 .6      82    .3662758   .0835336    4.3848      0.000    -8.9488      0.000
 .7     171    .4585444   .0534592    8.5775      0.000   -11.0412      0.000
------------------------------------------------------------------
```

Robinson's estimator of the series also clearly rejects its null hypothesis of $d = 0$ and shows Sea to be mean reverting but not covariance stationary.

```
. roblpr Sea, power(0.7 0.75:0.95)
Robinson estimates of fractional differencing parameter for Sea
-----------------------------------------------------------------
Power    Ords      Est d      Std Err   t(H0: d=0)    P>|t|
-----------------------------------------------------------------
  .7      171    .9237293    .0350966    26.3196      0.000
 .75      247    .9065404    .0309762    29.2657      0.000
  .8      359     .917925    .0235231    39.0223      0.000
 .85      517    .9001612    .0165281    54.4623      0.000
  .9      747    .8584385    .0127106    67.5373      0.000
 .95     1079    .7541746    .0111689    67.5247      0.000
-----------------------------------------------------------------

. roblpr Temp, power(0.7 0.75:0.95)
Robinson estimates of fractional differencing parameter for Temp
-----------------------------------------------------------------
Power    Ords      Est d      Std Err   t(H0: d=0)    P>|t|
-----------------------------------------------------------------
  .7      171     .663615    .0487188    13.6213      0.000
 .75      247     .670809    .0446191    15.0341      0.000
  .8      359    .6502788    .0386477    16.8258      0.000
 .85      517    .5918933    .0315372    18.7681      0.000
  .9      747     .529059    .0260836    20.2832      0.000
 .95     1079    .4386199    .0213949    20.5012      0.000
-----------------------------------------------------------------
```

In some circumstances, these frequency domain methods for estimating d start by detrending the data or first-differencing the data, so the problem is to estimate the fractional difference operator, $\tilde{d}$, for the series $\Delta^{\tilde{d}} x_t$. If the level of fractional integration of Δx_t is $\tilde{d}$, then x_t is integrated of order $1 + \tilde{d}$. It follows that if $\tilde{d} = 0$, then x_t has a unit root. The GPH estimator performed on the first difference of the series gives the following results:

```
. gphudak D.Sea, powers(0.4(0.1)0.7)
GPH estimate of fractional differencing parameter
---------------------------------------------------------------------------------
                                                        Asy.
Power   Ords    Est d     StdErr  t(H0: d=0)  P>|t|    StdErr  z(H0: d=0)  P>|z|
---------------------------------------------------------------------------------
 .4      19    .045245     .1828    0.2475    0.808     .194     0.2332    0.816
 .5      40   -.247045     .1253   -1.9710    0.056    .1194    -2.0690    0.039
 .6      83    -.38321     .07055  -5.4321    0.000   .07768    -4.9332    0.000
 .7     172   -.450413     .04986  -9.0333    0.000   .05206    -8.6524    0.000
---------------------------------------------------------------------------------
```

```
. gphudak D.Temp, powers(0.4(0.1)0.7)
GPH estimate of fractional differencing parameter
--------------------------------------------------------------------------------
                                                      Asy.
Power    Ords    Est d      StdErr    t(H0: d=0)  P>|t|   StdErr   z(H0: d=0)  P>|z|
--------------------------------------------------------------------------------
 .4       19   -.514832     .2786     -1.8476     0.083   .194     -2.6540     0.008
 .5       40   -.739729     .1408     -5.2550     0.000   .1194    -6.1952     0.000
 .6       83   -.617632     .08592    -7.1886     0.000   .07768   -7.9510     0.000
 .7      172   -.546858     .05236    -10.4443    0.000   .05206   -10.5052    0.000
--------------------------------------------------------------------------------
```

The estimates differ from those returned on the level of the series, but the test of $d = 0$ is still rejected. The estimate of d for the two series does not converge as nicely for the two series with ΔSea returning an estimate of approximately $1 - 0.45 = 0.55$, while that of ΔTemp is approximately $1 - 0.55 = 0.45$. These results suggest that Sea is mean reverting but not covariance stationary, while Temp is covariance stationary.[2]

15.5 Maximum likelihood estimation of the ARFIMA model

Consistent and asymptotically normal estimates of the fully specified ARFIMA(p, d, q) model are obtained by maximizing the log-likelihood function, including an estimate of the d parameter. Consider the ARFIMA(p, d, q) model in (15.1), where y_t is taken to be a mean-zero process. Estimation by maximum likelihood requires the construction of the covariance matrix of $\mathbf{y} = [\, y_1 \; y_2 \; \ldots \; y_T \,]'$ given by

$$E(\mathbf{yy'}) = \mathbf{V} = \begin{bmatrix} \gamma_0 & \gamma_1 & \cdots & \gamma_{T-1} \\ \gamma_1 & \gamma_0 & \ddots & \gamma_{T-2} \\ \vdots & \ddots & \ddots & \vdots \\ \gamma_{T-1} & \gamma_{T-2} & \cdots & \gamma_0 \end{bmatrix}$$

in which the autocovariances are often scaled by the variance of the disturbances (15.1), σ_v^2, so that $\gamma_k = E(y_t y_{t-k})/\sigma_v^2$. The matrix $\mathbf{V}$ is a symmetric Toeplitz matrix.

Based on the assumption that $\mathbf{y} \sim N(0, \mathbf{V})$, maximum likelihood estimators are obtained by maximizing the log-likelihood function

$$\log L = -\frac{T}{2} \log(2\pi) - \log |\mathbf{V}| - \frac{1}{2} \mathbf{y'} \mathbf{V}^{-1} \mathbf{y}$$

---

2. For brevity, an additional frequency-domain estimator, the local Whittle estimator (Whittle 1951, 1962) is not presented here. Full details can be found in Baum, Hurn, and Lindsay (2020).

Construction of the likelihood requires a practical method in which to compute the autocovariances of the fractionally integrated process, γ_k, to construct $\mathbf{V}$. The straightforward way to compute the values of γ_k is to recognize that because y_t is stationary, it can be represented as an infinite-order moving-average (MA) model

$$y_t = \sum_{j=0}^{\infty} \varphi_j v_t$$

where $\varphi_0 = 1$ and the remaining φ coefficients are complex combinations of the original ϕ and ψ parameters on the AR and MA components of the ARFIMA(p, d, q) model in (15.1). The autocovariances can then be constructed as

$$\gamma_k = \sum_{j=0}^{\infty} \varphi_j \varphi_{j+k} \tag{15.4}$$

In turns out that because of the persistence in the γ_k of a fractionally integrated process, this simple approach is infeasible because many terms are needed on the right-hand side of (15.4) to yield an accurate approximation of γ_k.

Based on the original work of Hosking (1981), Sowell (1992) achieved a major improvement in the speed of computation of the autocovariances, γ_k. Consequently, the official Stata command arfima (see [TS] **arfima**) implements the full maximum-likelihood estimation of the ARFIMA(p, d, q) model as proposed by Sowell. This approach requires choosing the values of p and q and fitting the full ARFIMA model conditional on those choices. This involves the challenge of choosing an appropriate ARMA specification.

For the Sea series, limited examination of the information criteria postestimation suggests that an ARFIMA$(2, d, 2)$ model might be appropriate. For example, the information criteria returned for ARFIMA$(2, d, 2)$ and ARFIMA$(4, d, 4)$ models are as follows, and both suggest that the former specification is to be preferred.

```
. quietly arfima Sea, ar(1/2) ma(1/2)
. estat ic
Akaike's information criterion and Bayesian information criterion
```

Model	N	ll(null)	ll(model)	df	AIC	BIC
.	1,558	.	-3604.306	7	7222.612	7260.07

Note: BIC uses N = number of observations. See [R] BIC note.

```
. quietly arfima Sea, ar(1/4) ma(1/4)
. estat ic
Akaike's information criterion and Bayesian information criterion
```

Model	N	ll(null)	ll(model)	df	AIC	BIC
.	1,558	.	-3601.792	11	7225.583	7284.446

Note: BIC uses N = number of observations. See [R] BIC note.

15.5 Maximum likelihood estimation of the ARFIMA model

The maximum likelihood estimates of the ARFIMA$(2, d, 2)$ model applied to the Sea series are

```
. arfima Sea, ar(1/2) ma(1/2) nolog vsquish
ARFIMA regression
Sample: 1880m1 - 2009m10                        Number of obs    =      1,558
                                                Wald chi2(5)     =   64319.30
Log likelihood = -3604.3061                     Prob > chi2      =     0.0000
```

Sea	Coef.	OIM Std. Err.	z	P>|z|	[95% Conf. Interval]	
Sea						
_cons	-66.08779	135.2249	-0.49	0.625	-331.1238	198.9482
ARFIMA						
ar						
L1.	.3229797	.0350799	9.21	0.000	.2542243	.3917351
L2.	.250393	.0352438	7.10	0.000	.1813165	.3194696
ma						
L1.	.6191094	.0259352	23.87	0.000	.5682773	.6699415
L2.	.6978272	.019849	35.16	0.000	.6589238	.7367306
d	.4985044	.0021135	235.86	0.000	.494362	.5026468
/sigma2	5.941244	.2129461	27.90	0.000	5.523878	6.358611

Note: The test of the variance against zero is one sided, and the two-sided
confidence interval is truncated at zero.

All the AR and MA terms are statistically significant. The estimate of d is 0.498 and is broadly in agreement with the GPH and Phillips frequency domain estimators. Note, however, that based on inspection of the 95% confidence interval, the hypothesis of $d < 0.5$ cannot be rejected.

The Temp series does not seem to support any lags being included in the ARFIMA model, and consequently an ARFIMA$(0, d, 0)$ model is fit.

```
. arfima Temp, nolog vsquish
ARFIMA regression
Sample: 1880m1 - 2009m10                        Number of obs    =      1,558
                                                Wald chi2(1)     =    5494.15
Log likelihood =  1208.2758                     Prob > chi2      =     0.0000
```

Temp	Coef.	OIM Std. Err.	z	P>|z|	[95% Conf. Interval]	
Temp						
_cons	.0064788	.5941636	0.01	0.991	-1.15806	1.171018
ARFIMA						
d	.49479	.0066753	74.12	0.000	.4817066	.5078733
/sigma2	.0123713	.0004434	27.90	0.000	.0115023	.0132403

Note: The test of the variance against zero is one sided, and the two-sided
confidence interval is truncated at zero.

The estimate of d is 0.495, and the hypothesis that $d < 0.5$ is also rejected in this instance.

The results of the frequency domain estimators and the maximum likelihood therefore suggest two conclusions for each of the Sea and Temp series. First, there is strong evidence that both of these series have long memory because the hypothesis that $d = 0$ is strongly rejected. This result is supportive of a widely held view that many environmental time series exhibit long-memory characteristics. Second, while the estimate of the fractional difference parameters of both of these series is in the vicinity of 0.5, there is no concrete evidence to support the hypothesis that $d < 0.5$. The series may therefore not be covariance stationary.

15.6 Fractional cointegration

The concept of cointegration discussed in chapter 7 may also be generalized to noninteger orders of integration. If two series are fractionally integrated of order $I(d)$, but a linear combination of the two is integrated of order $I(d - b)$ where $b > 0$, then the two series are fractionally cointegrated. There is therefore a natural extension of the Engle–Granger two-step procedure to the case of noninteger degrees of integration, as presented in Caporale and Gil-Alana (2004a,b). The first stage involves testing the individual series for the same order of fractional integration. At the second stage, a cointegrating regression is estimated, so the cointegrating parameter, β, is identified, along with the fractional differencing parameter, d. In the case where $d = 1$, the ordinary least-squares estimator, $\widehat{\beta}$ is super-consistent and converges at the rate T rather than at the usual rate of $\sqrt{T}$. The properties of the estimator in the case of fractional orders of integration are more difficult to establish. See, for example, Robinson and Marinucci (2001). There are methods for estimating β and d jointly (Nielsen 2007; Hualde and Robinson 2007) that are not considered here.

It is apparent from the results of the two-step estimators of d discussed previously that there is some doubt as to whether the mean sea level and mean global temperature share the same order of fractional integration. It will simply be assumed here that there is sufficient evidence to proceed with the analysis. The Robinson semiparametric estimate of the long-memory (fractional integration) parameter d computed by `roblpr` can be applied to a set of time series, with the fractional difference parameter for each series being estimated from a single log-periodogram regression, which allows the intercept and slope to differ for each series. It is also possible to constrain d to be the same for all the time series. Following this approach for Sea and Temp yields

15.6 Fractional cointegration

```
. constraint 1 Sea = Temp
. roblpr Sea Temp, constraints(1)
Robinson estimates of fractional differencing parameters
Power =   .9                              Ords       =      747
-----------------------------------------------------------------
Variable     |      Est d       Std Err      t       P>|t|
-------------+---------------------------------------------------
        Sea  |    .6937487     .0151171   45.8915    0.000
       Temp  |    .6937487     .0151171   45.8915    0.000
-----------------------------------------------------------------
```

These results indicate that the two series are nonstationary in a fractional sense. Furthermore, the estimate $\widehat{d} = 0.69$ is almost identical to the estimate reported by Ventosa-Santaulària, Heres, and Martínez-Hernández (2014), using a different estimation method. Estimating the cointegrating parameter β and testing the residuals for their fractional order of integration yields the following results:

```
. regress Sea Temp
      Source |       SS           df       MS      Number of obs   =     1,558
-------------+----------------------------------   F(1, 1556)      =   3502.68
       Model |  3675897.63         1   3675897.63  Prob > F        =    0.0000
    Residual |  1632949.22     1,556   1049.45322  R-squared       =    0.6924
-------------+----------------------------------   Adj R-squared   =    0.6922
       Total |  5308846.85     1,557    3409.664   Root MSE        =    32.395

         Sea |      Coef.   Std. Err.      t    P>|t|    [95% Conf. Interval]
        Temp |    171.1312   2.891539   59.18    0.000    165.4595    176.8029
       _cons |   -63.47391   .8262528  -76.82    0.000   -65.09459   -61.85322

. capture drop uhat
. predict double uhat, resid
. roblpr uhat
Robinson estimates of fractional differencing parameter for uhat
---------------------------------------------------------
Power   Ords      Est d      Std Err   t(H0: d=0)   P>|t|
---------------------------------------------------------
  .9    747     .4646748    .0248526    18.6972     0.000
---------------------------------------------------------
```

Taken at face value, these results confirm that fractional cointegration exists and that, because $\widehat{d} < 1/2$, the residuals are stationary. Based on this conclusion, the regression of Sea on Temp does not constitute a spurious regression. The estimated coefficient $\widehat{\beta}$ implies that a $1°C$ increase in mean global temperature is likely to increase mean sea levels by 171 mm or 1.31 mm/year per $°C$, an estimate that is much lower than that produced in chapter 9.

Ventosa-Santaulària, Heres, and Martínez-Hernández (2014) argue that if mean sea level and mean global temperature are nonstationary long-range dependent endogenous variables belonging to a system of simultaneous equations, the ordinary least-squares estimate of β will be biased and inconsistent. Accordingly, they suggest an instrumental-variables regression in which the contemporaneous value as well as two leads of a series measuring the extent of global sea ice serve as instruments for mean global temperature. Present and future levels of the extent of global sea ice may be considered valid instruments because they will not be correlated with sea level except indirectly through their relation with temperature. The time series of the global extent of sea ice is plotted in figure 15.4. It appears that the extent of sea ice has a strong negative correlation with mean global temperature. However, the behavior of the series appears to undergo a substantial change in the second half of the sample, indicative perhaps of a nonlinearity that does not appear to be present in the global temperature series.

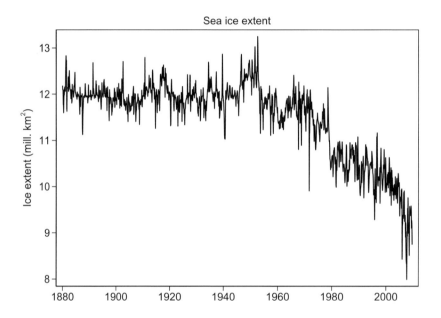

Figure 15.4. Time-series plot of the extent of global sea ice in millions of km^2. The data are monthly for the period January 1880 to December 2009.

Implementing once more the Robinson estimator in which all the series are constrained to have the same degree of fractional integration yields a similar estimate of d.

```
. constraint 2 Sea = IceSAt
. roblpr Sea Temp IceSAt, constraints(1/2)
Robinson estimates of fractional differencing parameters
Power =    .9                            Ords       =     747
------------------------------------------------------------
  Variable     |     Est d      Std Err      t       P>|t|
---------------+--------------------------------------------
  Sea          |   .6875769    .0127139   54.0807    0.000
  Temp         |   .6875769    .0127139   54.0807    0.000
  IceSAt       |   .6875769    .0127139   54.0807    0.000
------------------------------------------------------------
```

15.6 Fractional cointegration

Using sea ice and two leads as instruments for global mean temperature and estimating the fractional cointegrating regression by instrumental variables gives a slightly higher estimate of β.

```
. ivregress 2sls Sea (Temp = IceSAt IceSAtmas1 IceSAtmas2)
Instrumental variables (2SLS) regression      Number of obs   =     1,558
                                               Wald chi2(1)    =   2854.44
                                               Prob > chi2     =    0.0000
                                               R-squared       =    0.6424
                                               Root MSE        =    34.909
```

Sea	Coef.	Std. Err.	z	P>|z|	[95% Conf. Interval]
Temp	217.1425	4.064285	53.43	0.000	209.1766 225.1083
_cons	-61.95565	.8945288	-69.26	0.000	-63.7089 -60.20241

```
Instrumented: Temp
Instruments:  IceSAt IceSAtmas1 IceSAtmas2
. capture drop uhat
. predict double uhat
(option xb assumed; fitted values)
. roblpr uhat, power(0.7 0.75:0.95)
Robinson estimates of fractional differencing parameter for uhat
```

Power	Ords	Est d	Std Err	t(H0: d=0)	P>|t|
.7	171	.663615	.0487188	13.6213	0.000
.75	247	.670809	.0446191	15.0341	0.000
.8	359	.6502788	.0386477	16.8258	0.000
.85	517	.5918933	.0315372	18.7681	0.000
.9	747	.529059	.0260836	20.2832	0.000
.95	1079	.4386199	.0213949	20.5012	0.000

The estimated coefficient $\widehat{\beta}$ implies that a $1°C$ increase in mean global temperature is likely to increase mean sea levels by 217 mm in the long run or 1.67 mm/year per $°C$, which although larger than before is still lower than the figure in chapter 9.

Of course, the validity of this estimate depends upon the soundness of the estimation methodology. There are some concerns in this particular case that may be enumerated as follows:

1. The estimation of d simply enforces the requirement that all the series have the same order of fractional integration with the result that the actual estimate suggests fractional nonstationarity $(\widehat{d} > 1/2)$. The results of the two-step estimates call the first restriction into question, while the maximum-likelihood estimation of the full ARFIMA model raises doubts about the estimated value of d.

2. The instrumental-variables approach should eliminate inconsistency in the estimate of β, provided that the global extent of sea ice is a valid instrument. The nonlinearity evident in figure 15.4 raises doubts about the appropriateness of sea ice extent as an instrument.

3. Ventosa-Santaulària, Heres, and Martínez-Hernández (2014) present results from a finite-sample simulation experiment that shows that the instrumental estimates converge to the true parameter values as the sample size grows. It should be noted, however, that theoretical results relating to the instrumental-variables approach are not available and should therefore be treated with caution.

4. The estimate of d obtained from the instrumental-variables residuals is above 0.5 for many choices of the number of ordinates (`power`) in the log-periodogram regression.

5. The behavior of the residuals from the instrumental-variables regression plotted in figure 15.5 reflects the concerns raised in c and d and raises doubts about the conclusion that `Sea` and `Temp` are indeed fractionally cointegrated.

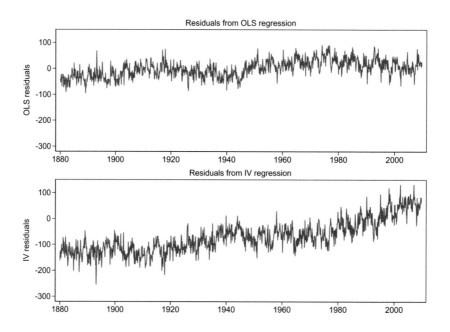

Figure 15.5. Time-series plots of the residuals obtained from fractional cointegrating regressions of mean sea level on global temperature estimated by ordinary least squares (top panel) and instrumental variables (bottom panel). The instrumental-variables regression uses global sea ice extent together with two lags as instruments for global temperature. The data are monthly for the period January 1880 to December 2009.

To conclude, fractional integration and cointegration undoubtedly provide interesting avenues for exploring the dynamics of environmental variables. The implementation of the econometrics needs care and attention to detail. The fact that the results are so sensitive to the choices made by the modeler is perhaps the reason why the method has not been widely used in the econometrics of the environment.

Exercises

1. One of the most well known examples of long memory is the time series of the minimal water level of the Nile River for the years 622–1284 AD measured at the Roda Gauge near Cairo (Beran 1994). The data in `nile.dta` represent annual minimum levels of the Nile River for the period 622–1286.

    a. Plot the data, and verify that there are long periods where the observations tend to stay at a high level and that there are long periods with low levels. Looking at short time periods, there seem to be cycles or local trends. However, looking at the whole series, there is no apparent persisting cycle.

    b. Plot the autocorrelations of the Nile data. Comment on the rate of decay of the autocorrelations.

    c. Estimate the fractional parameter d using the two-step methods described in the chapter. Comment on the similarity or otherwise of the estimates returned by these estimators.

    d. Fit an ARFIMA$(1, d, 2)$ model by maximum likelihood, and compare the estimate of d with that obtained in exercise 1c.

    e. Use the fitted ARFIMA model to predict the value of the series for the years 1285 to 1294. Plot the predictions and the 95% confidence interval. Comment on the results.

2. `maunaloa.dta` contains monthly levels of CO_2 emissions above Mauna Loa, Hawaii. Remove the seasonal effects by using the 12th seasonal difference (`S12.`) on the logarithm of the series. The "seasonal difference" in this case is the year-over-year change in the underlying variable; see [U] **11.4.4 Time-series varlists**.

    a. Fit an ARMA(1,2) model, and comment on the statistically significance of the estimated parameters and the degree of dependence.

    b. Estimate the long-range dependence using the modified Lo range statistic and the KPSS statistic. What do you conclude?

    c. Estimate the fractional parameter d using the two-step methods described in the chapter. Comment on the similarity or otherwise of the estimates returned by these estimators.

    d. Fit an ARFIMA$(1, d, 2)$ model by maximum likelihood. Compare the spectral density of this model with that of an ARFIMA$(1,0,2)$ model. Interpret the plots.

    e. Explore whether the fitted ARFIMA models enhance the ability to forecast the CO_2 emissions.

A   Using Stata

This appendix presents some of the basic concepts involved with using Stata effectively. Of necessity, the treatment is brief. For more detail, see Baum (2016).

A command, to Stata, is an instruction to perform some action. Commands that have been coded into the Stata kernel are known as "built in" commands. However, Stata is continually being expanded. Stata's official commands are updated fairly frequently and are distributed as bug fixes, enhancements to existing commands, and even entirely new commands during the lifetime of a given major release. Issuing the command `update query` from the command line will check for more recent versions of either Stata's executable (the kernel) or the ado-files. Issuing the command `version` shows the current version to which the command interpreter is set. Using `version #` sets the command interpreter and other features (such as random-number generators) to a particular version. This feature makes it possible for old programs to run correctly under more recent versions of Stata.[1]

In addition to "built in" commands, programs written in Stata's own programming language using an ado-file can be used to extend the functionality of official Stata. There are two particularly important websites to check:

- The *Stata Journal*, a quarterly refereed journal, is the primary method for distributing peer-reviewed user contributions. The Stata command `search` (see [R] **search**) will locate commands that have been documented in the *Stata Journal*, and with one click you may install them in your version of Stata.

- The Boston College Statistical Software Components (SSC) Archive contains many new community-contributed commands that extend Stata functionality.
  `ssc new` lists new packages.
  `ssc hot` reports on the most popular packages.
  `ado update, update` updates your installed packages.

Issuing the command `search` plus one or more keywords will present results from a keyword database and from the Internet, for instance, frequently asked questions from the Stata website, articles in the *Stata Journal* and its predecessor the *Stata Technical Bulletin*, and community-contributed commands from the SSC Archive and user sites. Stata's `help` (see [R] **help**) command can also be used in this manner because if it does not locate a help file, it passes the query to `search`.

---

[1]. The programs written for this book are compatible with Stata 16.

This appendix is arranged in four sections. The first discusses how to manage the myriad of files related to a project. The second is concerned with data and outlines the data types supported by Stata, how to input data, how to transfer data from various sources, and how to protect data in memory. The third section provides general hints on Stata programming. The final section is a smorgasbord of topics that are useful in the context of the kind of data encountered in environmental econometrics.

A.1 File management

Like most programs, Stata has a concept of the current working directory (CWD), and in interactive mode Stata displays the CWD in its status bar. If you save a file such as a .dta file or log file, it will be placed in the CWD unless a full file specification directing it to another directory. That is, `save myfile, replace` will save that file in the CWD. Likewise, attempting to use a file with the syntax `use myfile, clear` will return an error if the file is not located in the CWD.

One of the most common problems inexperienced users of Stata face is saving a data file and not knowing where it was saved. It is useful, therefore, to set up directories and folders dedicated to your projects and use the `cd` command to navigate to the specific directory associated with a project. You may change your current directory with the `cd` command, for instance, `cd "My Documents/project1"`.[2] The `pwd` command may be used to display the CWD at any time. Both `cd` and `pwd` are described in `cd` (see [D] **cd**).

You may want Stata to change the working directory to your preferred location automatically when you start Stata. This can be accomplished with `profile.do`. This file, placed in your home directory,[3] will execute a set of commands when Stata is invoked.

A.1.1 Locating important directories: adopath

The `adopath` command provides a list of six directories or folders that are important to Stata. The BASE directory contains the Stata program itself and the official ado-files that make up most of Stata. Do not tamper with the files in this directory. Stata's `update` (see [R] **update**) command will automatically modify the contents of the BASE directory. The SITE directory may reference a network drive in a university or corporate setting where a system administrator places ado-files to be shared by several users.

---

2. Even in a Windows environment, you can (and should) use the forward slash (/) as a directory separator. The backslash (\) usually used in Windows file specifications can be problematic in Stata do-files, because Stata interprets the backslash as an escape character, modifying its handling of the following character.
3. See `help profilew` (Windows), `help profilem` (Mac OS X), or `help profileu` (Unix/Linux) for operating system–specific details.

A.1.1 Locating important directories: adopath

The PERSONAL directory is, as its name suggests, personal, and personal ado-files may be placed in that directory. If you want to modify an official Stata ado-file, you should make a copy of it, change its name (for instance, rename sureg.ado to sureg2.ado), and place it in the PERSONAL directory.

The PLUS directory is automatically created when any community-contributed materials are downloaded. If you use search to locate and install community-contributed programs from the *Stata Journal* or *Stata Technical Bulletin*, their ado-files and help files will be located in a subdirectory of the PLUS directory.[4] If you use the ssc (see [R] **ssc**) command to download community-contributed materials from the SSC ("Boston College") Archive, or net install to download materials from a user's site, they will also be placed in a subdirectory of the PLUS directory.[5]

The adopath command, like sysdir, lists six directories:

```
. adopath
    [1]  (BASE)      "/Applications/Stata/ado/base/"
    [2]  (SITE)      "/Applications/Stata/ado/site/"
    [3]              "."
    [4]  (PERSONAL)  "~/Library/Application Support/Stata/ado/personal/"
    [5]  (PLUS)      "~/Library/Application Support/Stata/ado/plus/"
    [6]  (OLDPLACE)  "~/ado/"
```

The order of these directories is important because it defines how Stata will search for a command. It will attempt to find foo.ado in BASE, the location of Stata's official ado-files.[6] The third directory is ".", that is, the current working directory. The fifth is PERSONAL, and the sixth is PLUS. This pecking order implies that if foo.ado is not to be found among Stata's official ado-files or the SITE directory, Stata will examine the CWD. If that fails, it will look for foo.ado in PERSONAL (and its subdirectories). If that fails, it will look in PLUS (and its subdirectories) and as a last resort in OLDPLACE. If foo.ado is nowhere to be found, Stata will generate an unrecognized command error.

This search hierarchy indicates that an ado-file may be located in one of several places. In the next section, we discuss how you might choose to organize ado-files as well as do-files and data files related to a research project.

---

4. For instance, if sysdir shows that PLUS is ~/Library/Application Support/Stata/ado/plus/, foo.ado and foo.hlp will be placed in ~/Library/Application Support/Stata/ado/plus/f.
5. The OLDPLACE directory exists only for historical reasons and will be ignored.
6. As mentioned above, SITE may be ignored unless you are accessing Stata in a networked environment.

A.1.2 Organization of do-, ado-, and data files

It is crucially important that ado-files are placed on the adopath. You can place them in your CWD ([3] above in the adopath listing), but that is generally a bad idea because if you work in any other directory, those ado-files will not be found. If the ado-files are your own or have been written by a coworker, place them in PERSONAL. If you download ado-files from the SSC Archive, heed the advice that you should always use Stata, not a web browser, to perform the download and locate the files in the correct subdirectory in the PLUS directory.[7]

For do-files, data files, and log files, it makes sense to create a directory, or folder, in your home directory for each separate project and store all project-related files in that directory.[8] Referencing files in the same directory simplifies making a copy of that directory for a coworker or collaborator and makes it possible to run the do-files from an external drive such as a flash disk or from a shared storage location such as Dropbox or Google Drive. It is also a good idea to place a cd command at the top of each do-file referencing the CWD. Although this command would have to be altered if you moved the directory to a different computer, it will prevent the common mistake of saving data files or log files to a directory other than the project directory.

You might also have several projects that depend on one or two key data files. Rather than duplicating possibly large data files in each project directory, you can refer to them with a relative file specification. Say that your research directory is d:/data/research with subdirectories d:/data/research/project1 and d:/data/research/project2. Place the key data file master.dta in the research directory, and refer to it in the project directories with use ../master, clear. The double dot indicates that the file is to be found in the parent (enclosing) directory while allowing you to move your research directory to a different drive (or even to a Mac OS or Linux computer) without having to alter the use statement.

A.1.3 Editing Stata do- and ado-files

You should recognize that do- and ado-files are merely text files with filetypes of .do or .ado rather than .txt. As such, it is a poor idea to edit them in a word-processing application such as Microsoft Word. A word processor must read the text file and convert it into its own binary format. When the file is saved, it must reverse the process.[9] Furthermore, a word processor will usually present the file in a variable-width character format, which is harder to read. But the biggest objection to word processing a do-file or ado-file is the waste of your time; it is considerably faster to edit a file in the Do-file Editor and execute it immediately without the need to translate it back into text. Note also that the Do-file Editor has an advantage over most external text editors

---

7. If your installation's firewall prevents access to the Internet, you must add proxy server settings from IT staff to allow Stata's external access.
8. Recent versions of Stata provide a Project Manager, accessed by **File > New > Project...**.
9. This is the same translation process that takes place when you import delimited a text file and export delimited it back to text within Stata.

A.2.1 Data types 365

because it allows you to execute only a section of the file by selecting those lines and hitting the **Execute (do)** button.[10]

Stata's Do-file Editor supports syntax highlighting, automatic indentation, line numbering, bookmarks, and collapsible nodes, so there are few features provided by external text editors that are not readily available in the Do-file Editor. The Do-file Editor also supports autocompletion for words that are already in the text, Stata commands, quotes, parentheses, braces, and brackets.

A.2 Basic data management

A.2.1 Data types

Stata, as a programming language, supports more data types than many statistical packages. The major distinction to consider is between numeric and string data types. Data management tasks often involve conversions between numeric and string variables. For instance, data read from a text file (such as a .csv or tab-delimited file created by a spreadsheet) will often be considered to be a string variable by Stata even though most of its contents are numeric. The commands `destring` (see [D] **destring**) and `tostring` are helpful in this regard, as are `encode` (see [D] **encode**) and `decode`.

String variables may hold values up to 2,045 characters in length, one byte for each character. You usually need not declare their length, because Stata's string functions (`help string functions`) will generate a string variable long enough to hold the contents of any `generate` (see [D] **generate**) operation. They require as many bytes of storage per observation as their declaration. For instance, a `str20` variable requires 20 bytes per observation. If longer strings are required, use the `strL` datatype, which may hold up to two billion bytes.

Stata's numeric data types include `byte`, `int`, `long`, `float`, and `double`. The `byte`, `int`, and `long` data types can hold only integer contents. In summary,

Table A.1. Numeric data types

Storage type	Minimum	Maximum	Bytes
byte	-127	100	1
int	$-32,767$	$32,740$	2
long	$-2,147,483,647$	$2,147,483,620$	4
float	-1.701×10^{38}	1.701×10^{38}	4
double	-8.988×10^{307}	8.988×10^{307}	8

---

10. A useful compendium of information on text editors has been collected and organized by Nicholas J. Cox (2002a) and is available from the SSC Archive as `texteditors`, an HTML document.

The `long` integer data type can hold all signed 9-digit integers but only some 10-digit integers. Integers are held in their exact representation by Stata, so you may store a 9-digit integer (such as a U.S. Social Security number) as a `long`. However, lengthy identification numbers can also be stored as a `double` data type or as a string variable. In many cases, the latter will be a wiser choice because then you need not worry about possible truncation of values. You will also find it useful to use string variables when a particular identification code could contain characters. Storing these values as strings avoids later problems with numeric missing values.

As displayed above, the two floating-point data types, `float` and `double`, can hold very large numbers. A `float` contains approximately seven significant digits in its mantissa. This implies that if you read a set of nine-digit U.S. Social Security numbers into a `float`, they will not be held exactly. A `double` contains approximately 15 significant digits. For example, the residuals computed from a linear regression using `regress` and `predict eps, residual` should sum to exactly zero. In Stata's finite-precision computer arithmetic using the default `float` data type, residuals from such a regression will sum to a value in the range of 10^{-7} rather than 0.0. Thus, discussions of the **predict** (see [R] **predict**) command often advise using `predict double eps, residual` to compute more accurate residuals.

What are the implications of finite-precision arithmetic for working with Stata?

1. Store ID numbers with many digits as string variables, not as integers, floats, or doubles.

2. Do not rely on exact tests of a floating-point value against a constant, not even zero. The `reldif()` function (`help math functions`) may be used to test for approximate equality.[11]

3. As suggested above, use `double` floating-point values for any generated series where a loss of precision might be problematic, such as residuals, predicted values, scores, and the like.

4. You should be wary of variables' values having very different scales, particularly when a nonlinear estimation method is used. Any regression of `price` from the venerable `auto.dta` reference dataset on a set of regressors will display extremely large sums of squares in the analysis of variance table. Scaling `price` from dollars to thousands of dollars obviates this problem. The scale of this variable does not affect the precision of linear regression, but it could be problematic for nonlinear estimation techniques.

5. Use integer data types where it is appropriate to do so. Storing values as `byte` or `int` data types when feasible saves disk space and memory.

A useful command, particularly when working with datasets acquired from other statistical packages, is `compress` (see [D] **compress**). This command will examine each

---

11. Further details of this issue can be found in [U] **12.2.2 Numeric storage types**, Gould (2006) and Cox (2006).

variable in memory and determine whether it may be stored more efficiently. It is guaranteed never to lose information nor reduce the precision of measurements. The advantage of storing indicator (0/1) variables as `byte` data types rather than as four-byte `long` data types is substantial for many survey datasets with numerous indicator variables. It is an excellent idea to apply `compress` when performing the initial analysis of a new dataset.

A.2.2 Getting your data into Stata

Stata has the capability to read binary files in several formats other than its own binary `.dta` format. These include facilities described in [D] **import excel**, [D] **import sas**, [D] **import sasxport8**, [D] **import spss**, and [D] **import dbase**. Access to Structured Query Language and Microsoft Access databases is provided by `odbc` (see [D] **odbc**). Accessing text data usually involves `import delimited` (see [D] **import delimited**) for comma- or tab-delimited files, or `infile` or `infix` (see [D] **infile (fixed format)** or [D] **infix (fixed format)**) for fixed-format files, possibly with a dictionary (`.dct`) file.

Before carrying out statistical analysis, applied researchers face several thorny issues in converting their data into usable form. These range from the mundane (for example, a text-file dataset may have coded missing values as `99`) to the challenging (for example, a text-file dataset may be in a hierarchical format, with master records and detail records). Although a brief guide to these issues cannot possibly cover all the ways in which external data may be organized and transformed for use in Stata, several rules apply:

1. Familiarize yourself with the various Stata commands for data input. Reading the [U] **22 Entering and importing data** section is well worth the investment of your time.

2. When you need to manipulate a text file, use a text editor, not a word processor or spreadsheet.

3. Move the data into Stata as early in the process as possible, and perform all manipulations via well-documented do-files that may be edited and reexecuted if need be (or if a similar dataset is encountered).

4. Keeping track of multiple steps of the data input and manipulation process requires good documentation.

5. Working with anything but a simple rectangular data array will almost always require the use of `append` (see [D] **append**), `merge` (see [D] **merge**), or `reshape` (see [D] **reshape**). The command `append` simply adds Stata-format datasets stored on disk to the end of the dataset in memory, while the `merge` command joins corresponding observations from the dataset currently in memory with those from another Stata-format dataset matching on one or more key variables. The `reshape` command is used to change the format of the data in memory from wide form to long form and vice versa and is a crucial command to master when using

longitudinal or panel data (see chapter 12). Read the documentation for these commands and understand their capabilities.

Handling text files

Text files—often described as ASCII files—are a common source of raw data. Text files may have any file extension—they may be labeled .raw (as Stata would prefer), .txt, .csv, .tsv, or .asc. Every operating system supports a variety of text editors, many of which are freely available. A useful summary of text editors of interest to Stata users is edited by Nicholas J. Cox and is available as a webpage from the SSC as the package texteditors. You will find that a good text editor is much faster than a word processor when scrolling through a large data file. Many text editors colorize Stata commands, making them useful for Stata program development. Text editors are also extremely useful when working with large survey datasets that are accompanied with machine-readable codebooks, which are often many megabytes in size. Searching those codebooks for particular keywords with a powerful text editor is efficient.

Text files may be free format or fixed format. A free-format file contains several fields per record separated by delimiters, which are characters not to be found within the fields. A purely numeric file (or one with simple string variables such as U.S. state codes) may be space delimited; that is, successive fields in the record are separated by one or more space characters. The columns in the file need not be aligned. These data may be read from a text file (by default with extension .raw) with Stata's infile command, which must assign names (and if necessary, data types) to the variables.

If string variables with embedded spaces are to be used in a space-delimited file, they themselves must be delimited, usually with quotation marks in the text file. So what should you do if your text file is space-delimited and contains string variables with embedded spaces? That is a difficult question because no mechanical transformation will generally solve this problem. If the data are downloadable from a webpage that offers formatting choices, you should choose a tab-delimited rather than a space-delimited format. The other option, comma-delimited text, or comma-separated values (.csv), has its own difficulties. Tab-delimited text avoids most problems.

The import delimited command

If tab-delimited text files are to be read, the infile command no longer is the correct tool for the job, and import delimited should be used instead.[12] The import delimited command reads a tab-delimited or comma-delimited (CSV) text file whether or not a spreadsheet program was involved in its creation. For instance, most database programs contain an option to generate a tab-delimited or comma-delimited export file, and many datasets available for web download are in one of these formats.

---

12. Prior to Stata 13, the command with this functionality was insheet.

A.2.2 Getting your data into Stata

The `import delimited` command is handy. As long as one observation in your target Stata dataset is contained on a single record with tab or comma delimiters, this is the command to use. Stata will automatically try to determine which delimiter is in use (but options `tab` and `comma` are available), or any ASCII character may be specified as a delimiter with the `delimiter(`*char*`)` option. For instance, some European database exports use semicolon (;) delimiters because standard European numeric formats use a comma as the decimal separator. If the first line (or a specified line) of the file contains variable names, they will be used and translated into valid Stata variable names unless they contain invalid characters such as spaces. This is very useful because if data are being extracted from a spreadsheet, they will often have that format. The issue of embedded spaces or commas no longer arises in tab-delimited data, and you can rely on the first line of the file to define the variable names. However, the names must be distinct and follow naming rules. For instance, a variable name cannot start with a numeral.

It is particularly important to heed any informational or error messages produced by the data input commands. If you know how many observations are present in the text file, check to see if the number Stata reports is correct. Likewise, the `summarize` or `codebook` commands should be used to discern whether the number of observations—minimum and maximum—for each numeric variable is sensible. Data entry errors can often be detected by noting that a particular variable takes on nonsensical values, usually denoting the omission of one or more fields on that record. Such an omission may also trigger one or more error messages. For instance, leaving out a numeric field on a particular record will move an adjacent string field into that variable. Stata will then complain that it cannot read the string as a number. A distinct advantage of the tab- or comma-delimited formats is that missing values may be coded with two successive delimiters. The command `assert` (see [D] `assert`) can be used to good advantage to ensure that reasonable values appear in the data.

An additional distinction exists between `infile` and `import delimited`, that is, `infile` may be used with `if` *exp* and `in` *range* qualifiers to selectively input data. For instance, with a large text-file dataset, you could use `in 1/1000` to read only the first 1,000 observations and verify that the input process is working properly. These qualifiers may not be used with `import delimited`, but you may specify a range of rows to read with the `rowrange()` option to read less than the entire file.

Accessing data stored in spreadsheets

Copy-and-paste techniques should not be used to transfer data from another application directly to Stata, because this process cannot be reliably replicated. If the data are presently in a spreadsheet, the appropriate portion of that spreadsheet should be copied and pasted (in Excel, **Paste Special...** to ensure that only values are stored) into a new blank sheet. If Stata variable names are to be added, leave the first row blank so that they may be filled in. Save only that sheet as *Text Only – Tab delimited* to a new filename. If you use the file extension `.raw`, it will simplify reading the file into Stata using the `import delimited` command.

The command `import excel` command may be used to read the contents of an Excel or Excel-compatible worksheet, in either `.xls` or `xlsx` format, into Stata. You may specify the worksheet to be input and optionally provide a cell range from which data are to be read. The first row may be used to provide Stata variable names. Note that if a column of the worksheet contains at least one cell with nonnumerical text (such as `NA`), the entire column is imported as a string variable. Therefore, you should be familiar with Stata's string-to-numeric conversion capabilities.

There are two caveats regarding dates. Both Excel and Stata work with the notion that calendar dates are successive integers from an arbitrary starting point. To read the dates into a Stata date variable, they must be formatted with a four-digit year, preferably in a format with delimiters (for example, 12/6/2004 or 6-Dec-2004). It is much easier to make these changes in the spreadsheet program before reading the data into Stata. Second, Mac OS users of Excel should note that Excel's default is the 1904 Date System. If the spreadsheet was produced in Excel for Windows and the steps above are used to create a new sheet with the desired data, the dates will be off by four years (the difference between Excel for Mac and Excel for Windows defaults). Uncheck the preference *Use the 1904 date system* before saving the file as text.

Importing data from other package formats

In many cases, the foreign data are already in the format of some other statistical package or application. As mentioned above, Stata can handle importing and exporting from several other applications. Of course, the mapping between packages is not always one-to-one. In Stata, a value label stands alone and may be attached to any variable or set of variables, whereas in other packages it is generally an attribute of a variable and must be duplicated for similar variables.

An important distinction between Stata on one hand and SAS and SPSS on the other is Stata's flexible set of data types. As already noted, Stata, like the C language in which its core code is written, offers five numeric data types, while some other packages do not have this broad array of data types and resort to storing all numeric data in a single data type. This simplicity bears a sizable cost because an indicator variable only requires a single byte of storage, whereas a double-precision floating-point variable requires eight bytes to hold up to 15 decimal digits of accuracy. Stata allows the user to specify the data type based on the contents of each variable, which can result in considerable savings in terms of both disk space and execution time when reading or writing those variables to disk.

An alternative solution for data transfer between databases uses some flavor of structured query language (SQL). Stata can perform Open Data Base Connectivity (ODBC) operations with databases accessible via that protocol (see [D] **odbc** for details). Because most SQL databases as well as non-SQL data structures such as Excel and Microsoft Access support ODBC, this is often suggested as a workable solution to dealing with foreign data. It does require that the computer system on which you are running Stata is equipped with ODBC drivers. These are installed by default on Windows systems with Microsoft Office but may require the purchase of a third-party product for Mac OS or Linux systems. If the necessary database connectivity is available, Stata's `odbc` is a full-featured solution. It allows for both the query of external databases and the insertion or update of records in those databases.

A.2.3 Other data issues

Protecting the data in memory

Several Stata commands replace the data in memory with a new dataset. For instance, the `collapse` (see [D] **collapse**) command makes a dataset of summary statistics, while `contract` (see [D] **contract**) makes a dataset of frequencies or percentages. In a program, you may want to invoke one of these commands but may want to retain the existing contents of memory for further use in the do-file. You need the `preserve` and `restore` (see [P] **preserve**) commands, which will allow you to set aside the current contents of memory in a temporary file and bring them back when needed. The ability to set the current dataset aside (without having to explicitly `save` it) and bring it back into memory when needed is a useful feature. Even greater flexibility for handling these operations is provided by `frames` (see [D] **frames**), which enables Stata to work with more than one dataset in memory.

Missing data handling

Stata possesses 27 numeric missing-value codes: the system missing value . and 26 others from .a through .z. They are treated as large positive values and sort in that order, with plain . being the smallest missing value (see [U] **12.2.1 Missing values**). This allows qualifiers such as `if` *variable* `<.` to exclude all possible missing values. To make your code as readable as possible, use `missing()`.

Stata's standard practice for missing-data handling is to omit those observations from any computation. For `generate` or `replace`, missing values are typically propagated so that any function of missing data is missing.[13] In univariate statistical computations (such as `summarize`), computing a mean or standard deviation, only nonmissing cases are considered. For multivariate statistical commands, Stata generally practices casewise deletion, which is when an observation in which any variable is missing is deleted from the computation. The `missing(`*y1,y2,...,yn*`)` function returns 1 if any of the arguments is missing and 0 otherwise. That is, it provides the user with a casewise deletion indicator.

Stata also provides for missing values in string variables. The empty or null string (`""`) is taken as missing. To Stata, there is an important difference between a string variable containing one or more spaces and a string variable containing no spaces (al-

---

[13]. There is one important exception. If you use a command such as `generate hx = x > 10000`, the resulting indicator (0,1) variable will be set to 1 for missing values of `x`. This command is properly written as `generate hx = x > 10000 if !mi(x)`, which ensures that missing values of `x` will be missing in `hx` as well.

though they will appear identical to the naked eye). That suggests that you should not include one or more spaces as a possible value of a string variable.

Recoding missing values: the mvdecode and mvencode commands

When importing data from another statistical package or a spreadsheet or database, differing notions of missing-data codes may hinder the proper rendition of the data within Stata. Likewise, if the data are to be used in another program that does not use the . notation for missing-data codes, there may be a need to use an alternative representation of Stata's missing data. The `mvdecode` and `mvencode` commands (see [D] **mvencode**) may be useful in those circumstances. The `mvdecode` command permits you to recode various numeric values to missing, as would be appropriate when missing data have been represented as -99, -999, 0.001, and so on. Stata's full set of 27 numeric missing-data codes may be used, so -9 may be mapped to `.a`, -99 to `.b`, etc. The `mvencode` command provides the inverse function, allowing Stata's missing values to be revised to numeric form. Like `mvdecode`, `mvencode` can map each of the 27 numeric missing-data codes to a different numeric value.

A.2.4 String-to-numeric conversion and vice versa

Quite commonly, a variable imported from an external source will be misclassified as string when it should be considered as numeric. For instance, if the first value read by `import delimited` is NA, that variable will be classified as a string variable. Stata provides several methods for converting string variables to numeric:

1. If the variable has merely been incorrectly classified as a string, you may apply the brute force approach of the `real()` function. This will create missing values for any observations that cannot be interpreted as numeric.

2. A more subtle approach is given by the `destring` command, which can transform variables in place (with the `replace` option) and may be used with a *varlist* to apply the same transformation to an entire set of variables with a single command. This is useful if there are several variables that may require conversion. However, `destring` should be used only for variables that have genuine numeric content but happen to have been misclassified as string variables.

3. If the variable truly has string content and you need a numeric equivalent, you may use the `encode` command. You should not apply `encode` to a string variable that has purely numeric content (for instance, one which has been misclassified as a string variable) because `encode` will attempt to create a value label for each distinct value of the variable.

A.3 General programming hints

As you move away from interactive use of Stata and make greater use of do-files and ado-files in your research, the style of the contents of those files becomes more important. The concepts dealt with in this section are all crucial to writing efficient Stata code.

Variable names

Stata variable names must be distinct and follow certain rules of syntax. For instance, they may not contain embedded spaces, nor hyphens (-), nor characters outside the sets A-Z, a-z, 0-9, _. In particular, a full stop or period (.) cannot appear within a variable name. Variable names must start with a letter or an underscore. Most importantly, case matters—STATE, State, and state are three different variables to Stata. The Stata convention is to use lowercase names by default for all variables to avoid confusion and to use uppercase only for some special reason.

Observation numbering: _n and _N

The observations in the dataset are numbered. When you refer to an observation, you may do so with its observation number. The highest observation number, corresponding to the total number of observations, is known as _N, while the current observation number is known as _n. Under some circumstances, the meanings of these two symbols will be altered. The observation numbers will change if a sort (see [D] sort) command alters the order of the dataset in memory.

The varlist

Many Stata commands accept a *varlist*, which is a list of one or more variables to be used. A *varlist* can contain the variable names, or a wild card (*) such as in *id may be used. The * will stand in for an arbitrary set of characters. A *varlist* can also contain a hyphenated list such as x1-x4. That refers to all variables in the dataset between x1 and x4, including those two, in the order that the variables appear in the dataset. The dataset order is that provided by describe (see [D] describe) and is shown in the Variables window. It can be modified by the order command.

The numlist

Many Stata commands require use of a list of numeric arguments known as a numlist. A numlist can be provided in several ways. It may be spelled out explicitly, for example, as 0.5 1.0 1.5. It may involve a range of values, such as 1/4 or -3/3. These lists would include the integers between those limits. You could also specify 10 15 to 30, which would count from 10 to 30 by 5s, or use a colon to say the same thing—10 15:30. You may count by steps, as 1(2)9, which is a list of the first five odd integers, or 9(-2)1, which is the same list in reverse order. Square brackets may be used in place of parentheses.

One thing that generally should not appear in a numlist is a comma. A comma in a numlist will usually cause a syntax error. Loop constructs in other programming often spell out a range with an expression such as 1,10. In Stata, such an expression will involve a numlist of 1/10. One of the primary uses of the *numlist* is in the `forvalues` (see [P] **forvalues**) statement (but not all valid numlists are acceptable in `forvalues`).

The if exp and in range qualifiers

Stata commands operate on all the observations in memory by default. Almost all Stata commands accept the qualifiers `if` *exp* and `in` *range*. These qualifiers restrict the command to a subset of the observations. In many problems, the desired subset of the data is not defined in terms of observation numbers (as specified with `in` *range*) but in terms of some logical condition. In that case, it is more useful to use the `if` *exp* qualifier. Of course, `if` *exp* can be used to express an `in` *range* condition, but the most common use of `if` *exp* involves the transformation of data or specification of a statistical procedure for a subset of data identified by `if` *exp* as a logical condition.

Local macros

If you are familiar with lower-level programming languages such as Fortran or C, you may find Stata's terminology for various objects rather confusing. In those languages, you refer to a variable with statements such as x = 2. Although you might have to declare x before its use—for instance, as `integer` or `float`—the notion of a variable in those languages refers to an entity that can be assigned a single value, either numeric or string. In contrast, the Stata variable refers to one column of the dataset which contains _N values, one per *observation*.

So what corresponds to a Fortran or C variable in Stata's command language? Stata's local macro is a container which can hold either a single object, such as a number or variable name, or a set of objects such as a list of variable names. The Stata macro is just an alias that has both a name and a value. When its name is dereferenced, it returns its value. That operation may be carried out at any time and does not alter its value. Alternatively, the macro's value can be easily modified with an additional command.

A local macro is created in a do-file or in an ado-file and ceases to exist when that do-file terminates either normally or abnormally. The Stata command to define a local macro is `local` (see [P] **macro**). In most cases, the `local` statement is written without an equal sign (=). It is acceptable syntax to use an equal sign following the macro's name, but it is a bad idea to get in the habit of using it unless it is required. The equal sign causes the remainder of the expression to be evaluated rather than merely aliased to the macro's name. Defining a macro with an equal sign will cause evaluation of the remainder of the command as a numeric expression or as a string expression.

To work with the value of the macro, it must be dereferenced. To dereference and access the value of the macro, the macro's name is preceded by the left single-quote

A.3 General programming hints

character (`` ` ``) and followed by the right single-quote character (`'`).[14] Stata uses different opening and closing quote characters to signify the beginning and end of a macro name because macro references may be nested. It is important to understand that if macros are nested, they are evaluated from the inside out.[15] To dereference the macro, the correct punctuation is vital. Defining `local alpha 1.5`, we refer to its content as `` `alpha' ``.

Stata's macros are useful in constructing lists or as counters and loop indices. They are that, but they play a much larger role in Stata do-files and ado-files and in the return values of many Stata commands. Macros are one of the key elements of Stata's programming language that allow you to avoid repetitive commands and the retyping of computed results. Macros allow you to change the performance of your do-file by merely altering the contents of a local macro. In this manner, your do-file can be made quite general, and that set of Stata commands may be reused or adapted for use in similar tasks with a minimum of effort.

Global macros

Global macros are distinguished from local macros by their manner of creation (with the `global` statement; see [P] **macro**) and their means of reference. The value of the global macro is obtained by dereferencing it with the dollar sign, `$`, in front of the name taking the place of the punctuation surrounding the local macro's name when it is dereferenced. A global macro exists for the duration of the Stata program or interactive session.

Unless there is an explicit need for a global macro—a symbol with global scope—it is usually preferable to use a local macro. It is easy to forget that a global symbol was defined in do-file *A*. By the time you run the do-file *G* or *H* in that session of Stata, you may find that they do not behave as expected, because they now pick up the value of the global symbol. Such problems are quite difficult to debug. Authors of Fortran or C programs have always been encouraged to keep definitions local unless they absolutely must be visible outside the module. This is good advice for Stata programmers as well.

Scalars

In addition to Stata's variables and local and global macros, there are two additional entities related to every analysis command: scalars and matrices. Scalars, like macros, can hold either numeric or string values, but a scalar can hold only a single value. Most analysis commands return one or more results as numeric scalars. For instance, `describe` returns the scalars `r(N)` and `r(k)`, corresponding to the number of observations and variables in the dataset. A scalar is also much more useful for storing a single numeric result, such as the mean of a variable, rather than storing that value in a Stata

---

14. These characters are found in different places on different languages' keyboards. The left single quote, or left tick, is the French accent grave. The right single quote is commonly known as the apostrophe. These symbols cannot be copied and pasted from Stata manuals.
15. For a thorough discussion of these issues, see Cox (2002c).

variable containing _N copies of the same number. A scalar may be referred to in any subsequent Stata command by its name. The distinction between a macro and a scalar appears when it is referenced. The macro must be dereferenced to refer to its value, while the scalar is merely named.

Stata's scalars are typically used in a numeric context. When a numeric quantity is stored in a macro, it must be converted from its internal (binary) representation into a printable form. By storing the result of a computation—for instance, a variable's mean or standard deviation—in a scalar, no conversion of its value need take place. However, a scalar can only appear in an expression where a Stata variable or a numeric expression could be used. For instance, one cannot specify a scalar as part of an in *range* qualifier, because its value will not be extracted. It may be used in an if *exp* qualifier because that contains a numeric expression. Most of Stata's statistical and estimation commands return various numeric results as scalars.

Stata is capable of working with scalars of the same name as Stata variables. As the manual suggests, Stata will not become confused, but you well may. So you should avoid using the same names for both entities; see Kolev (2006). Scalars and matrices share the same namespace, so you cannot have both a scalar named gamma and a matrix named gamma.

Stata's scalars play a useful role in do-files. By defining scalars at the beginning of the do-file and referring to them throughout the code, you make the do-file parametric. This avoids the difficulties of changing various constants in the do-file's statements everywhere where they appear. You may often need to repeat a complex data transformation task for a different category.

Matrices

Stata supports a broad range of matrix operations on real matrices, as described in matrix (see [P] **matrix**). Stata also provides a dedicated matrix language, Mata, that operates in a separate environment within Stata.

Stata's estimation commands typically create both scalars and Stata matrices, in particular, the matrix e(b), containing the set of estimated coefficients, and the matrix e(V), containing the estimated variance–covariance matrix of the coefficients. These matrices may be manipulated by Stata's matrix commands, and their contents used in later commands. Like all Stata estimation commands, regress (see [R] **regress**) produces matrices e(b) and e(V) as the row vector of estimated coefficients (a $1 \times k$ matrix) and the estimated variance–covariance matrix of the coefficients (a $k \times k$ symmetric matrix), respectively. These matrices may be examined with the matrix list command, or they may be copied for use in do-files using the matrix statement. The command matrix beta = e(b) will create a matrix beta in your program as a copy of the last estimation command's coefficient vector.

A.3 General programming hints

For those Stata users who are writing do-files, Stata matrices are likely to be useful in two particular contexts: stored results as described above and a way of organizing information for presentation. References to matrix elements appear in square brackets. Because Stata does not have a vector data type, all Stata matrices have two subscripts, and both subscripts must be given in any reference. A range of rows or a range of columns may be specified in an expression; see `matrix` for details. Stata's traditional matrices are distinctive in that their elements may be addressed both conventionally by their row and column numbers (counting from 1, not 0) as well as by their row and column names.

Stata's matrices are often useful devices for housekeeping purposes such as the accumulation of results that are to be presented in tabular form. The `tabstat` command may generate descriptive statistics for a set of `by`-groups. Likewise, `statsmat` can be used to generate a matrix of descriptive statistics for a set of variables or for a single variable over `by`-groups. In summary, judicious use of Stata's traditional `matrix` commands eases the burden of many housekeeping tasks and makes it feasible to update material in tabular form without retyping.

Looping

Two of Stata's most useful commands are `forvalues` and `foreach`. These versatile tools have essentially supplanted other mechanisms in Stata for looping. You could also use `while` (see [P] **while**) to construct a loop, most commonly when you are unsure how many times to repeat the loop contents. This is a common task when seeking convergence of a numeric quantity to some target value. For instance,

```
while reldif(newval, oldval) > 0.001 {
    ...
}
```

would test for the *relative difference* between successive values of a criterion and would exit the loop when that difference was less than 0.1%. Conversely, if the computational method is not guaranteed to converge, this could become an infinite loop.

In contrast, when you have a defined set of values over which to iterate, the commands `forvalues` and `foreach` are the tools of choice. These commands are followed by a left brace ({), one or more following command lines, and a terminating line containing only a right brace (}). You can place as many commands in the loop body as you wish. One of the most important uses of `forvalues` is looping over variables where the variables have been given names with an integer component. This avoids the need for separate statements to transform each of the variables. The integer component need not be a suffix. A nested loop is readily constructed with two `forvalues` statements.

The `foreach` command is especially useful in constructing efficient do-files. This command interacts perfectly with some of Stata's most common constructs: the macro, the varlist, and the numlist. As with `forvalues`, a local macro is defined as the loop index. Rather than cycling through a set of numeric values, `foreach` specifies that the loop index iterates through the elements of a local (or global) macro, the variable names of a varlist, or the elements of a numlist. The list can also be an arbitrary list of elements on the command line or a new varlist of valid names for variables not present in the dataset. This syntax allows `foreach` to be used flexibly with any set of items, regardless of pattern.

The generate command

The fundamental commands for data transformation are `generate` and `replace`. These function in the same way, but two rules govern their use. The `generate` function may be used only to create a new variable whose name is not currently in use. On the other hand, `replace` can be used only to revise the contents of an existing variable. Unlike other Stata commands whose names can all be abbreviated, `replace` must be spelled out.

The egen command

While the functions available for use with `generate` or `replace` are limited to those listed in *exp* (see [U] **13 Functions and expressions**), Stata's `egen` command provides an open-ended list of capabilities. Just as Stata's command `set` can be extended by placing additional .ado and .sthlp files on the ado-path, the functions that may be invoked from `egen` are those defined by ado-files whose names start with _g and are stored on the ado-path. Several of those functions are part of official Stata, as documented by [D] **egen**. But your copy of Stata may include additional `egen` functions—either those you have written yourself or those downloaded from the SSC Archive.

If you seek spreadsheet-like functionality in Stata's data transformations, you should become acquainted with the rowwise `egen` functions. Like the equivalent spreadsheet functions, the rowwise functions support the calculation of sums, averages, standard deviations, extrema, and counts across several Stata variables. Wildcards may be used. For example, if you have state-level U.S. Census variables pop1890, pop1900, ..., pop2000, you may use `egen nrCensus = rowmean(pop*)` to compute the average population of each state over those decennial censuses. As discussed in the treatment of missing values in section A.2.3, the rowwise functions operate in the presence of missing values. The mean will be computed for all 50 states, although several were not part of the United States in 1890.

A.3 General programming hints

Several `egen` rowwise functions (`rowmax()`, `rowmean()`, `rowmedian()`, `rowmin()`, `rowsd()`, `rowtotal()`) all ignore missing values. For example, `rowmean(y1,y2,y3)` will compute the mean of three, two, or one of the variables, returning missing only if all three variables' values are missing for that observation. The `egen` functions `rownonmiss()` and `rowmiss()` return, respectively, the number of nonmissing and missing elements in their varlists. Although `correlate` (see [R] **correlate**) *varlist* uses casewise deletion to remove any observation containing missing values in any variable of the varlist from the computation of the correlation matrix, the alternative command `pwcorr` computes pair-wise correlations using all available data for each pair of variables.

The number of nonmissing elements in the rowwise varlist may be computed with `rownonmiss()` with `rowmiss()` as the complementary value. Other official rowwise functions include `rowmax()`, `rowmin()`, `rowtotal()`, and `rowsd()` (row standard deviation). The functions `rowfirst()` and `rowlast()` give the first (last) nonmissing values in the *varlist*. You may find this useful if the variables refer to sequential items. For instance, wages earned per year over several years, with missing values when unemployed. `rowfirst()` would return the earliest wage observation, and `rowlast()` the most recent.

Official `egen` also provides several statistical functions that compute a statistic for specified observations of a variable and place that constant value in each observation of the new variable. Because these functions generally allow the use of `by` *varlist*:, they may be used to compute statistics for each by-group of the data, as discussed in section A.3. This facilitates computing statistics for each household for individual-level data or each industry for firm-level data. The `count()`, `mean()`, `min()`, `max()`, and `total()`[16,17] functions are especially useful in this context.

Computation for by-groups

One of Stata's most useful features is the ability to transform variables or compute statistics over by-groups. By-groups are defined with the `by` *varlist*: prefix and are often useful in data transformations using `generate`, `replace`, and `egen`. Using `by` *varlist*: with one or more categorical variables, a command will be repeated automatically for each value of the `by` *varlist*:. However, it also has its limitations: `by` *varlist*: can only execute a single command.[18]

---

16. Before Stata 9, the `egen` `sum()` function performed this task but was often confused with `generate`'s `sum()` function. Hence it was renamed.
17. Note that `egen`'s `total()` function treats missing values as zeros, so the `total()` of a variable with all missing values is computed as zero rather than missing. You can change this behavior with this function's `missing` option.
18. Because only a subset of `egen` functions allow a `by` *varlist*: prefix or `by(`*varlist*`)` option, the documentation should be consulted to determine whether a particular function is "byable", in Stata parlance.

Under the control of a by-group, the meanings of _n and _N are altered. Those markers usually refer to the current observation and last defined observation in the dataset, respectively. Within a by-group, _n is the current observation of the group and _N is the last observation of the group.

Although `egen`'s statistical functions can be handy, creating variables with constant values or constant values over by-groups in a large dataset will consume a great deal of Stata's available memory. If the constant values are needed only for a subsequent transformation such as computing each state population's deviation from average size and will not be used in later analyses, you should `drop` those variables at the earliest opportunity. Alternatively, consider other Stata commands that can provide this functionality. Ben Jann's (2004) `center` command, available from the SSC Archive, can transform a variable into deviation from mean form, and it works with by-groups.

A.4 A smorgasbord of important topics

There are several programming topics that are of particular importance to environmental econometrics. Because many of the datasets contain higher frequency time-series data, important time-series tools like dates, times, and time-series operators are essential elements of the programmer's toolkit.

Date and time handling

Stata does not have a separate data type for calendar dates. Dates are represented, because they are in a spreadsheet program by numbers known as %t values measuring the time interval from a reference date or epoch. For example, the epoch for Stata and SAS is midnight on January 1, 1960. Days following that date have positive integer values, while days prior have negative integer values. These are known as %td values. Other calendar frequencies are represented by the number of weeks, months, quarters, or half-years since that reference date: %tw, %tm, %tq, and %th values, respectively. The year is represented as a %ty value, ranging from 100 to 9999 A.D. You may also use consecutive integers and the *generic* form, as %tg.

Stata also supports business-day calendars via the command `bcal` (see [D] **bcal**). These calendars are custom calendars specifying the dates to be included and excluded in the calendar specification. For instance, most financial markets are closed on weekends and are also closed on holidays. A business-day calendar allows you to set up the data in "trading time", which is crucial for many time-series commands that require that there are no gaps in the sequence of daily values. Stata can construct a business-day calendar from the dates represented in a variable.

A.4 A smorgasbord of important topics

Stata provides support for accurate intradaily measurements of time, down to the millisecond. A date-and-time variable is known as a %tc (clock) value, and it may be defined to any intraday granularity: hours, minutes, seconds, and milliseconds.[19] For more information, see [U] **12.3 Dates and times**. The tsset (see [TS] tsset) command has a delta() option, by which you can specify the frequency of data collection. For instance, you may have annual data collected only at five-year intervals, or high-frequency financial markets transactions data timestamped by day, hour, minute, and second, collected every 5 or 10 minutes.

It is important when working with variables containing dates and times to ensure that the proper Stata data type is used for their storage. Weekly and lower-frequency values (including generic values) may be stored as data type int or as data type float. Daily (%td) values should be stored as data type long or as data type float. If the int data type is used, dates more than 32,740 days from January 1, 1960 (that is, beyond August 21, 2049) cannot be stored.

Much more stringent requirements apply to clock (date-and-time) values. These values *must* be stored as data type double to avoid overflow conditions. Clock values, like other time values, are integers, and there are 86,400,000 milliseconds in a day. The double data type is capable of precisely storing date and time measurements within the range of years (AD 100–9999) defined in Stata.

Although it is important to use the appropriate data type for date-and-time values, you should avoid using a larger data type than needed. The int data type requires only two bytes per observation; long and float data types require four bytes; and the double data type requires eight bytes. Although every date-and-time value could be stored as a double, that would be wasteful of memory and disk storage, particularly in a dataset with many observations.

A suite of functions (see help datetime functions) is available to handle the definition of date variables and date-and-time arithmetic. Display of date variables in calendar formats (such as 08 Nov 2016) and date and time variables with the desired intraday precision is handled by the definition of proper formats. As with numeric variables, you should distinguish between the content or value of a date-and-time variable and the format in which it will be displayed.

If you are preparing to move data from a spreadsheet into Stata with the import delimited command, make sure that any date variables in the spreadsheet display as four-digit years. It is possible to deal with two-digit years such as 11/08/16 in Stata, but it is easier to format the dates with four-digit years (for example, 11/08/2016) before reading those data into Stata.

---

[19]. There are also %tC values, which account for *leap seconds* for very precise measurement of time intervals.

Time-series operators

Stata provides the time-series operators L., F., D., and S. which allow the specification of lags, leads (forward values), differences, and seasonal differences, respectively. The time-series operators make it unnecessary to create a new variable to use a lag, difference, or lead. When combined with a *numlist*, they allow the specification of a set of these constructs in a single expression. Consider the lag operator, L., which when prepended to a variable name refers to the (first-)lagged value of that variable: L.x. A number may follow the operator, so L4.x would refer to the fourth lag of x. More generally, a numlist may be used, so L(1/4).x refers to the first through fourth lags of x, and L(1/4).(x y z) defines a list of four lagged values of each of the variables x, y and z. These expressions may be used anywhere that a *varlist* is required.

Similar to the lag operator, the lead operator, F., allows specification of future values of one or more variables. Strictly speaking, the lead operator is unnecessary because a lead is a negative lag, and an expression such as L(-4/4).x will work, labeling the negative lags as leads. The difference operator, D., may be used to generate differences of any order. The first difference, D.x, is Δx or $x_t - x_{t-1}$. The second difference, D2.x, is not $x_t - x_{t-2}$ but rather $\Delta(\Delta x_t)$; that is, $\Delta(x_t - x_{t-1})$, or $x_t - 2x_{t-1} + x_{t-2}$. You can also combine the time-series operators so that LD.x is the lag of the first difference of x (that is, $x_{t-1} - x_{t-2}$) and refers to the same expression as DL.x. The seasonal difference, S., is used to compute the difference between the value in the current period and the period one year ago. For quarterly data, S.x would generate $x_t - x_{t-4}$, and S2.x generates $x_t - x_{t-8}$.

In addition to being easy to use, time-series operators will never misclassify an observation. You could refer to a lagged value as x[_n-1] or a first difference as x[_n] - x[_n-1], but that construction is not only cumbersome but also dangerous. Consider an annual time-series dataset in which the 1981 and 1982 data are followed by the data for 1984, 1985, ..., with the 1983 data not appearing in the dataset (that is, not recorded as missing values but physically absent). The observation-number constructs above will misinterpret the lagged value of 1984 to be 1982, and the first difference for 1984 will incorrectly span the two-year gap. The time-series operators will not make this mistake. Because tsset has been used to define year as the time-series calendar variable, the lagged value or first difference for 1984 will be properly coded as missing whether the 1983 data are stored as missing in the dataset. It is therefore preferable always to use time-series operators when referring to past or future values or computing differences in a time-series dataset.

The time-series operators also provide an important benefit in the context of longitudinal or panel datasets (see [XT] **xt**) in which each observation, $x_{i,t}$, is identified with both an i and a t subscript. If those data are xtset (see [XT] **xtset**) or tsset, using the time-series operators will ensure that references will not span panels. For instance, z[_n-1] in a panel context will allow you to reference $z_{1,T}$ (the last observation of panel 1) as the prior value of $z_{2,1}$ (the first observation of panel 2). In contrast, L.z (or D.z) will never span panel boundaries. Panel data should always be xtset or tsset, and any time-series references should make use of the time-series operators.

A.5 Factor variables and operators

Stata allows you to use any integer variable with values in the range (0, 32740) as a factor variable (see [U] **11.4.3 Factor variables**). Such a variable, when preceded with the i. operator, will produce a set of indicator (0/1) variables for all but the base level of the factor variable. The base level is usually the lowest observed value of the variable, but it may be specified as a particular value or the largest value. Just as the time-series operators create specified variables "on the fly", the factor variable operator will cause the set of indicator variables to be temporarily created and usable in most estimation and postestimation commands.

One of the most common uses of factor variables is in the construction of interaction terms. For instance, consider the variables ratify, indicating whether an environmental treaty on pollution abatement has been ratified and taking on two values (1, 2), and corruption, an index of corruption taking on five values which need not be consecutive integers, for example, (11, 21, 32, 44, 56). The expression i.ratify#i.corruption specifies the full set of interactions between the nonbase-level values, which in this case is the interaction of ratify=2 with corruption=(21, 32, 44, 56). The interaction operator # separates the two factor variables and may be repeated. For instance, we might have i.ratify#i.corruption#i.ets, where ets indicates whether an emissions trading scheme is in place.

If an interaction with a continuous variable is required, the c. operator must be used to specify that the variable is to be treated as continuous rather than as a set of indicators. The expression i.corruption#c.gdp would produce a set of interaction terms and allow the effects of gross domestic product to differ by level of corruption in a regression context. The c. operator can also be used to specify interactions among continuous variables or indeed powers of continuous variables; for instance, c.gdp#c.co2 or c.gdp#c.gdp. Using this notation, rather than creating extra variables in the dataset, is essential for the computation of marginal means, predictive margins, and marginal effects via the margins (see [R] **margins**) command.

The factorial interaction operator, ##, can be used as a shorthand to specify a variable list containing both main effects and interactions. For instance, i.ratify##c.gdp would be expanded to the *varlist* 2.ratify gdp 2.ratify × gdp.

A.6 Circular variables

As discussed in chapter 2, wind direction is measured in compass degrees, $0° - 360°$. It is an example of a *circular statistic* (Cox 2009) and must therefore be treated with care. Although it is a continuous variable, it would make no sense to use its value (in terms of degrees of a circle) without recognizing the innate nature of the measurement.[20]

The wind direction variable in the Cerrillos dataset has been transformed by applying Nicholas Cox's (2004) command `circsummarize` from the `circular` package, available from the SSC Archive. To collapse the hourly Cerrillos data to weekly averages, we use the following code:

```
. generate days = dofc(datevec)
. format days %td
. summarize days, mean
. local ll = `r(min)´
. local ul = `r(max)´
. generate winddir_mu = .
(52,584 missing values generated)
. forvalues w=`ll´/`ul´ {
  2.         quietly circsummarize dv if days ==`w´, ci
  3.         quietly replace winddir_mu = `r(vecmean)´ if days ==`w´
  4. }
. collapse co temperature vv pm25 winddir_mu, by(days)
. rename winddir_mu dv
. tsset days, daily
        time variable:  days, 01jan2009 to 31dec2014
                delta:  1 day
. capture drop winddir
. generate winddir = int((dv-1)/45) + 1
```

In the hourly dataset, `datevec` is a Stata "clocktime" variable, which is a numeric variable that must be stored as a `double` that contains both the calendar date and time of day. We first compute the calendar day corresponding to each observation with the `dofc()` function and then compute the week in which that day falls using the `wofd()` function. See [D] **Datetime** for further details on date and time conversions.

The `circsummarize` command is "byable", but it returns the result only for the last element of the by-list. To store two of its return values for each week, use a `forvalues` loop over calendar weeks to execute that command for each week in turn. We are then ready to use the `collapse` command, with its default statistical function of the group mean, to generate a weekly version of the dataset for the variables we will use. The `tsset` command is then used to define the data as a weekly time series.

---

20. For instance, the simple average of a wind direction of 10° and 350° is 180°, representing almost precisely the opposite direction!

References

Abrigo, M. R. M., and I. Love. 2016. Estimation of panel vector autoregression in Stata. *Stata Journal* 16: 778–804. https://doi.org/10.1177/1536867X1601600314.

Akaike, H. 1974. A new look at the statistical model identification. *IEEE Transactions on Automatic Control* 19: 716–723. https://doi.org/10.1109/TAC.1974.1100705.

———. 1976. Canonical correlation analysis of time series and the use of an information criterion. In *System Identification: Advances and Case Studies*, ed. R. K. Mehra and D. G. Lainiotis, 27–96. New York: Academic Press. https://doi.org/10.1016/S0076-5392(08)60869-3.

Alogoskoufis, G., and R. Smith. 1991. On error correction models: Specification, interpretation, estimation. *Journal of Economic Surveys* 5: 97–128. https://doi.org/10.1111/j.1467-6419.1991.tb00128.x.

Anderson, T. W., and C. Hsiao. 1981. Estimation of dynamic models with error components. *Journal of the American Statistical Association* 76: 598–606. https://doi.org/10.2307/2287517.

Andrews, D. W. K., and P. Guggenberger. 2003. A bias-reduced log-periodogram regression estimator for the long-memory parameter. *Econometrica* 71: 675–712. https://doi.org/10.1111/1468-0262.00420.

Angrist, J. D., and J.-S. Pischke. 2009. *Mostly Harmless Econometrics: An Empiricist's Companion*. Princeton, NJ: Princeton University Press.

Anselin, L. 1988. *Spatial Econometrics: Methods and Models*. Boston: Kluwer.

———. 1995. SpaceStat, A software program for the analysis of spatial data. Unpublished manuscript, Regional Research Institute, West Virginia University.

Arellano, M., and S. Bond. 1991. Some tests of specification for panel data: Monte Carlo evidence and an application to employment equations. *Review of Economic Studies* 58: 277–297. https://doi.org/10.2307/2297968.

Arellano, M., and O. Bover. 1995. Another look at the instrumental variable estimation of error-components models. *Journal of Econometrics* 68: 29–51. https://doi.org/10.1016/0304-4076(94)01642-D.

Bacon, D. W., and D. G. Watts. 1971. Estimating the transition between two intersecting straight lines. *Biometrika* 58: 525–534. https://doi.org/10.2307/2334387.

Baillie, R. T. 1996. Long memory processes and fractional integration in econometrics. *Journal of Econometrics* 73: 5–59. https://doi.org/10.1016/0304-4076(95)01732-1.

Bates, D. M., and D. G. Watts. 1988. *Nonlinear Regression Analysis and Its Applications*. New York: Wiley. https://www.doi.org/10.1002/9780470316757.

Battaglia, F., and L. Orfei. 2002. Outlier detection and estimation in nonlinear time series. *Journal of Time Series Analysis* 26: 107–121. https://doi.org/10.1111/j.1467-9892.2005.00392.x.

Baum, C. F. 2000. kpss: Stata module to compute Kwiatkowski–Phillips–Schmidt–Shin test for stationarity. Statistical Software Components S410401, Department of Economics, Boston College. https://ideas.repec.org/c/boc/bocode/s410401.html.

———. 2003. dmariano: Stata module to calculate Diebold–Mariano comparison of forecast accuracy. Statistical Software Components S433001, Department of Economics, Boston College. https://ideas.repec.org/c/boc/bocode/s433001.html.

———. 2006. *An Introduction to Modern Econometrics Using Stata*. College Station, TX: Stata Press.

———. 2007. urcovar: Stata module to perform Elliott–Jansson test for unit roots with stationary covariates. Statistical Software Components S456863, Department of Economics, Boston College. https://ideas.repec.org/c/boc/bocode/s456863.html.

———. 2009. levpredict: Stata module to compute log-linear level predictions reducing retransformation bias. Statistical Software Components S457001, Department of Economics, Boston College. https://ideas.repec.org/c/boc/bocode/s457001.html.

———. 2016. *An Introduction to Stata Programming*. 2nd ed. College Station, TX: Stata Press.

———. 2017. fcstats: Stata module to compute time series forecast accuracy statistics. Statistical Software Components S458358, Department of Economics, Boston College. https://ideas.repec.org/c/boc/bocode/s458358.html.

Baum, C. F., S. Hurn, and K. Lindsay. 2020. Local Whittle estimation of the long-memory parameter. *Stata Journal* 20: 565–583. https://doi.org/10.1177/1536867X20953569.

Baum, C. F., and T. Rõõm. 2000. lomodrs: Stata module to perform Lo R/S test for long range dependence in timeseries. Statistical Software Components S412601, Department of Economics, Boston College. https://ideas.repec.org/c/boc/bocode/s412601.html.

———. 2001. sts18: A test for long-range dependence in a time series. *Stata Technical Bulletin* 60: 2–3. Reprinted in *Stata Technical Bulletin Reprints*. Vol. 10, pp. 37–39. College Station, TX: Stata Press.

Baum, C. F., and M. E. Schaffer. 2004. actest: Stata module to perform Cumby–Huizinga general test for autocorrelation in time series. Statistical Software Components S457668, Department of Economics, Boston College. https://ideas.repec.org/c/boc/bocode/s457668.html.

Baum, C. F., M. E. Schaffer, and S. Stillman. 2003. Instrumental variables and GMM: Estimation and testing. *Stata Journal* 3: 1–31. https://doi.org/10.1177/1536867X0300300101.

———. 2007. Enhanced routines for instrumental variables/generalized method of moments estimation and testing. *Stata Journal* 7: 465–506. https://doi.org/10.1177/1536867X0800700402.

Baum, C. F., and V. Wiggins. 2000. sts16: Tests for long memory in a time series. *Stata Technical Bulletin* 57: 39–44. Reprinted in *Stata Technical Bulletin Reprints*. Vol. 10, pp. 362–368. College Station, TX: Stata Press.

Beran, J. 1994. *Statistics for Long-Memory Processes*. New York: Chapman & Hall/CRC.

Berndt, E. R., B. H. Hall, R. E. Hall, and J. A. Hausman. 1974. Estimation and inference in nonlinear structural models. In *Annals of Economic and Social Measurement, Volume 3, number 4*, ed. S. V. Berg, 653–665. Cambridge, MA: National Bureau of Economic Research.

Bloomfield, P. 1992. Trends in global temperatures. *Climate Change* 21: 1–16. https://doi.org/10.1007/BF00143250.

Blundell, R., and S. Bond. 1998. Initial conditions and moment restrictions in dynamic panel data models. *Journal of Econometrics* 87: 115–143. https://doi.org/10.1016/S0304-4076(98)00009-8.

Bollerslev, T. 1986. Generalized autoregressive conditional heteroskedasticity. *Journal of Econometrics* 31: 307–327. https://doi.org/10.1016/0304-4076(86)90063-1.

Boot, T., and A. Pick. 2018. Optimal forecasts from Markov switching models. *Journal of Business & Economic Statistics* 36: 628–642. https://doi.org/10.1080/07350015.2016.1219264.

Borsky, S., and P. A. Raschky. 2015. Intergovernmental interaction in compliance with an International Environmental Agreement. *Journal of the Association of Environmental and Resource Economists* 2: 161–203. https://doi.org/10.1086/679666.

Box, G. E. P., and G. M. Jenkins. 1970. *Time Series Analysis: Forecasting and Control*. San Francisco: Holden Day.

Breusch, T. S. 1978. Testing for autocorrelation in dynamic linear models. *Australian Economic Papers* 17: 334–355. https://doi.org/10.1111/j.1467-8454.1978.tb00635.x.

Breusch, T. S., and A. R. Pagan. 1979. A simple test for heteroscedasticity and random coefficient variation. *Econometrica* 47: 1287–1294. https://doi.org/10.2307/1911963.

Brockwell, P. J., and R. A. Davis. 1991. *Time Series: Theory and Methods.* 2nd ed. New York: Springer.

Cameron, A. C., and P. K. Trivedi. 2010. *Microeconometrics Using Stata.* Rev. ed. College Station, TX: Stata Press.

———. 2013. *Regression Analysis of Count Data.* 2nd ed. Cambridge: Cambridge University Press.

Campbell, S. D., and F. X. Diebold. 2005. Weather forecasting for weather derivatives. *Journal of the American Statistical Association* 100: 6–16. https://doi.org/10.1198/016214504000001051.

Caporale, G. M., and L. A. Gil-Alana. 2004a. Fractional cointegration and real exchange rates. *Review of Financial Economics* 13: 327–340. https://doi.org/10.1016/j.rfe.2003.12.001.

———. 2004b. Fractional cointegration and tests of present value models. *Review of Financial Economics* 13: 245–258. https://doi.org/10.1016/j.rfe.2003.09.009.

Carson, R. T. 2010. The environmental Kuznets curve: Seeking empirical regularity and theoretical structure. *Review of Environmental Economics and Policy* 4: 3–23. https://doi.org/10.1093/reep/rep021.

Chan, K. S., and H. Tong. 1986. On estimating thresholds in autoregressive models. *Journal of Time Series Analysis* 7: 178–190. https://doi.org/10.1111/j.1467-9892.1986.tb00501.x.

Cifuentes, L. 2010. Relación de la Norma de Calidad Primaria PM2.5 con la Norma de Calidad Primaria de PM10. Technical report, Comision Nacional del Medio Ambiente, Region Metropolitana de Santiago.

Clemente, J., A. Montañés, and M. Reyes. 1998. Testing for a unit root in variables with a double change in the mean. *Economics Letters* 59: 175–182. https://doi.org/10.1016/S0165-1765(98)00052-4.

CME Group. 2015. CBOT soybeans vs. DCE soybean mean and soybean oil—crush spread. https://www.cmegroup.com/trading/agricultural/files/pm374-cbot-soybeans-vs-dce-soybean-meal-and-soybean-oil.pdf.

Cook, R. D., and S. Weisberg. 1983. Diagnostics for heteroskedasticity in regression. *Biometrika* 70: 1–10. https://doi.org/10.2307/2335938.

Cox, N. J. 2002a. texteditors: Stata module—some notes on text editors for Stata users. Statistical Software Components S423801, Department of Economics, Boston College. https://ideas.repec.org/c/boc/bocode/s423801.html.

———. 2002b. tsspell: Stata module for identification of spells or runs in time series. Statistical Software Components S426901, Department of Economics, Boston College. https://ideas.repec.org/c/boc/bocode/s426901.html.

———. 2002c. Speaking Stata: How to face lists with fortitude. *Stata Journal* 2: 202–222. https://doi.org/10.1177/1536867X0200200208.

———. 2003. extremes: Stata module to list extreme values of a variable. Statistical Software Components S430801, Department of Economics, Boston College. http://econpapers.repec.org/software/bocbocode/s430801.htm.

———. 2004. circular: Stata module for circular statistics. Statistical Software Components S436601, Department of Economics, Boston College. https://ideas.repec.org/c/boc/bocode/s436601.html.

———. 2006. Stata tip 33: Sweet sixteen: Hexadecimal formats and precision problems. *Stata Journal* 6: 282–283. https://doi.org/10.1177/1536867X0600600211.

———. 2009. To the vector belong the spoils: Circular statistics in Stata. UK Stata Users Group meeting proceedings. https://ideas.repec.org/p/boc/usug09/04.html.

Cumby, R. E., and J. Huizinga. 1992. Testing the autocorrelation structure of disturbances in ordinary least squares and instrumental variables regressions. *Econometrica* 60: 185–195. https://doi.org/10.2307/2951684.

Davidson, J. E. H., D. F. Hendry, F. Srba, and S. Yeo. 1978. Econometric modelling of the aggregate time-series relationship between consumer's expenditure and income in the United Kingdom. *Economic Journal* 88: 661–692. https://doi.org/10.2307/2231972.

Davies, R. B. 1987. Hypothesis testing when a nuisance parameter is present only under the alternative. *Biometrika* 74: 33–44. https://doi.org/10.2307/2336019.

Davies, R. B., and H. T. Naughton. 2014. Cooperation in environmental policy: A spatial approach. *International Tax and Public Finance* 21: 923–954. https://doi.org/10.1007/s10797-013-9280-1.

De Jong, P. 1988. The likelihood for a state space model. *Biometrika* 75: 165–169. https://doi.org/10.2307/2336450.

———. 1991. The diffuse Kalman filter. *Annals of Statistics* 19: 1073–1083. https://doi.org/10.1214/aos/1176348139.

Dempster, A. P., N. M. Laird, and D. B. Rubin. 1977. Maximum likelihood from incomplete data via the EM algorithm. *Journal of the Royal Statistical Society, Series B* 39: 1–22. https://doi.org/10.1111/j.2517-6161.1977.tb01600.x.

Dickey, D. A., and W. A. Fuller. 1979. Distributions of the estimators for autoregressive time series with a unit root. *Journal of the American Statistical Association* 74: 427–431. https://doi.org/10.2307/2286348.

———. 1981. Likelihood ratio statistics for autogressive time series with a unit root. *Econometrica* 49: 1057–1072. https://doi.org/10.2307/1912517.

Diebold, F. X., and R. S. Mariano. 1995. Comparing predictive accuracy. *Journal of Business & Economic Statistics* 13: 253–263. https://doi.org/10.1080/07350015.1995.10524599.

Dolado, J. J., T. Jenkinson, and S. Sosvilla-Riverso. 1990. Cointegration and unit roots. *Journal of Economic Surveys* 4: 249–273. https://doi.org/10.1111/j.1467-6419.1990.tb00088.x.

Doornik, J. A. 2009. Autometrics. In *The Methodology and Practice of Econometrics: A Festschrift in Honour of David F. Hendry*, ed. J. L. Castle and N. Shephard, 88–121. Oxford: Oxford University Press.

Doornik, J. A., and H. Hansen. 2008. An omnibus test for univariate and multivariate normality. *Oxford Bulletin of Economics and Statistics* 70: 927–939. https://doi.org/10.1111/j.1468-0084.2008.00537.x.

Drukker, D. M., H. Peng, I. R. Prucha, and R. Raciborski. 2013. Creating and managing spatial-weighting matrices with the spmat command. *Stata Journal* 13: 242–286. https://doi.org/10.1177/1536867X1301300202.

Durbin, J. 1959. Efficient estimation of parameters in moving-average models. *Biometrika* 46: 306–316. https://doi.org/10.2307/2333528.

Elliott, G., and M. Jansson. 2003. Testing for unit roots with stationary covariates. *Journal of Econometrics* 115: 75–89. https://doi.org/10.1016/S0304-4076(03)00093-9.

Elliott, G., T. J. Rothenberg, and J. H. Stock. 1996. Efficient tests for an autoregressive unit root. *Econometrica* 64: 813–836. https://doi.org/10.2307/2171846.

Engle, R. F. 1982. Autoregressive conditional heteroskedasticity with estimates of the variance of United Kingdom inflation. *Econometrica* 50: 987–1008. https://doi.org/10.2307/1912773.

———. 2002. Dynamic conditional correlation: A simple class of multivariate generalized autoregressive conditional heteroskedasticity models. *Journal of Business & Economic Statistics* 20: 339–350. https://doi.org/10.1198/073500102288618487.

Engle, R. F., and C. W. J. Granger. 1987. Cointegration and error correction: Representation, estimation, and testing. *Econometrica* 55: 251–276. https://doi.org/10.2307/1913236.

Florax, R. J. G. M., H. Folmer, and S. J. Rey. 2003. Specification searches in spatial econometrics: The relevance of Hendry's methodology. *Regional Science & Urban Economics* 33: 557–579. https://doi.org/10.1016/S0166-0462(03)00002-4.

Frances, P. H., and N. Haldrup. 1994. The effects of additive outliers on tests for unit roots and cointegration. *Journal of Business & Economic Statistics* 12: 471–478. https://doi.org/10.2307/1392215.

Fredriksson, P. G., and D. L. Millimet. 2002. Strategic interaction and the determination of environmental policy across U.S. states. *Journal of Urban Economics* 51: 101–122. https://doi.org/10.1006/juec.2001.2239.

Fredriksson, P. G., E. Neumayer, and G. Ujhelyi. 2007. Kyoto Protocol cooperation: Does government corruption facilitate environmental lobbying? *Public Choice* 133: 231–251. https://doi.org/10.1007/s11127-007-9187-4.

Geweke, J., and S. Porter-Hudak. 1983. The estimation and application of long memory time series models. *Journal of Time Series Analysis* 4: 221–238. https://doi.org/10.1111/j.1467-9892.1983.tb00371.x.

Ghaddar, D. K., and H. Tong. 1981. Data transformation and self-exciting threshold autoregression. *Journal of the Royal Statistical Society, Series C* 30: 238–248. https://doi.org/10.2307/2346347.

Giacomini, R., and B. Rossi. 2010. Forecast comparisons in unstable environments. *Journal of Applied Econometrics* 25: 595–620. https://doi.org/10.1002/jae.1177.

Gilbert, C. L. 1986. Professor Hendry's econometric methodology. *Oxford Bulletin of Economics and Statistics* 48: 283–307. https://doi.org/10.1111/j.1468-0084.1986.mp48003007.x.

Glosten, L. R., R. Jagannathan, and D. E. Runkle. 1993. On the relation between the expected value and the volatility of the nominal excess return on stocks. *Journal of Finance* 48: 1779–1801. https://doi.org/10.1111/j.1540-6261.1993.tb05128.x.

Godfrey, L. G. 1978. Testing against general autoregressive and moving average error models when the regressors include lagged dependent variables. *Econometrica* 46: 1293–1301. https://doi.org/10.2307/1913829.

Goldfeld, S. M., and R. E. Quandt. 1972. *Nonlinear Methods in Econometrics*. Amsterdam: North-Holland.

Gould, W. 2006. Mata Matters: Precision. *Stata Journal* 6: 550–560. https://doi.org/10.1177/1536867X0600600407.

Granger, C. W. J. 1969. Investigating causal relations by econometric models and cross-spectral methods. *Econometrica* 37: 424–438. https://doi.org/10.2307/1912791.

Granger, C. W. J., and A. P. Anderson. 1978. *An Introduction to Bilinear Time Series Models*. Gottingen: Vandenhoeck & Ruprecht.

Granger, C. W. J., and R. Joyeux. 1980. An introduction to long-memory time series models and fractional differencing. *Journal of Time Series Analysis* 1: 15–29. https://doi.org/10.1111/j.1467-9892.1980.tb00297.x.

Granger, C. W. J., and R. Ramanathan. 1984. Improved methods for combining forecasts. *Journal of Forecasting* 3: 197–204. https://doi.org/10.1002/for.3980030207.

Grassi, S., E. Hillebrand, and D. Ventosa-Santaulària. 2013. The statistical relation of sea-level and temperature revisited. *Dynamics of Atmospheres and Oceans* 64: 1–9. https://doi.org/10.1016/j.dynatmoce.2013.07.001.

Gregory, A. W., and B. E. Hansen. 1996a. Tests for cointegration in models with regime and trend shifts. *Oxford Bulletin of Economics and Statistics* 58: 555–560. https://doi.org/10.1111/j.1468-0084.1996.mp58003008.x.

———. 1996b. Residual-based tests for cointegration in models with regime shifts. *Journal of Econometrics* 70: 99–126. https://doi.org/10.1016/0304-4076(69)41685-7.

Grossman, G. M., and A. B. Krueger. 1993. Environmental impacts of a North American Free Trade Agreement. In *The Mexico-U.S. Free Trade Agreement*, ed. P. M. Garber, 13–56. Cambridge, MA: MIT Press.

———. 1995. Economic growth and the environment. *Quarterly Journal of Economics* 110: 353–377. https://doi.org/10.2307/2118443.

Haggan, V., and T. Ozaki. 1981. Modelling nonlinear random vibrations using an amplitude-dependent autoregressive time series model. *Biometrika* 68: 189–196. https://www.doi.org/10.2307/2335819.

Hall, A. D., J. Skalin, and T. Teräsvirta. 2001. A nonlinear time series model of El Niño. *Environmental Modelling & Software* 16: 139–146. https://doi.org/10.1016/S1364-8152(00)00077-3.

Hamilton, J. D. 1989. A new approach to the economic analysis of nonstationary time series and the business cycle. *Econometrica* 57: 357–384. https://doi.org/10.2307/1912559.

———. 1994. *Time Series Analysis*. Princeton, NJ: Princeton University Press.

Hannan, E. J. 1980. The estimation of the order of an ARMA process. *Annals of Statistics* 8: 1071–1081. https://doi.org/10.1214/aos/1176345144.

Hannan, E. J., and B. G. Quinn. 1979. The determination of the order of an autoregression. *Journal of the Royal Statistical Society, Series B* 41: 190–195. https://doi.org/10.1111/j.2517-6161.1979.tb01072.x.

Hansen, B. E. 1995. Rethinking the univariate approach to unit root testing: Using covariates to increase power. *Econometric Theory* 11: 1148–1171. https://doi.org/10.1017/S0266466600009993.

———. 1996. Inference when a nuisance parameter is not identified under the null hypothesis. *Econometrica* 64: 413–430. https://doi.org/10.2307/2171789.

Hansen, J., R. Ruedy, M. Sato, and K. Lo. 2010. Global surface temperature change. *Reviews of Geophysics* 48(4). https://doi.org/10.1029/2010RG000345.

Hansen, L. P. 1982. Large sample properties of generalised method of moments estimators. *Econometrica* 50: 1029–1054. https://doi.org/10.2307/1912775.

Hansen, P. R., and A. Lunde. 2005. A forecast comparison of volatility models: Does anything beat a GARCH(1,1)? *Journal of Applied Econometrics* 20: 873–889. https://doi.org/10.1002/jae.800.

Hardin, J. W., and J. M. Hilbe. 2018. *Generalized Linear Models and Extensions*. 4th ed. College Station, TX: Stata Press.

Harris, D., D. I. Harvey, S. J. Leybourne, and A. M. R. Taylor. 2009. Testing for a unit root in the presence of a possible break in trend. *Econometric Theory* 25: 1545–1588. https://doi.org/10.1017/S0266466609990259.

Harvey, A. C. 1989. *Forecasting, Structural Time Series Models and the Kalman Filter*. Cambridge: Cambridge University Press.

Harvey, A. C., and S. Thiele. 2016. Testing against changing correlation. *Journal of Empirical Finance* 38: 575–589. https://doi.org/10.1016/j.jempfin.2015.09.003.

Harvey, D. I., S. J. Leybourne, and A. M. R. Taylor. 2009. Unit root testing in practice: Dealing with uncertainty over trend and initial condition. *Econometric Theory* 25: 587–636. https://doi.org/10.1017/S026646660809018X.

Haslett, J., and A. E. Raftery. 1989. Space-time modelling with long memory dependence: Assessing Ireland's wind power resource. *Applied Statistics* 38: 1–50. https://doi.org/10.2307/2347679.

Hausman, J. A. 1978. Specification tests in econometrics. *Econometrica* 46: 1251–1271. https://doi.org/10.2307/1913827.

Hayashi, F. 2000. *Econometrics*. Princeton, NJ: Princeton University Press.

Hendry, D. F. 1979. Predictive failure and econometric modelling in macroeconomics: The transactions demand for money. In *Economic Modelling: Current Issues and Problems in Macroeconomic Modelling in the UK and the US*, ed. P. Ormerod, 217–242. London: Heinemann.

———. 1995. *Dynamic Econometrics*. Oxford: Oxford University Press.

Hendry, D. F., and G. J. Anderson. 1977. Testing dynamic specfications in small simultaneous systems: An application to a model of building society behaviour in the United Kingdom. In *Frontiers of Quantitative Economics*, ed. M. D. Intriligator, 361–383. Amsterdam: North-Holland.

Hodrick, R. J., and E. C. Prescott. 1997. Postwar U.S. business cycles: An empirical investigation. *Journal of Money, Credit and Banking* 24: 1–16. https://doi.org/10.2307/2953682.

Holtz-Eakin, D., W. K. Newey, and H. S. Rosen. 1988. Estimating vector autoregressions with panel data. *Econometrica* 56: 1371–1395. https://doi.org/10.2307/1913103.

Hosking, J. R. M. 1981. Fractional differencing. *Biometrika* 68: 165–176. https://doi.org/10.1093/biomet/68.1.165.

Hoti, S., M. McAleer, and L. L. Pauwels. 2007. Measuring risk in environmental finance. *Journal of Economic Surveys* 21: 970–998. https://doi.org/10.1111/j.1467-6419.2007.00526.x.

Hualde, J., and P. M. Robinson. 2007. Root-n-consistent estimation of weak fractional cointegration. *Journal of Econometrics* 140: 450–484. https://doi.org/10.1016/j.jeconom.2006.07.004.

Hurn, S., V. L. Martin, P. C. B. Phillips, and J. Yu. 2020. *Financial Econometric Modeling*. New York: Oxford University Press.

Hurst, H. 1951. Long term storage capacity of reservoirs. *Transactions of the American Society of Civil Engineers* 116: 770–799.

Jackson, J. A., A. Zerbini, P. Clapham, R. Constantine, C. Garrigue, N. Hauser, M. M. Poole, and C. S. Baker. 2008. Progress on a two-stock catch allocation model for reconstructing population histories of east Australia and Oceania. Technical Report No. SC/60/SH14, International Whaling Commission.

Jaffe, A. B., S. R. Peterson, P. R. Portney, and R. N. Stavins. 1995. Environmental regulation and the competitiveness of U.S. manufacturing: What does the evidence tell us? *Journal of Economic Literature* 33: 132–163.

Jann, B. 2004. center: Stata module to center (or standardize) variables. Statistical Software Components S444102, Department of Economics, Boston College. https://ideas.repec.org/c/boc/bocode/s444102.html.

Jarque, C. M., and A. K. Bera. 1987. A test for normality of observations and regression residuals. *International Statistical Review* 2: 163–172. https://doi.org/10.2307/1403192.

Johansen, S. 1988. Statistical analysis of cointegration vectors. *Journal of Economic Dynamics & Control* 12: 231–254. https://doi.org/10.1016/0165-1889(88)90041-3.

———. 1991. Estimation and hypothesis testing of cointegration vectors in Gaussian vector autoregressive models. *Econometrica* 59: 1551–1580. https://doi.org/10.2307/2938278.

———. 1995. *Likelihood-Based Inference in Cointegrated Vector Autoregressive Models*. Oxford: Oxford University Press.

Johansen, S., R. Mosconi, and B. Nielsen. 2000. Cointegration analysis in the presence of structural breaks and deterministic trend. *Econometrics Journal* 3: 216–249. https://doi.org/10.1111/1368-423X.00047.

Kaufmann, R. K., and D. I. Stern. 1997. Evidence for human influence on climate from hemispheric temperature relations. *Nature* 388: 39–44. https://doi.org/10.1038/40332.

Keller, W., and A. Levinson. 2002. Pollution abatement costs and foreign direct investment inflows to U.S. states. *Review of Economics and Statistics* 84: 691–703. https://doi.org/10.1162/003465302760556503.

Khan, H., C. R. Knittel, K. Metaxoglou, and M. Papineau. 2015. How do carbon emissions respond to business-cycle shocks? Unpublished manuscript.

Kolev, G. I. 2006. Stata tip 31: Scalar or variable? The problem of ambiguous names. *Stata Journal* 6: 279–280. https://doi.org/10.1177/1536867X0600600209.

Kuznets, S. 1955. Economic growth and income inequality. *American Economic Review* 45: 1–28.

Kwiatkowski, D. P., P. C. B. Phillips, P. Schmidt, and Y. Shin. 1992. Testing the null hypothesis of stationarity against the alternative of a unit root: How sure are we that economic series have a unit root? *Journal of Econometrics* 54: 159–178. https://doi.org/10.1016/0304-4076(92)90104-Y.

Lanne, M., and M. Liski. 2004. Trends and breaks in per-capita carbon dioxide emissions, 1870–2028. *Energy Journal* 25(4): 41–65. https://doi.org/10.5547/ISSN0195-6574-EJ-Vol25-No4-3.

Lee, T.-H., H. White, and C. W. J. Granger. 1993. Testing for neglected nonlinearity in time-series models: A comparison of neural network methods and standard tests. *Journal of Econometrics* 56: 269–290. https://doi.org/10.1016/0304-4076(93)90122-L.

LeSage, J., and R. K. Pace. 2009. *Introduction to Spatial Econometrics*. Boca Raton, FL: Chapman & Hall/CRC.

Li, M., S. J. Koopman, R. Lit, and D. Petrova. 2020. Long-term forecasting of El Niño events via dynamic factor simulations. *Journal of Econometrics* 214: 46–66. https://doi.org/10.1016/j.jeconom.2019.05.004.

Ljung, G. M., and G. E. P. Box. 1978. On a measure of lack of fit in time series models. *Biometrika* 65: 297–303. https://doi.org/10.2307/2335207.

Lo, A. W. 1991. Long-term memory in stock market prices. *Econometrica* 59: 1279–1313. https://doi.org/10.2307/2938368.

Long, L. S., and J. Freese. 2000. fitstat: Stata module to compute fit statistics for single equation regression models. Statistical Software Components S407201, Department of Economics, Boston College. https://ideas.repec.org/c/boc/bocode/s407201.html.

Luukkonen, R., P. Saikkonen, and T. Teräsvirta. 1988. Testing linearity against smooth transition autoregressive models. *Biometrika* 75: 491–499. https://doi.org/10.2307/2336599.

Maddala, G. S. 1992. *Introduction to Econometrics*. 2nd ed. New York: Macmillan.

Mandelbrot, B. 1972. Statistical methodology for nonperiodic cycles: From the covariance to R/S analysis. *Annals of Economic and Social Measurement* 1: 259–290.

Mann, H. B., and A. Wald. 1943. On the statistical treatment of linear stochastic difference equations. *Econometrica* 11: 173–220. https://doi.org/10.2307/1905674.

Martin, V. L., S. Hurn, and D. Harris. 2013. *Econometric Modelling with Time Series: Specification, Estimation and Testing*. New York: Cambridge University Press.

Mayer, T., and S. Zignago. 2011. Notes on CEPII's distances measures: the GeoDist database. CEPII Working Paper No. 2011-25. http://www.cepii.fr/PDF_PUB/wp/2011/wp2011-25.pdf.

McCracken, J. P., J. Schwartz, A. Diaz, N. Bruce, and K. R. Smith. 2013. Longitudinal relationship between personal CO and personal PM2.5 among women cooking with woodfired cookstoves in Guatemala. *PLOS ONE* 8: e55670. https://doi.org/10.1371/journal.pone.0055670.

MMA. 2011. Norma primaria de calidad ambiental para material particulado fino respirable MP2,5. Reporte Oficial. Technical report, Ministerio del Medio Ambiente.

Motanari, A., R. Rosso, and M. S. Taqqu. 1996. Some long-run properties of rainfall records in Italy. *Journal of Geophyiscal Research* 101: 431–438. https://doi.org/10.1029/96JD02512.

Mundlak, Y. 1978. On the pooling of time series and cross section data. *Econometrica* 46: 69–85. https://doi.org/10.2307/1913646.

Murdoch, J. C., T. Sandler, and W. P. M. Vijverberg. 2003. The participation decision versus the level of participation in an environmental treaty: A spatial probit analysis. *Journal of Public Economics* 87: 337–362. https://doi.org/10.1016/S0047-2727(01)00152-9.

Muscatelli, V. A., and S. Hurn. 1992. Cointegration and dynamic time sereis models. *Journal of Economic Surveys* 6: 1–43. https://doi.org/10.1111/j.1467-6419.1992.tb00142.x.

Nelson, D. B. 1991. Conditional heteroskedasticity in asset returns: A new approach. *Econometrica* 59: 347–370. https://doi.org/10.2307/2938260.

Newey, W. K., and K. D. West. 1987. A simple, positive semi-definite, heteroscedasticity and autocorrelation consistent covariance matrix. *Econometrica* 55: 703–708. https://doi.org/10.2307/1913610.

Ng, S., and P. Perron. 2001. Lag length selection and the construction of unit root tests with good size and power. *Econometrica* 69: 1519–1554. https://doi.org/10.1111/1468-0262.00256.

Nickell, S. 1981. Biases in dynamic models with fixed effects. *Econometrica* 49: 1417–1426. https://doi.org/10.2307/1911408.

Nielsen, M. Ø. 2007. Local Whittle analysis of stationary fractional cointegration and the implied-realized volatility relation. *Journal of Business & Economic Statistics* 25: 427–446. https://doi.org/10.1198/073500106000000314.

Ord, K. 1975. Estimation methods for models of spatial interaction. *Journal of the American Statistical Association* 70: 120–126. https://doi.org/10.2307/2285387.

Ouliaris, S., A. R. Pagan, and J. Restrepo. 2016. Quantitative macroeconomic modeling with structural vector autoregressions—An EViews implementation. http://www.eviews.com/StructVAR/structvar.html.

Percival, D., J. Overland, and H. Morfjeld. 2001. Interpretation of North Pacific variability as a short and long memory process. Technical Report Series No. 065, NRCSE.

Pérez, J. E. P. 2002. ghansen: Stata module to perform Gregory–Hansen test for cointegration with regime shifts. Statistical Software Components S457327, Department of Economics, Boston College. https://ideas.repec.org/c/boc/bocode/s457327.html.

Perron, P. 1990. Testing for a unit root in a time series with a changing mean. *Journal of Business & Economic Statistics* 8: 153–162. https://doi.org/10.2307/1391977.

Perron, P., and T. J. Vogelsang. 1992. Nonstationarity and level shifts with an application to purchasing power parity. *Journal of Business & Economic Statistics* 10: 301–320. https://doi.org/10.2307/1391544.

———. 1993. Erratum: The Great Crash, the oil price shock, and the unit root hypothesis. *Econometrica* 61: 248–249. https://doi.org/10.2307/2951792.

Phillips, P. C. B. 1999. Discrete fourier transforms of fractional processes. Discussion Paper No. 1243, Cowles Foundation. https://papers.ssrn.com/sol3/papers.cfm?abstract_id=216308.

———. 2007. Unit root log periodogram regression. *Journal of Econometrics* 138: 104–124. https://doi.org/10.1016/j.jeconom.2006.05.017.

Phillips, P. C. B., and B. E. Hansen. 1990. Statistical inference in instrumental variables regressions with I(1) errors. *Review of Economic Studies* 57: 99–125. https://doi.org/10.2307/2297545.

Phillips, P. C. B., and P. Perron. 1988. Testing for a unit root in time series regression. *Biometrika* 75: 335–346. https://doi.org/10.2307/2336182.

Pitcher, T. J., D. Kalikoski, K. Short, D. Varkey, and P. Ganapathiraju. 2009. An evaluation of progress in implemting ecosystem-based management of fisheries in 33 countries. *Marine Policy* 33: 223–232. https://doi.org/10.1016/j.marpol.2008.06.002.

Poskitt, D. S., and A. R. Tremayne. 1986. The selection and use of linear and bilinear time series models. *International Journal of Forecasting* 2: 101–114. https://doi.org/10.1016/0169-2070(86)90033-6.

Pretis, F. 2020. Econometric models of climate systems: The equivalence of energy balance models and cointegrated vector autoregressions. *Journal of Econometrics* 214: 256–273. https://doi.org/10.1016/j.jeconom.2019.05.013.

Quandt, R. E. 1958. The estimation of the parameters of a linear regression system obeying two separate regimes. *Journal of the American Statistical Association* 53: 873–880. https://doi.org/10.2307/2281957.

Rahmstorf, S. 2007. A semi-empirial approach to projecting future sea-level rise. *Science* 315: 368–370. https://doi.org/10.1126/science.1135456.

Ramsey, J. B. 1969. Tests for specification errors in classical linear least-squares regression analysis. *Journal of the Royal Statistical Society, Series B* 31: 350–371. https://doi.org/10.1111/j.2517-6161.1969.tb00796.x.

Robinson, P. M. 1995. Log-periodogram regression of time series with long range dependence. *Annals of Statistics* 23: 1048–1072. https://doi.org/10.1214/aos/1176324636.

Robinson, P. M., and D. Marinucci. 2001. Narrow-band analysis of nonstationary processes. *Annals of Statistics* 29: 947–986. https://doi.org/10.1214/aos/1013699988.

Roodman, D. 2009. How to do xtabond2: An introduction to difference and system GMM in Stata. *Stata Journal* 9: 86–136. https://doi.org/10.1177/1536867X0900900106.

Rossi, B., and T. Sekhposyan. 2016. Forecast rationality tests in the presence of instabilities, with applications to Federal Reserve and survey forecasts. *Journal of Applied Econometrics* 31: 507–532. https://doi.org/10.1002/jae.2440.

Rossi, B., and M. Soupre. 2017. Implementing tests for forecast evaluation in the presence of instabilities. *Stata Journal* 17: 850–865. https://doi.org/10.1177/1536867X1801700405.

Saikkonen, P. 1991. Asymptotically efficient estimation of cointegration regressions. *Econometric Theory* 7: 1–21. https://doi.org/10.1017/S0266466600004217.

Saikkonen, P., and R. Luukkonen. 1988. Lagrange multiplier tests for testing non-linearities in time series models. *Scandinavian Journal of Statistics* 15: 55–68.

Sarafidis, V., and T. Wansbeek. 2020. Celebrating 40 years of panel data analysis: past, present and future. Working Paper 06/20, Monash University. https://www.monash.edu/business/ebs/research/publications/ebs/wp06-2020.pdf.

Sargan, J. D. 1964. Wages and prices in the United Kingdom: A study in econometric methodology. In *Econometric Analysis for National Economic Planning*, ed. P. Hart, G. Mills, and J. Whitaker. Vol. 16 of *Colston Papers*. London: Butterworth Co.

Schaffer, M. E. 2010. egranger: Stata module to perform Engle–Granger cointegration tests and 2-step ECM estimation. Statistical Software Components S457210, Department of Economics, Boston College. http://ideas.repec.org/c/boc/bocode/s457210.html.

Schreck, S., J. Lundquist, and W. Shaw. 2008. Reseach needs for wind resource characterization. Technical Report No. NREL/TP-500-43521, National Renewable Energy Laboratory.

Schunck, R. 2013. Within and between estimates in random-effects models: Advantages and drawbacks of correlated random effects and hybrid models. *Stata Journal* 13: 65–76. https://doi.org/10.1177/1536867X1301300105.

Schunck, R., and F. Perales. 2017. Within- and between-cluster effects in generalized linear mixed models: A discussion of approaches and the xthybrid command. *Stata Journal* 17: 89–115. https://doi.org/10.1177/1536867X1701700106.

Schwarz, G. 1978. Estimating the dimension of a model. *Annals of Statistics* 6: 461–464. https://doi.org/10.1214/aos/1176344136.

Seber, G. A. F., and C. J. Wild. 1989. *Nonlinear Regression*. New York: Wiley.

Shi, S., S. Hurn, and P. C. B. Phillips. 2020. Causal change detection in possibly integrated systems: Revisiting the money–income relationship. *Journal of Financial Econometrics* 18: 158–180. https://doi.org/10.1093/jjfinec/nbz004.

Shi, S., P. C. B. Phillips, and S. Hurn. 2018. Change detection and the causal impact of the yield curve. *Journal of Time Series Analysis* 39: 966–987. https://doi.org/10.1111/jtsa.12427.

Silvennoinen, A., and T. Teräsvirta. 2016. Testing constancy of unconditional variance in volatility models by misspecification and specification tests. *Studies in Nonlinear Dynamics & Econometrics* 20: 347–364. https://doi.org/10.1515/snde-2015-0033.

Silvennoinen, A., and S. Thorp. 2016. Crude oil and agricultural futures: an analysis of correlation dynamics. *Journal of Futures Markets* 36: 522–544. https://doi.org/10.1002/fut.21770.

Sims, C. A. 1980. Macroeconomics and reality. *Econometrica* 48: 1–48. https://doi.org/10.2307/1912017.

Sowell, F. 1992. Maximum likelihood estimation of stationary univariate fractionally integrated time series models. *Journal of Econometrics* 53: 165–188. https://doi.org/10.1016/0304-4076(92)90084-5.

Stephenson, D. B., V. Pravan, and R. Bojariu. 2000. Is the North Atlantic oscillation a random walk. *International Journal of Climatology* 20: 1–18. https://doi.org/10.1002/(SICI)1097-0088(200001)20:1⟨1::AID-JOC456⟩3.0.CO;2-P.

Stern, D. I., and R. K. Kaufmann. 2014. Anthropogenic and natural causes of climate change. *Climatic Change* 122: 257–269. https://doi.org/10.1007/s10584-013-1007-x.

Stock, J. H., and M. W. Watson. 1993. A simple estimator of cointegration vectors in higher order integrated systems. *Econometrica* 61: 783–820. https://doi.org/10.2307/2951763.

Swanson, N. R. 1998. Money and output viewed through a rolling window. *Journal of Monetary Economics* 41: 455–474. https://doi.org/10.1016/S0304-3932(98)00005-1.

Tai, A. P. K., L. J. Mickley, and D. J. Jacob. 2010. Correlations between fine particulate matter ($PM_{2.5}$) and meteorological variables in the United States: Implications for the sensitivity of $PM_{2.5}$ to climate change. *Atmospheric Environment* 44: 3987–3984. https://doi.org/10.1016/j.atmosenv.2010.06.060.

Teräsvirta, T. 1994a. Testing linearity and modelling nonlinear time series. *Kybernetika* 30: 319–330.

———. 1994b. Specification, estimation and evaluation of smooth transition autoregressive models. *Journal of the American Statistical Association* 89: 208–218. https://doi.org/10.2307/2291217.

Teräsvirta, T., C.-F. Lin, and C. W. J. Granger. 1993. Power of the neural network linearity test. *Journal of Time Series Analysis* 14: 209–220. https://doi.org/10.1111/j.1467-9892.1993.tb00139.x.

Theil, H. 1966. *Applied Economic Forecasting*. New York: Rand McNally.

Thoma, M. A. 1994. Subsample instability and asymmetries in money-income causality. *Journal of Econometrics* 64: 279–306. https://doi.org/10.1016/0304-4076(94)90066-3.

Tobin, J. 1958. Estimation of relationships for limited dependent variables. *Econometrica* 26: 24–36. https://doi.org/10.2307/1907382.

Tong, H., and K. S. Lim. 1980. Threshold autoregression, limit cycles and cyclical data. *Journal of the Royal Statistical Society, Series B* 42: 245–292. https://doi.org/10.1111/j.2517-6161.1980.tb01126.x.

Tse, Y. K., and A. K. C. Tsui. 2002. A multivariate generalized autoregressive conditional heteroscedasticity model with time-varying correlations. *Journal of Business & Economic Statistics* 20: 351–362. https://doi.org/10.1198/073500102288618496.

Ventosa-Santaulària, D., D. R. Heres, and L. C. Martínez-Hernández. 2014. Long-memory and the sea level-temperature relationship: A fractional cointegration approach. *PLOS One* 9: e113439. https://doi.org/10.1371/journal.pone.0113439.

Volterra, V. 1930. *The Theory of Functionals and of Integro-Differential Equations.* New York: Dover.

Vuong, Q. H. 1989. Likelihood ratio tests for model selection and non-nested hypotheses. *Econometrica* 57: 307–333.

Wagner, M. 2012. The Phillips unit root tests for polynomials of integrated processes. *Economics Letters* 114: 299–303. https://doi.org/10.1016/j.econlet.2011.11.006.

———. 2015. The environmental Kuznets curve, cointegration and nonlinearity. *Journal of Applied Econometrics* 30: 948–967. https://doi.org/10.1002/jae.2421.

White, H. 1980. A heteroskedasticity-consistent covariance matrix estimator and a direct test for heteroskedasticity. *Econometrica* 48: 817–838. https://doi.org/10.2307/1912934.

White, W. B., and Z. Liu. 2008. Non-linear alignment of El Niño to the 11-yr solar cycle. *Geophysical Research Letters* 35: L19607. https://doi.org/10.1029/2008GL034831.

Whittle, P. 1951. *Hypothesis Testing in Time Series Analysis.* Uppsala, Sweden: Almqvist & Wiksells.

———. 1954. On stationary processes in the plane. *Biometrika* 41: 434–449. https://doi.org/10.2307/2332724.

———. 1962. Gaussian estimation in stationary time series. *Bulletin of the International Statistical Institute* 39: 105–129.

Wiener, N. 1958. *Nonlinear Problems in Random Theory.* London: Wiley.

Wooldridge, J. M. 2010. *Econometric Analysis of Cross Section and Panel Data.* 2nd ed. Cambridge, MA: MIT Press.

———. 2016. *Introductory Econometrics: A Modern Approach.* 6th ed. New York: Cengage Learning.

Zakoïan, J.-M. 1994. Threshold heteroskedastic models. *Journal of Economic Dynamics and Control* 18: 931–955. https://doi.org/10.1016/0165-1889(94)90039-6.

Zellner, A. 1962. An efficient method of estimating seemingly unrelated regressions and tests of aggregation bias. *Journal of the American Statistical Association* 57: 348–368. https://doi.org/10.2307/2281644.

Author index

A

Abrigo, M. R. M. 281
Akaike, H. 64, 84
Alogoskoufis, G. 131
Anderson, A. P. 203
Anderson, G. J. 131
Anderson, T. W. 280
Andrews, D. W. K. 349
Angrist, J. D. 323
Anselin, L. 283, 304
Arellano, M. 280

B

Bacon, D. W. 212
Baillie, R. T. 339
Baker, C. S. 310
Bates, D. M. 217
Battaglia, F. 205, 211
Baum, C. F. .... xxvii,
        23, 46, 49, 118, 121, 159, 170,
        344, 347, 349, 351, 361
Bera, A. K. 25
Beran, J. 195, 359
Berndt, E. R. 61
Bloomfield, P. 339
Blundell, R. 280
Bojariu, R. 339
Bollerslev, T. 232
Bond, S. 280
Boot, T. 226
Borsky, S. 284, 290, 307
Bover, O. 280
Box, G. E. P. 23, 55
Breusch, T. S. 21, 23
Brockwell, P. J. 347
Bruce, N. 17

C

Cameron, A. C. 43, 309
Campbell, S. D. 253
Caporale, G. M. 354
Carson, R. T. 147
Chan, K. S. 213
Cifuentes, L. 11
Clapham, P. 310
Clemente, J. 124, 127
CME Group 130
Constantine, R. 310
Cook, R. D. 21
Cox, N. J. 13, 52, 174, 235, 365, 366,
        375, 384
Cumby, R. E. 23

D

Davidson, J. E. H. 131
Davies, R. B. 209, 284
Davis, R. A. 347
De Jong, P. 178
Dempster, A. P. 224
Diaz, A. 17
Dickey, D. A. 112
Diebold, F. X. 159, 253
Dolado, J. J. 129
Doornik, J. A. 26, 304
Drukker, D. M. 290
Durbin, J. 60, 205

E

Elliott, G. 119, 120
Engle, R. F. 129, 133, 135, 229, 232,
        248

F

Florax, R. J. G. M. 303, 304
Folmer, H. 303, 304
Frances, P. H. 127
Fredriksson, P. G. 284, 337
Freese, J. 324
Fuller, W. A. 112

G

Ganapathiraju, P. 284
Garrigue, C. 310
Geweke, J. 347
Ghaddar, D. K. 215
Giacomini, R. 159
Gil-Alana, L. A. 354
Gilbert, C. L. 304
Glosten, L. R. 239
Godfrey, L. G. 23
Goldfeld, S. M. 212
Gould, W. 366
Granger, C. W. J. . . . 85, 129, 133, 135,
    174, 201, 203, 339
Grassi, S. 191, 193
Gregory, A. W. 143
Grossman, G. M. 77
Guggenberger, P. 349

H

Haggan, V. 217
Haldrup, N. 127
Hall, A. D. 195
Hall, B. H. 61
Hall, R. E. 61
Hamilton, J. D. 35, 177, 220, 347
Hannan, E. J. 64, 84
Hansen, B. E. 120, 136, 143, 209
Hansen, H. 26
Hansen, J. 5
Hansen, L. P. 45, 46
Hansen, P. R. 233
Hardin, J. W. 171
Harris, D. 35, 41, 127, 137, 178
Harvey, A. C. . . 177, 178, 184, 189, 251
Harvey, D. I. 120, 127
Haslett, J. 339

Hauser, N. 310
Hausman, J. A. 49, 61, 273
Hayashi, F. 15, 35
Hendry, D. F. 131, 304
Heres, D. R. 340, 355, 358
Hilbe, J. M. 171
Hillebrand, E. 191, 193
Hodrick, R. J. 78
Holtz-Eakin, D. 280, 281
Hosking, J. R. M. 339, 341, 352
Hoti, S. 229
Hsiao, C. 280
Hualde, J. 354
Huizinga, J. 23
Hurn, S. 35, 41, 87, 129, 137, 178,
    231, 351
Hurst, H. 339, 343

J

Jackson, J. A. 310
Jacob, D. J. 17
Jaffe, A. B. 257
Jagannathan, R. 239
Jann, B. 380
Jansson, M. 120
Jarque, C. M. 25
Jenkins, G. M. 23, 55
Jenkinson, T. 129
Johansen, S. 137, 146
Joyeux, R. 339

K

Kalikoski, D. 284
Kaufmann, R. K. 101
Keller, W. 255
Khan, H. 97, 100
Knittel, C. R. 97, 100
Kolev, G. I. 376
Koopman, S. J. 195
Krueger, A. B. 77
Kuznets, S. 147
Kwiatkowski, D. P. 117

L

Laird, N. M. 224
Lanne, M. 106
Lee, T-H. 203
LeSage, J. 283
Levinson, A. 255
Leybourne, S. J. 120, 127
Li, M. 195
Lim, K. S. 208, 211
Lin, C.-F. 201
Lindsay, K. 351
Liski, M. 106
Lit, R. 195
Liu, Z. 198
Ljung, G. M. 23
Lo, A. W. 344
Lo, K. 5
Long, L. S. 324
Love, I. 281
Lunde, A. 233
Lundquist, J. 152
Luukkonen, R. 214, 215

M

Maddala, G. S. 303, 304
Mandelbrot, B. 343
Mann, H. B. 77
Mariano, R. S. 159
Marinucci, D. 354
Martin, V. L. 35, 41, 137, 178, 231
Martínez-Hernández, L. C. 340, 355, 358
Mayer, T. 289
McAleer, M. 229
McCracken, J. P. 17
Metaxoglou, K. 97, 100
Mickley, L. J. 17
Millimet, D. L. 284
MMA 11
Montañés, A. 124, 127
Morfjeld, H. 339
Mosconi, R. 146
Motanari, A. 339
Mundlak, Y. 274

Murdoch, J. C. 284
Muscatelli, V. A. 129

N

Naughton, H. T. 284
Nelson, D. B. 240
Neumayer, E. 337
Newey, W. K. 28, 280, 281
Ng, S. 114
Nickell, S. 279
Nielsen, B. 146
Nielsen, M. Ø. 354

O

Ord, K. 294
Orfei, L. 205, 211
Ouliaris, S. 95
Overland, J. 339
Ozaki, T. 217

P

Pace, R. K. 283
Pagan, A. R. 21, 95
Papineau, M. 97, 100
Pauwels, L. L. 229
Peng, H. 290
Perales, F. 274, 275
Percival, D. 339
Pérez, J. E. P. 144
Perron, P. 114, 116, 121, 123, 124
Peterson, S. R. 257
Petrova, D. 195
Phillips, P. C. B. 87, 116, 117, 136, 231, 348
Pick, A. 226
Pischke, J.-S. 323
Pitcher, T. J. 284
Poole, M. M. 310
Porter-Hudak, S. 347
Portney, P. R. 257
Poskitt, D. S. 208
Pravan, V. 339
Prescott, E. C. 78
Pretis, F. 148
Prucha, I. R. 290

Q

Quandt, R. E. 208, 212
Quinn, B. G. 64, 84

R

Raciborski, R. 290
Raftery, A. E. 339
Rahmstorf, S. 193
Ramanathan, R. 174
Ramsey, J. B. 201
Raschky, P. A. 284, 290, 307
Restrepo, J. 95
Rey, S. J. 303, 304
Reyes, M. 124, 127
Robinson, P. M. 349, 354
Roodman, D. 280, 281
Rööm, T 344
Rosen, H. S. 280, 281
Rossi, B. 159, 160
Rosso, R. 339
Rothenberg, T. J. 119
Rubin, D. B. 224
Ruedy, R. 5
Runkle, D. E. 239

S

Saikkonen, P. 136, 214, 215
Sandler, T. 284
Sarafidis, V. 255
Sargan, J. D. 131
Sato, M. 5
Schaffer, M. E. 23, 46, 49, 140
Schmidt, P. 117
Schreck, S. 152
Schunck, R. 274, 275
Schwartz, J. 17
Schwarz, G. 64, 84
Seber, G. A. F. 217
Sekhposyan, T. 160
Shaw, W. 152
Shi, S. 87
Shin, Y. 117
Short, K. 284
Silvennoinen, A. 143, 251
Sims, C. A. 77, 89

Skalin, J. 195
Smith, K. R. 17
Smith, R. 131
Sosvilla-Riverso, S. 129
Soupre, M. 160
Sowell, F. 352
Srba, F. 131
Stavins, R. N. 257
Stephenson, D. B. 339
Stern, D. I. 101
Stillman, S. 46, 49
Stock, J. H. 119, 136
Swanson, N. R. 87

T

Tai, A. P. K. 17
Taqqu, M. S. 339
Taylor, A. M. R. 120, 127
Teräsvirta, T. ...201, 213, 214, 217, 219,
    220, 226, 251
Theil, H. 159
Thiele, S. 251
Thoma, M. A. 87
Thorp, S. 143
Tobin, J. 332
Tong, H. 208, 211, 213, 215
Tremayne, A. R. 208
Trivedi, P. K. 43, 309
Tse, Y. K. 248
Tsui, A. K. C. 248

U

Ujhelyi, G. 337

V

Varkey, D. 284
Ventosa-Santaulària, D. ...191, 193, 340,
    355, 358
Vijverberg, W. P. M. 284
Vogelsang, T. J. 121, 123
Volterra, V. 208
Vuong, Q. H. 53

W

Wagner, M. 147, 148
Wald, A. 77
Wansbeek, T. 255
Watson, M. W. 136
Watts, D. G. 212, 217
Weisberg, S. 21
West, K. D. 28
White, H. 22, 27, 203
White, W. B. 198
Whittle, P. 294, 351
Wiener, N. 208
Wiggins, V. 347, 349
Wild, C. J. 217
Wooldridge, J. M. 15, 274

Y

Yeo, S. 131
Yu, J. 231

Z

Zakoïan, J.-M. 239
Zellner, A. 81, 102
Zerbini, A. 310
Zignago, S. 289

Subject index

A

ACF . . . . *see* autocorrelation function
additive outlier model 123
ADF test . . . . *see* augmented Dickey–Fuller test
`Adjusted R-squared` label 18
ado-file 361
`adopath` command 362, 364
AIC . . . *see* Akaike information criterion
air pollution 11
Akaike information criterion 64, 84
AMEs . . . . . . *see* average marginal effects
analysis of variance F statistic 20
Anderson–Hsiao model 280
AO model . . . *see* additive outlier model
ARCH . . . . *see* autoregressive conditional heteroskedasticity
`arch` command 234
Arellano–Bond model 280
`arfima` command 351
ARFIMA model . . . . . . *see* autoregressive fractional integration moving-average model
`arima` command 61, 180
ARMA model . . . . *see* autoregressive moving-average model
ARMA-X model . . . . *see* autoregressive moving-average model that includes additional explanatory variables
$AR(p)$ model 59
ASCII text file 368
`assert` command 369
asymmetric GARCH model 239
asymptotic efficiency 38
asymptotic normality 36
asymptotic properties 35

asymptotic unbiasedness 35
augmented Dickey–Fuller test 113
    lag length selection 114
autocorrelation 23
autocorrelation function 61, 65
autoregressive conditional heteroskedasticity 232
autoregressive fractional integration moving-average model 351
autoregressive model 59
autoregressive moving-average model 55, 61
autoregressive moving-average model that includes additional explanatory variables 61
average marginal effects 321

B

backslash 362
`bcal` command 380
bilinear time-series model 203
binomial logit model 312
binomial probit model 312
biofuels 143
Blundell–Bond model 280
Box–Jenkins models 55
Box–Pierce Q test 23
Breusch–Godfrey test 23
Breusch–Pagan test 21
built-in command 361
byable 379

C

capital asset pricing model 242
CAPM . . *see* capital asset pricing model
carbon monoxide 13
casewise deletion 371

cd command 362
censored regression model 313
center command 380
Cerrillos 12
Chile 11
choropleth map 9, 284
circsummarize command 14, 384
circular 384
circular package 14
circular statistics 13
classical decomposition 2
clemao1 command 124
clemao2 command 124
Clemente–Montañés–Reyes test 124
clemio1 command 124
clemio2 command 124
cluster–robust standard errors 29
clustering 231
CO_2 8
cointegration 129
collapse command 371
comma-delimited file 368
comma-separated values file 368
compress command 366
consistency 35
contiguity weights 288
contract command 371
convergence in probability 35
correlate command 379
correlated random-effects model ... 274
Cramér–Rao lower bound 38
CRE model .... see correlated random-effects model
crush spread 130
.csv files .. see comma-separated values file
Cumby–Huizinga test 23
current working directory 362
CWD ..... see current working directory
cyclical component 189

D
D. difference operator 382
Dalton minimum 225
delimiters 368

delta method 42, 301
    marginal effect of probit model ...
    322
dereferencing 374
describe command 373
destring command 372
deterministic trend 6
DF-GLS test .... see generalized least-squares version of the augmented Dickey–Fuller test
dfgls command 119
dfuller command 113
Dickey–Fuller distribution 113
Dickey–Fuller test 112
dictionary file 367
Diebold–Mariano test 159
difference operator .... see D. difference operator
discrete data 9
discrete Fourier transform 346
dmariano command 159
Doornik–Hansen test 25
DPD model .... see dynamic panel-data model
Durbin–Wu–Hausman test 49
dynamic forecast 154
dynamic model 55
dynamic panel-data model 279, 280

E
egen
    command 378
    count() function 379
    egen max() function 379
    mean() function 379
    min() function 379
    egen rowfirst() function 379
    rowlast() function 379
    rowmax() function 379
    rowmean() function 378
    rowmin() function 379
    egen rowmiss() function 379
    egen rownonmiss() function .. 379
    rowsd() function 379

egen, *continued*
    `rowtotal()` function..........379
    `total()` function.............379
`egranger` command...............140
El Niño..............................4
El Niño Southern Oscillation.........4
Elliott–Rothenberg–Stock test.....119
embedded spaces..................368
`encode` command..................260
Engle–Granger regression..........135
environmental Kuznets curve.......78
environmental risk................229
ESTAR model .. *see* exponential smooth transition model
estat
    `archlm` command..............233
    `aroots` command...............71
    `duration` command...........225
    `hettest` command..............22
    `imtest` command...............22
    `ovtest` command..............201
    `period` command..............190
    `transition` command........225
`estimates store` command.......273
ethanol...........................143
Euclidian distance................287
ex ante forecast..................154
ex post forecast..................154
exact identification...............45
excess return.....................231
exogeneity........................45
expectation maximization algorithm...
   224
exponential smooth transition model..
   213

F
`F.` lead operator...................382
factor variables...................16
`fcast compute` command.........157
first-order conditions..............40
fixed effects......................264
fixed format......................368
`foreach` command................377
forecast accuracy.................158

forecast error variance decomposition..
   94
foreign direct investment..........256
`forvalues` command..............377
forward slash.....................362
fractional cointegration...........354
fractional integration.........339, 340
free format.......................368

G
gamma distribution................172
gamma function...................237
GARCH .. *see* generalized autoregressive conditional heteroskedasticity
generalized autoregressive conditional heteroskedasticity........232
generalized error distribution......237
generalized least-squares version of the augmented Dickey–Fuller test
   119
generalized linear model..........171
generalized method of moments.....45
`generate` command...............378
Geweke–Porter-Hudak estimator...347
`ghansen` package.................144
`giacross` command..............159
GLM...... *see* generalized linear model
`glm` command....................172
`global` command.................375
global macro.....................378
GMM........ *see* generalized method of moments
GPH estimator.......................
    ... *see* Geweke–Porter-Hudak estimator
`gphudak` command................347
Granger causality..................85
green bonds......................229

H
HAC standard errors................
    .... *see* heteroskedasticity- and autocorrelation-consistent standard errors
Hannan information criterion.......84

Hannan–Quinn information criterion 64
Hansen–Sargan J test 46
`hausman` command 272
Hausman test 49, 272
`help` command 361
Hessian matrix 40
heteroskedasticity 21
heteroskedasticity tests 21
heteroskedasticity- and autocorrelation-consistent standard errors 28
HIC .. *see* Hannan information criterion
`histogram` command 2, 34
HQIC .. *see* Hannan–Quinn information criterion
humpback whale 310
hybrid model 274

I

I(1) 108
`if` qualifier 374
`import`
    `dbase` command 367
    `delimited` command 368, 369, 372, 381
    `excel` command 367, 370
    `sas` command 367
    `sasxport8` command 367
    `spss` command 367
`in` qualifier 374
in-sample forecast 154
incidental parameter problem 266
`infile` command 368
information matrix 41
innovational outlier model 123
integrated GARCH 233
interval forecast 154
intracluster correlation 29
IO model .... *see* innovational outlier model
`ivreg2` command 47
`ivregress` command 28

J

J statistic 46
Jarque–Bera test 25
Johansen algorithm 138
joint hypothesis test 19

K

Kalman filter 177
Kalman gain 178
`kpss` command 345
KPSS test ... *see* Kwiatkowski, Phillips, Schmidt, and Shin test
Kwiatkowski, Phillips, Schmidt, and Shin test 117, 345

L

`L.` lag operator 382
lag operator .... *see* `L.` lag operator
Lagrange multiplier test 44
lead operator ..... *see* `F.` lead operator
leptokurtosis 237
`levpredict` command 170
likelihood-ratio test 43
linear probability model 312
linear regression 11
linear regression model diagnostics 20–26
    explanatory variable diagnostics 18–20
link function 171
load-weighted price 56
`local` command 374
local level model 185
local linear trend 185
local macro 378
local Whittle estimator 351
log price relative 231
log-likelihood function 39
logistic smooth transition model ... 213
`logit` command 318
lognormal distribution 33
`lomodrs` command 344
long memory 339
longitudinal data 7, 255

Subject index

loop constructs 377
LSTAR model .... *see* logistic smooth transition model

M

Mac OS X spreadsheet dates 370
macro 374
macro evaluation 374
MA(q) model .... *see* moving-average model
marginal effects 321
`margins` command 383
Markov switching model 220
Mata 376
`matrix list` command 376
`matrix` command 376
Maunder minimum 198
maximum likelihood estimator 38
mean absolute error 159
mean absolute percentage error 159
mean squared error 159
method of moments 44
MGARCH ..... *see* multivariate generalized autoregressive conditional heteroskedasticity
`mgarch dcc` command 248
`mgarch dvech` command 246
`missing()` function 371
missing string value 371
`mkspline` command 121
`mlexp` command 42
`modlpr` command 349
Moran scatterplot 292
moving-average model 60
`mswitch` command 222
`mswitch ar` command 222
`mswitch dr` command 222
multivariate generalized autoregressive conditional heteroskedasticity 242
multivariate GARCH models vech 246
`mvdecode` command 372
`mvencode` command 372
`mvtest normality` command 26

N

nested loop 377
nested macros 375
`net install` command 363
`newey` command 28
nonlinearity 3
nonstationarity 6
normal equations 15, 40
normality 24
numlist 373

O

observation equation 181
ODBC .... *see* Open Data Base Connectivity
`odbc` command 367, 370
odds ratio 323
`ologit` command 327
OLS .... *see* ordinary least squares
Open Data Base Connectivity 370
`oprobit` command 327
option pricing 7
`order` command 373
order of integration 108
ordered logit model 313
ordered probit model 313
ordinary least squares 15
out-of-sample forecast 154
outer product of gradient 42
overidentification 45

P

p-value 18
PACF .... *see* partial autocorrelation function
panel data 7, 255
    Nickell bias 279
partial autocorrelation function .. 63, 65
periodicity 2
periodogram 346
Phillips modified log-periodogram test 349
Phillips–Perron test 116
platykurtosis 237
plim 35

PM2.5 11
point forecast 154
pollution abatement costs 256
pollution haven hypothesis 255
pooled ordinary least-squares model...
.... 262
population regression function 14
`pperron` command 116
`predict` command 366
premature deaths 11
`preserve` command 371
`probit` command 318
productivity 99
`profile.do` do-file 362
`pwcorr` command 379

Q

quantile plot 26
quasi–maximum likelihood estimator ..
.... 43

R

R/S statistic 343
R^2 17
random effects 264
random walk model 185
range over scale statistic 343
`real()` function 372
recursive forecast 155
`regress` command 16
regression disturbances 14
regression residual 15
regulatory compliance 284
relative difference 377
`reldif()` function 377
`replace` command 378
RESET test 201
`reshape` command 259
`restore` command 371
retransformation bias 166
risk free rate 231
Robinson log-periodogram regression
estimator 349
`roblpr` command 349, 354
robust standard errors 27

root mean squared error 17, 159
`rosssekh` command 160
rowwise functions 378

S

S. seasonal difference 382
sample regression function 15
sandwich estimator 43
Santiago 12
`sarima()` option 73
`save` command 362
SBIC .... see Schwarz or Bayesian
information criterion
scalars 375
Schwarz or Bayesian information criterion 64, 84
sea level rise 176
`search` command 361, 363
seasonal component 188
seasonal difference operator ..... see S. seasonal difference
seasonal fluctuations 2
seasonal ARMA model 71
seemingly unrelated regression 83
self exciting threshold autoregressive
model 208
senstivity 325
SETAR model ....
.... see self exciting threshold
autoregressive model
shapefile 288
smooth transition model 212
SO_2 8
`sort` command 373
Southern Oscillation 4
Southern Oscillation Index 4
space-delimited files 368
spatial
    autoregressive model 294
    dependency parameter 287
    error model 295
    lag 292
    lag model 295
    model 283

Subject index

spatial, *continued*
    relationships 8
    weighting matrix 283
`spbalance` command 257
specificity 325
spillover effects 300
`spmatrix` command 288
`spregress` command 297
`spset` command 297
SQL ..... *see* structured query language
`ssc` command 363, 378
SSC Archive 361
`sspace` command 178, 179
`sspace postestimation` command ...
    179
standard error of regression 17
standard errors
    cluster–robust 29
    independently and identically distributed 27
    Newey–West 28
    White 27
Stata Journal 361, 363
Stata Technical Bulletin 363
state equation 181
state-space form 175
stationarity 55
`statsmat` command 377
stochastic trend 6
string missing values 371
`strL` datatype 365
structural breaks 121
structural vector autoregressive model
    94–101
structured query language 367, 370
sunspots 3, 198
SUR .... *see* seemingly unrelated regression
`sureg` command 83
sustainability index 229
SVAR model ...... *see* structural vector autoregressive model

T

t test 18
t test of significance 18
tab-delimited file 368
`tabstat` command 377
TAR model .... *see* threshold autoregressive model
temperature 2, 13
temperature anomaly 5
`testparm` command 268
text files 368
`texteditors` command 368
Theil's U 159
thermal inversion 12
threshold autoregressive model 208
time-series operators 382
time-varying conditional covariance ...
    244
`tobit` command 331
Tobit model 313
trace test 142
trend, seasonal, cyclical, and irregular components 2
`tsset` command 257, 381–383

U

`ucm` command 184
unbalanced panel 255
unit root 108
unit-root tests
    structural breaks 124–126
univariate GARCH model
    forecasting 235
unobserved component time-series models 184
unobserved heterogeneity 7, 264
`update` command 362
`urcovar` command 121
`use` command 362

V

`v23` command 203
V23 test 201
VAR .... *see* vector autoregression
`var` command 83

variance–covariance estimators......16
varlist............................373
VARMA model...............*see* vector autoregressive moving-average model
`varsoc` command...............66, 84
VCE...........*see* variance–covariance estimators
`vec` command.....................138
VECM.......*see* vector error-correction model
`vecrank` command.................142
vector autoregression...............77
vector autoregressive forecast......156
vector autoregressive model.....80–94, 281
vector autoregressive models
    forecast error variance decomposition......................92
    Granger causality test..........85
    impulse–response analysis......87
vector autoregressive moving-average model..............77, 79, 80
vector error-correction model......129
    forecast......................157
`version` command.................361
volatility..........................6
volatility models
    dynamic conditional correlation...
   248–252
    exponential generalized autoregressive conditional heteroskedasticity....................240
    threshold ARCH...............239

W
Wald test...........................44
weak stationarity..................105
weather derivatives..................7
`while` command...................377
White test.........................22
wide and long forms...............258
wildcards....................373, 378
willingness to pay.................311
wind direction.....................13
wind speed....................13, 152
WTP............*see* willingness to pay

X
`xi` command......................275
`xtabond2` command...............280
`xthybrid` command...............275
`xtreg, be` command..............266
`xtreg, fe` command..............266
`xtset` command.............257, 383
`xtsum` command..................265

Z
z statistic..........................48
zero conditional mean..............15